AN INTRODUCTION TO
STOCHASTIC PROCESSES

AN INTRODUCTION TO
STOCHASTIC PROCESSES

WITH SPECIAL REFERENCE TO
METHODS AND APPLICATIONS

THIRD EDITION

BY

M. S. BARTLETT, F.R.S.

Emeritus Professor of Biomathematics
University of Oxford

CAMBRIDGE UNIVERSITY PRESS

CAMBRIDGE

LONDON · NEW YORK · MELBOURNE

Published by the Syndics of the Cambridge University Press
The Pitt Building, Trumpington Street, Cambridge CB2 1RP
Bentley House, 200 Euston Road, London NW1 2DB
32 East 57th Street, New York, NY 10022, USA
296 Beaconsfield Parade, Middle Park, Melbourne 3206, Australia

First published 1955
Reprinted 1956, 1960, 1961, 1962
Second edition 1966
Third edition 1978

Printed in Great Britain at the
University Press, Cambridge

Library of Congress Cataloguing in Publication Data
Bartlett, Maurice Stevenson.
An introduction to stochastic processes, with special reference to
methods and applications.
Bibliography: p.
Includes Index.
1. Stochastic processes. I. Title.
QA274.B37   1978   519.2   76-57094
ISBN 0 521 21585 4
(Second edition ISBN 0 521 04116 3)

TO MY FRIENDS

Fate, Time, Occasion, Chance and Change?—To these
All things are subject....

(P. B. SHELLEY, *Prometheus Unbound*,
Act II, sc. iv, ll. 119–20)

CONTENTS

PREFACE TO THE FIRST EDITION

The theory of stochastic processes may be regarded as the 'dynamic' part of statistical theory, with a multiplicity of applications. Nevertheless, in spite of the importance and breadth of the subject, which has shown an accelerated progress in the last twenty-five years, there was at the time J. E. Moyal and I first planned (in 1946) to write a book on stochastic processes no general work available, as distinct from more specialized monographs. Our original plan was to survey the general theory with especial reference to its uses and applications, both in physics and statistics. However, the many new developments, with their tremendous range from fundamental theory to specific applications and techniques, both delayed the completion of this project and made it difficult to confine it within one book. We finally felt it more useful to split our contributions into separate volumes. Mr Moyal, who has been actively interested in the development of the basic mathematical theory, would be responsible for dealing with this. A much more elementary discussion of mathematical methods and statistical techniques, addressed to the applied mathematician and statistician, would be attempted in the present introductory work. Although some reference could be made here to physical applications, a systematic discussion of stochastic processes in physics would require yet a third volume (which I hope Mr Moyal will write after the completion of his book on the mathematical theory).

My own book is now much closer in aim and content to my North Carolina lecture notes on stochastic processes (circulated in mimeographed form in 1947), although of course considerably amplified and brought more up to date. Any attempt, even at this level, to survey the whole field must necessarily risk the criticism of omission or patchiness, but there nevertheless has seemed to me to be a strong case for referring to so many topics in the same volume, in order to stress unifying principles, and to demonstrate the frequent value of the same technique in different applications. My theoretical approach is at times admittedly left rather formal and incomplete. In some sections mathematical results have been summarized without proof; this is not only for reasons of space but also because the proofs would not always fit in with the limited mathematical aim and scope of this volume. Those who require a more complete treatment from the mathematical point of view may be referred to Mr Moyal's forthcoming book, or to other theoretical publications which have appeared in the last few years (for example, Lévy (1948), Doob (1953)).† It would, however, be a pity if applied mathematicians or statisticians were put off from using some of the mathematical and statistical techniques available because they did not feel able to absorb all the more pure mathematical theory. As a statistician I find it at times rather exasperating when the mathematics of stochastic processes tends to become so abstract; time spent in wrestling with it can hardly be spared unless, as of course mathematics is best fitted to do, it deepens one's perception of the over-all theoretical picture in the probabilistic and statistical sense.

The placing of the examples and applications is somewhat uneven, as they may be found with the relevant theory and methods, or separately. In particular, if specifically physical applications are mentioned there is no separation; thus cascade showers are referred to as examples of multiplicative chains in Chapter 3, whilst the closely related theory of population growth is deferred until Chapter 4. Some compromise seemed essential, and a strictly logical and uniform pattern was considered impracticable. As no single application could be discussed at any great length, any that appeared to require an excessive prior familiarity or explanation has been omitted or merely mentioned. The summary of communication theory included in Chapter 7 is hardly an exception to such rules, as its principles are quite general and not dependent on particular physical communication systems. Some of the random walk and Markov chain applications referred to in the earlier chapters have now been rather fully treated by W. Feller in his recent book *An Introduction to Probability Theory and its Applications* (vol. 1), but the restriction to discrete probabilities somewhat limited the methods and class of problems he included.

In the concluding chapters I have attempted to survey methods

† See Bibliography (for § 1·3) at end.

of statistical inference for stochastic processes, and in particular for stationary time-series. Genuine examples here are more scanty than I would like. This is partly a reflection of the more particularized character of stochastic process data, as emphasized in the text, so that suitable 'stock examples' are not so readily available. Many of these statistical methods of analysis are, moreover, comparatively new and still unfamiliar. It is hoped that they will, like the mathematical techniques developed in the earlier part of the book, be of interest to a wide class of reader.

In spite of numerous references to original sources, it would be impossible in this wide survey to indicate in all cases the names of those first responsible for the various developments. Important theoretical contributions have been made in particular by American, French, Russian and Swedish writers on probability, the fundamental work of Kolmogorov perhaps calling most for explicit mention. In addition, however, to the systematic theory, the variety of individual and historical applications in physics, biology, medicine, economics or other fields, in many cases preceding and stimulating the general theory, should not be forgotten. Often in the text—for example, if more than one writer has contributed—general references have been deliberately omitted, but those most relevant are listed at the end of the book. This rule has also usually been adopted in regard to any references to my own work. A reference to Moyal (M) denotes the separate volume on the mathematical theory of stochastic processes mentioned above and not published at the time of writing.

The 'decimal system' of numbering the chapter sections and subsections has been used. Equations have been numbered separately for each subsection. A Glossary of some of the standard types of stochastic processes appears just before the general Index.

Finally, grateful acknowledgments are made to D. G. Kendall and J. E. Moyal, with whose work in recent years I am fortunate enough to have been in close contact, as will be evident in some parts of the book. Explicit acknowledgments have usually been inserted in the text for any unpublished work quoted—for example, work by research students. I am greatly indebted for a number of helpful comments or corrections to A. M. Walker,† who very kindly read through most of the final draft, to J. E. Moyal of course, but particularly in regard to §§ 3·5, 5·1 and 5·11; also to D. V. Lindley, in § 1·2. However, all the contents, including its limitations and remaining slips,‡ are my own responsibility. My sincere thanks are offered to Miss Barbara Appleby for her careful preparation of the final typescript; to Mrs G. W. Walls for typing an earlier draft; and to Mrs A. Linnert for assistance with some of the tables and figures. Acknowledgments are made to the Editor and publishers of *Biometrika*, for kind permission to reproduce figs. 12, 13 and 14 from my paper in vol. 37, pp. 1–16, and of *Applied Statistics*, for figs. 1, 6 and 7 from my paper in vol. 2, pp. 44–64.

M.S.B.

August 1953

† Mr Walker, with Dr H. C. Gupta and P. A. Wallington, also kindly assisted with proof-reading.

‡ A few slips have been corrected in the second impression. Further corrections will always be gratefully received.

PREFACE TO THE SECOND EDITION

The reception given to the first edition has, apart from the occasional pure-minded mathematical critic, been more gratifying than I had dared hope. Several readers, however, while sympathetic, have told me that they found the text rather condensed for easy reading. Unfortunately this could hardly be remedied, if at all, without complete re-writing, about which, as I explain below, neither my publishers nor I would have been very enthusiastic; so I decided that the book must be accepted (or not) with a limited revision. A major problem was what to add from the avalanche of further developments. It seemed most sensible, once it had been decided to maintain the book's identity and keep additions to a minimum, to refer on the whole to specific developments of interest to me, such as further progress in spectral analysis, including the spectral analysis of point processes.

A further problem was how to fit the additions in with the original text. It was convenient, both for the publishers and myself, to do this mainly by the insertion of extra sections. This means that the original text has sometimes acquired a slightly historical flavour, as, for example, with the numerical examples in § 9·2; but, together with the additions, I hope that a reasonable perspective is maintained. The new sections are §§ 2·23, 4·22, 5·22, 6·13, 6·52, 6·53, 7·12, 7·2 and 7·21 (in place of the old 7·2), 9·21 (the old 9·21 being now 9·22), 9·23, 9·4 and 9·41. The main further additions are to §§ 3·3, 3·5, 4·21, 4·4, 9·22 and an amendment to the last example in § 6·31 (V. Ramesám pointing out to me the non-stationarity of the process as originally discussed).

In this edition references to J. E. Moyal's intended companion volume have regretfully had to be deleted. However, as I had already mentioned in my first preface, other books on the mathematical theory fill some of the gaps in my own discussion, which I nevertheless believe to be still a reasonably self-contained introduction to the whole subject of stochastic processes. Further relevant references have of course been added to the Bibliography: but for a really comprehensive list the reader may now be referred to the Bibliography on time series and stochastic processes just issued by the International Statistical Institute.

Figs. 15, 16 and 17, new to this edition, are reproduced with the kind permission of the Society from my paper in vol. 25, pp. 264–296 of the *J. R. Statist. Soc.*, Series B.

M.S.B.

January, 1964.

PREFACE TO THE THIRD EDITION

The policy adopted for the second edition of incorporating new material by *adding* new sections wherever feasible has been repeated for this third edition, with the continued constraint that such new material should be fairly severely limited to maintain a reasonable size for the book as a whole. New material has been largely restricted to some recent developments on the following topics:

(i) extinction probabilities for variable environments (new section §2·31)

(ii) first passage-times for diffusion processes (end of §3·5)

(iii) multidimensional diffusion equations with radial symmetry (new section §3·51)

(iv) approximating and limiting solutions dependent on a parameter (§5·3)

(v) lattice models and Markov fields (new section §6·54)

(vi) extension of the χ^2 technique of §8·2 (new section §8·22)

(vii) use of the likelihood function for stochastic models (new section §8·31)

(viii) use of the phase of periodogram components in the estimation and separation of discrete harmonic components (end of §§9·22 and 9·23)

(ix) the spectral analysis of line processes (end of §9·23)

(x) analysis of multidimensional lattice models (new section §9·42).

Fig. 18 is reproduced with kind permission from my paper in vol. 130, pp. 457–77 of *J. R. Statist. Soc.*, Series A.

M.S.B.

January, 1976.

Chapter 1

GENERAL INTRODUCTION

1·1 Preliminary remarks

In this book we are going to consider a subject which in particular applications has arisen since the beginnings of probability theory, but the systematic treatment of which has only recently begun to receive the attention it deserves. We may, roughly speaking, think of this subject as the 'dynamic' part of statistical theory, or the statistics of 'change', in contrast with the 'static' statistical problems which have hitherto been the more systematically studied. By a *stochastic process* we shall in the first place mean some possible actual, e.g. physical, process in the real world, that has some random or stochastic element involved in its structure. It will be convenient, however, also to use the same phrase for the mathematical representation as well as the physical concept, just as with the word 'probability', especially here where we shall be mainly interested in the mathematical theory in its role as a theory of statistical phenomena.

Many obvious examples of such processes are to be found in various branches of science and technology, for example, the phenomenon of Brownian motion, the growth of a bacterial colony, or the fluctuating numbers of electrons and photons in a cosmic-ray shower. In many of these examples the statistical or random variables under study, such as the coordinates of a Brownian particle, are changing with *time*, but change involving any other parameter may arise; for example, a stochastic process involving space parameters as well as time is the 'velocity field' of a turbulent fluid.

The mathematical theory which is the starting point of the theoretical developments is the theory of probability, as this is the basis of all statistical theory. In view of this central position of the mathematical theory of probability, its elements are summarized in the next section; but in view of the many controversial discussions over its interpretation, it may be as well to stress at once that we shall always use it as a theory about *statistical* phenomena. There are many situations where observations on particular phenomena can be repeated under similar conditions, but where, however closely one attempts to control the conditions under which the observations are made, there are irregular or random variations between the results of different trials. Nevertheless, a survey of all the trials often indicates regularities which stabilize as the number of trials is increased; such regularities are called *statistical* properties. (The word 'trials' is of course used here in a broad sense; thus in cointossing experiments we may consider either repeated tossing of the same coin or simultaneous tossing of many similar coins.)

It is important to remember that while we make use of the idea of the probability of an event at a single trial, the concept of probability, at least in this statistical context, has a physical meaning only in relation to some aggregate or *ensemble* of trials. It may in fact be measured, with a margin of error, by the relative frequency of times the event has occurred after a reasonably large number of repetitions. No confusion should arise from this last remark if its dual content is realized; it can only be a rough statement if empirical and referring to actual events, but it can be given a precise mathematical meaning if referring to a hypothetical or conceptual model for such events.

After the elements of probability theory have been summarized, we can begin to classify and specify stochastic processes from the probabilistic point of view.

1·2   Elements of probability theory†

We first consider a finite number k mutually exclusive 'elementary events' A_s, one of which must occur on a 'trial'. Now in practice if we had the results of n actual trials, we should have empirical frequency ratios r_s/n, such that the sum $\Sigma_s r_s/n = 1$. Correspondingly in our conceptual and axiomatic model we postulate probability numbers p_s each between 0 and 1 such that $\Sigma_s p_s = 1$. Symbolically, we can write for the entire *probability distribution*

$$\Sigma_s p_s A_s \equiv p_1 A_1 + p_2 A_2 + \ldots + p_k A_k. \tag{1}$$

† § 1·2 (and § 1·21) may be omitted by a reader already familiar with probability theory.

Further, for the derived probability p, say, of 'either A_1 or A_2', we must necessarily, from the correspondence with frequency, postulate the *addition law* for mutually exclusive events

$$p = p_1 + p_2, \tag{2}$$

or in general probability notation (using the symbol $A_1 + A_2$ for A_1 *or* A_2)

$$P\{A_1 + A_2\} = P\{A_1\} + P\{A_2\}. \tag{3}$$

Suppose now the elementary events A_s are grouped in two distinct ways. In the first we have resulting mutually exclusive and exhaustive classes or sets B_i, say, such that B_1 is the sum of some of the A_s, B_2 the sum of some A_s not included in B_1, and so on. Then by use of the addition law we easily see that we have a new distribution $\Sigma_i p_i . B_i$, where $p_i . = P\{B_i\}$. Similarly in the second grouping let the new distribution be $\Sigma_j p_{.j} C_j$. We will next consider the probability p_{ij} to be attached to the symbolic product $B_i C_j$, by which we mean the set of all A_s common to both B_i and C_j. It is evident from their construction that the composite events $B_i C_j$, over all i and j, are also mutually exclusive and exhaustive, so that we have a third distribution $\Sigma_{i,j} p_{ij} B_i C_j$. We have further by the addition law

$$p_i . = \Sigma_j p_{ij}, \quad p_{.j} = \Sigma_i p_{ij}. \tag{4}$$

The identity $$p_{ij} = p_i . (p_{ij}/p_i .), \tag{5}$$

known as the *multiplication law*, defines (for $p_i . \neq 0$) a new quantity $p_{ij}/p_i .$ which obviously lies between 0 and 1 and from (4) satisfies the relation $\Sigma_j (p_{ij}/p_i .) = 1$. Now for actual frequency ratios r_{ij}/n, $r_i ./n$, $r_{.j}/n$, the corresponding identity would read

$$r_{ij}/n = (r_i ./n)(r_{ij}/r_i .), \tag{6}$$

where the second factor denotes the frequency ratio of the composite event $B_i C_j$ *for those trials in which the event B_i occurred*. Hence we call the quantity $p_{ij}/p_i .$ the *conditional* probability of $B_i C_j$ for given B_i (or more simply of C_j for given B_i). In terms of the notation (3) this is written

$$P\{B_i C_j\} = P\{B_i\} P\{C_j \mid B_i\}. \tag{7}$$

The probabilities $P\{B_i\}$ and $P\{C_j\}$ are called *independent* when $P\{C_j \mid B_i\} = P\{C_j\}$, whence it readily follows that $P\{B_i \mid C_j\} = P\{B_i\}$. The relation (5) then becomes simply

$$p_{ij} = p_{i.}\,p_{.j}. \tag{8}$$

Two entire distributions are independent if (8) is true for all i and j, whence

$$\Sigma_{i,j}\,p_{ij}\,B_i\,C_j = (\Sigma_i\,p_{i.}\,B_i)\,(\Sigma_j\,p_{.j}\,C_j). \tag{9}$$

If a value x_i is associated with the event B_i we say that the *random variable*† X has a probability $P\{X = x_i\} = p_{i.}$ of having the realized value x_i. If another random variable Y is associated with C_j, then the simultaneous or joint distribution of X and Y is specified by the p_{ij}. If (8) holds, X and Y are called independent.

1·21　Distribution functions and their properties.

The above elementary rules contain in essence the whole of probability theory, but they have the limitation that they cannot be applied without extension when the number of events A_s is no longer finite. This extension is needed not only for dealing with random variables that have a continuous range of possible values, but is convenient even for demonstrating the consistency and relevance of the axiomatic theory we have just set up when we compare its theoretical predictions, for example, the so-called 'laws of large numbers', with our intuitive ideas about the stability of frequency ratios obtained from actual trials.‡

In this generalization, based on the mathematical theory of measure and due to Kolmogorov, the additive function $P\{A_s\}$ becomes a *completely* additive set function, such that the probability of any sum of a finite *or enumerable* sequence of sets is uniquely defined with a meaning consistent with the elementary

† We distinguish the random variable X from an ordinary variable x, but of course (as in the general notation P for a probability) a capital letter does not necessarily denote a random quantity.

‡ It might be noticed that such theoretical laws, which refer of course to results from a conceptual set of trials, require some appropriate specification of the probability relations for these trials as a whole. It is simplest and usually pertinent to assume that a result from any trial is independent of any set of results from the other trials, and the independence assumption has thus played a rather important role in statistical theory, though the development of stochastic process theory has shown that some relaxation of this assumption is permissible.

finite theory. We shall not consider this generalization in detail here, but merely note some of the definitions and results that will be useful. For a rigorous discussion the reader is referred to Kolmogorov (1933), or Cramér (1937, 1946).

If a random variable X has possible (real) values x in the continuous range from $-\infty$ to ∞, we must postulate from our elementary theory probabilities $P\{x_i < X \leqslant x_{i+1}\}$ for some finite set of intervals $x_i < x \leqslant x_{i+1}$. Our more general starting point is that such probabilities given for *all* such finite sets define a *unique* probability measure $P\{B_r\}$ for a completely additive class of sets B_r containing such intervals (the smallest such completely additive class of sets is known as the class of Borel sets, and provides a sufficiently wide generalization). The most important values of the set function $P\{B_r\}$ are those corresponding to the sets defined by $X \leqslant x$. These are identical with the values of an ordinary function $F(x)$, say, of the single value x; conversely, $F(x)$ uniquely specifies $P\{B_r\}$. We call

$$P\{X \leqslant x\} = F(x) \tag{1}$$

the cumulative probability or distribution function of X. The properties of $F(x)$ are as follows. It is a never decreasing positive function of x (continuous everywhere to the right), tending to 0 as $x \to -\infty$ and to 1 as $x \to \infty$. It has at most an enumerable set of discontinuities, and is differentiable 'almost everywhere', i.e. except on a set of (Lebesgue) measure zero. $F(x)$ can always be expressed as the sum of three positive components

$$F(x) = c_1^2 F_1(x) + c_2^2 F_2(x) + c_3^2 F_3(x), \tag{2}$$

where $c_1^2 + c_2^2 + c_3^2 = 1$, and $F_1(x)$, $F_2(x)$, $F_3(x)$ have the following properties:

(i) $F_1(x)$ is a 'step' function equal to the sum of the jumps p_s of $F(x)$ at all points of discontinuity less than or equal to x, i.e.

$$F_1(x) = \sum_{x_s \leqslant x} \{F(x_s + 0) - F(x_s - 0)\} = \sum_{x_s \leqslant x} p_s. \tag{3}$$

(ii) $F_2(x)$ is an absolutely continuous function

$$F_2(x) = \int_{-\infty}^{x} f(x)\, dx, \tag{4}$$

where $f(x)$ is the probability density or frequency function corresponding to $F_2(x)$.

(iii) $F_3(x)$ is the 'singular' component, which is continuous but has zero derivative almost everywhere. We shall assume that this third component is absent unless the contrary is stated, as it does not arise in practice. Many distributions (though by no means all) are, moreover, either of the type $F_1(x)$ or $F_2(x)$, and may then be classified without ambiguity as 'discrete' or 'continuous' respectively. In such cases it will sometimes be convenient to write in the 'discrete' case

$$P\{X=x\} \equiv p(x) = \begin{cases} p_s & (x=x_s), \\ 0 & (x \neq \text{any } x_s); \end{cases} \tag{5}$$

and in the 'continuous' case

$$P\{X \text{ in } x, x+dx\} \equiv p(x) = f(x)\,dx. \tag{6}$$

In both these cases this amounts to a formal identification of $p(x)$ with $dF(x)$, where $F(x)$ is the Stieltjes integral

$$F(x) = \int_{-\infty}^{x} dF(x) \tag{7}$$

(the integral, like those below, is a Lebesgue-Stieltjes integral, corresponding to the probability measure postulated, though the distinction between Riemann and Lebesgue integration does not arise in most practical problems).

There is no fundamental distinction between scalar or vector random variables. For the latter we write

$$P\{\mathbf{X} \leqslant \mathbf{x}\} \equiv P\{X_i \leqslant x_i \text{ for } i=1, ..., n\} = F(x_1, x_2, ..., x_n)$$
$$\equiv F(\mathbf{x}).$$

For vector variables we can have distributions with pure probability densities $f(\mathbf{x})$ or purely discrete distributions, but other cases, such as a two-dimensional probability density in a three-dimensional space, may also arise.

The average or expected value of an ordinary function $w(X)$ of the random variable X is defined as the (absolutely convergent) integral

$$E\{w(X)\} \equiv \int_{-\infty}^{\infty} w(x)\,dF(x), \tag{8}$$

and similarly for functions $w(\mathbf{X})$ of a vector random variable. In particular, we have the moment formulae

$$\text{Mean } m = E\{X\}, \quad \text{Mean square } E\{X^2\},$$
$$\text{Variance } \sigma^2 = E\{(X-m)^2\},$$

and for a vector variable

> Product or bilinear moment $E\{X_i X_j\}$,
> Covariance $w_{ij} = \rho_{ij}\sigma_i\sigma_j = E\{(X_i - m_i)(X_j - m_j)\}$,
> Covariance matrix $\mathbf{V} = E\{(\mathbf{X} - \mathbf{m})(\mathbf{X} - \mathbf{m})'\}$,

where in the last formula $\mathbf{m} = E\{\mathbf{X}\}$ and $\mathbf{X}$ is written as a column vector or matrix with $\mathbf{X}'$ its transpose. The standard deviation σ is the positive square root of the variance and the correlation coefficient ρ_{ij} is defined as the bilinear moment between the two variables $(X_i - m_i)/\sigma_i$ and $(X_j - m_j)/\sigma_j$.

The moment-generating function (m.g.f.) of X, when it exists, is given by

$$M(\theta) = E\{e^{\theta X}\}, \tag{9}$$

where θ is an auxiliary variable. This function always exists and is, moreover, a uniformly continuous function of θ when θ is imaginary, say $i\phi$. It is then a function $C(\phi)$ of the real variable ϕ, and is known as the characteristic function. The rth derivative (if it exists) of $C(\phi)$ at $\phi = 0$ gives the quantity $i^r E\{X^r\}$. $C(\phi)$ and $F(x)$ are determined uniquely one from the other, the general inversion formula from $C(\phi)$ to $F(x)$ being

$$\Delta F(x) = \lim_{T \to \infty} \frac{1}{2\pi} \int_{-T}^{T} \frac{\Delta e^{-i\phi x}}{-i\phi} C(\phi)\, d\phi, \tag{10}$$

where $\Delta\xi(x) \equiv \xi(x+h) - \xi(x)$ for any function $\xi(x)$, and $F(x)$ is assumed continuous at x and $x+h$. For a vector variable X we define

$$M(\boldsymbol{\theta}) = E\{e^{\boldsymbol{\theta}' \mathbf{x}}\}. \tag{11}$$

The logarithm of $M(\theta)$ defines the *cumulant or semi-invariant* function $K(\theta)$, whose rth derivative (when it exists) at $\theta = 0$ is the rth cumulant or semi-invariant κ_r. In particular,

$$\kappa_1 = m = E\{X\}, \quad \kappa_2 = \sigma^2,$$

and in the case of a vector variable

$$\kappa_{ij} = w_{ij} = \rho_{ij}\sigma_i\sigma_j.$$

The *normal or Gaussian distribution* for a vector variable $\mathbf{X}$ may be most simply defined in terms of its cumulant function, which is

$$K(\boldsymbol{\theta}) = \boldsymbol{\theta}'\mathbf{m} + \tfrac{1}{2}\boldsymbol{\theta}'\mathbf{V}\boldsymbol{\theta}. \tag{12}$$

Its distribution function, which may be obtained from $C(\boldsymbol{\phi})$ by inversion, is equivalent to a density function

$$f(\mathbf{x}) = (2\pi)^{-\frac{1}{2}n} |\mathbf{V}|^{-\frac{1}{2}} \exp\{-\tfrac{1}{2}(\mathbf{x}-\mathbf{m})'\mathbf{V}^{-1}(\mathbf{x}-\mathbf{m})\}, \quad (13)$$

where $|\mathbf{V}|$ denotes the determinant of $\mathbf{V}$. Any linear transformation on $\mathbf{X}$ still yields a normal distribution function.

In the special case of a discrete random variable X with possible values $0, 1, 2, \ldots$, it is convenient also to define the probability-generating function (p.g.f.) by

$$\Pi(z) = E\{z^X\} = M(\log z), \quad (14)$$

the probability $X = r$ being the coefficient of z^r in $\Pi(z)$ (this is equivalent to the *partition function* of statistical mechanics). Two well-known distributions of this type are the binomial distribution for which

$$\Pi(z) = [1 + p(z-1)]^n, \quad (15)$$

and the Poisson distribution, a limiting case of (15) but extremely important in its own right in the theory of stochastic processes,

$$\Pi(z) = e^{m(z-1)}. \quad (16)$$

If two random variables X and Y are independent, we have $F(x, y) = F_X(x) F_Y(y)$ for all x and y. It follows that

$$C(\phi_1, \phi_2) = C_X(\phi_1) C_Y(\phi_2).$$

Putting $\phi_1 \equiv \phi_2 \equiv \phi$, we obtain the characteristic function of the random variable $Z = X + Y$. The distribution of Z when X and Y are independent is called the 'convolution' of the separate distributions $F_X(x)$ and $F_Y(y)$, and is given more directly by

$$F_Z(z) = \int_{-\infty}^{\infty} F_X(z-y) \, dF_Y(y). \quad (17)$$

An important theorem known as the Central Limit Theorem is concerned with the tendency of the distribution of a sum of independent random variables $S_n \equiv X_1 + X_2 + \ldots + X_n$ to become normal as n increases. In the particular case when the X's have the same distribution, the relevant result is that the random variable $(S_n - nm)/(\sigma\sqrt{n})$ tends to normality with zero mean and unit variance provided m and $\sigma^2 \neq 0$ exist as finite quantities. In the general case, a *sufficient* condition which is useful is that in addition to the existence of m_i and $\sigma_i \geqslant \sigma \ (\neq 0)$ in the 'standardized' random variable $(S_n - \Sigma_i m_i)/\sqrt{\Sigma_i \sigma_i^2}$, we have $E\{|\ X_i\ |^k\} < A$, say, for all i and some $k > 2$. The method of proof usually makes use of the 'continuity' theorem for characteristic functions, that if (i) a sequence of characteristic functions $C_n(\phi)$ converges to $C(\phi)$ for all ϕ, and moreover (ii) the limit $C(\phi)$ is continuous at $\phi = 0$ (*or* (ii) the convergence is uniform within $|\ \phi\ | < \alpha$ for some $\alpha > 0$), then the corresponding sequence of distribution functions $F_n(x)$ tends to a valid limiting distribution $F(x)$ (at all points of continuity of $F(x)$), where $C(\phi)$ is the characteristic function of $F(x)$.

In the case where X and Y are not independent, we define the conditional probability $G(y\ |\ x)$ of $Y \leqslant y$ when $X = x$ from the relation

$$F(x, y) = \int_{-\infty}^{x} G(y\ |\ x)\, dF_X(x), \tag{18}$$

where the marginal distribution $F_X(x) \equiv F(x, \infty)$. Notice that although for density functions $P\{X = x\} = 0$, the relation (18) still permits the definition of $G(y\ |\ x)$. In such a case, (18) is equivalent to a density relation

$$f(x, y) = g(y\ |\ x) f_X(x). \tag{19}$$

A conditional average of $w(Y)$, given $X = x$, is defined as

$$E\{w(Y)\ |\ x\} = \int_{-\infty}^{\infty} w(y)\, dG(y\ |\ x). \tag{20}$$

Conditional probabilities and averages will be of fundamental importance in the theory of stochastic processes, as we shall be interested in the future behaviour of random variables, given the realized values of variables already observed in the past.

1·3   Theoretical classification and specification of stochastic processes

We have noted that a random variable X may often be confined to a finite or enumerable discrete set of values, and if we consider some stochastic process characterized by a random variable changing with the time t, it is sometimes convenient to consider separately cases where the random variable X is 'discrete', for example, the size of a population, with possible values $0, 1, 2, \ldots$. However, it is now important to consider also the nature of the parameter t. If this is 'discrete', e.g. $t = \ldots, -2, -1, 0, 1, 2, \ldots$, our stochastic process must be specified by the *simultaneous* probability distribution of any vector set of X's, $\mathbf{X} \equiv X(t_1), X(t_2), \ldots, X(t_n)$, say. While this involves in the limit an extension to vector random variables with an enumerably infinite set of components, this creates no new difficulty, being already encountered in the classical study of sequences of independent events. We shall sometimes refer to such stochastic processes as *random sequences*.

For t taking values over a continuous range (e.g. 0 to ∞, or $-\infty$ to ∞), the problem is more complicated, for a random function $X(t)$ involves the notion of a vector random variable $\mathbf{X}$ defined in a space of a non-enumerable infinity of dimensions. The further extension of probability theory required for this will not be discussed here. It is sufficient for our purpose to make two remarks:

(i) If such a stochastic process $X(t)$ is properly specified from the theoretical point of view, it follows that for any arbitrary finite set of possible values t_r ($r = 1, \ldots, n$) of the parameter t, we shall have a vector random variable $\mathbf{R}_n \equiv (X_1, X_2, \ldots, X_n)$ with a valid distribution function $F_n(\mathbf{r})$. Validity here implies more than the validity of $F_n(\mathbf{r})$ for a fixed set of values t_r; for example, $F_3(x_1, x_2, x_3)$ is a function also of t_1, t_2, t_3, and may be written more fully $F_3(x_1, x_2, x_3; t_1, t_2, t_3)$, say. We may obtain $F_2(x_1, x_2; t_1, t_2)$ as $F_3(x_1, x_2, \infty; t_1, t_2, t_3)$, and this implies that the latter function should not depend any longer on t_3.

(ii) A specification as in (i) is *necessary*, but it has been shown (for example, by Doob (1937)) that without some restriction on the regularity of the possible realized functions $x(t)$ to be admitted, it is not *sufficient* to answer all questions about $X(t)$ we may ask, such as the probability that $X(t) > 0$ for all t in a given interval. But such a restriction is no real limitation from the practical point of view, for it is obvious that no realized function can be in effect measured at more than an enumerable set of points, and we should endeavour to specify our theoretical models and processes accordingly. Such a specification in effectively an enumerable set of variables $X_1, X_2, \ldots$ will be relevant also when we later consider problems of statistical inference for stochastic processes and may sometimes be more conveniently accomplished by a change in the variables considered; for example, if events are occurring independently and randomly in time, a specification of the random function $N(t)$, representing the total number of events at time t, could be given in terms of the random *times* $T_1, T_2, \ldots$ at which $N = 1, 2, \ldots$.

In some cases such a change of variable may show that the effective number of dimensions or 'degrees of freedom' of the process is still finite. For example, the stochastic process defined by the harmonic terms

$$X(t) = A \cos \lambda t + B \sin \lambda t, \tag{1}$$

where λ is a constant, and A and B normal or Gaussian random variables with zero means, unit variances and zero correlation coefficient, has only two degrees of freedom. It is evident from the property that linear combinations of A and B still give a joint normal distribution (though here a degenerate one) that any set of X's is jointly normal, a property which characterizes what is called a *normal stochastic process*. Such a process has a cumulant function for the X's involving only their means and covariance matrix. The means are here zero and further

$$E\{X_r X_s\} = \cos \lambda (t_r - t_s). \tag{2}$$

It may be noted that the function (2) (and the mean of X) involve t only through the differences $t_r - t_s$. Any distribution function $F_n(\mathbf{r})$ will thus in this example also involve t only through the differences $t_r - t_s$, and not depend on our time origin. This property characterizes an important type of process called a *stationary process*, to be discussed further in Chapter 6. Processes which are not stationary may sometimes be called *evolutionary*. For example, if for the same random constants A and B we define

$$Y(t) = A + Bt, \tag{3}$$

then the properties of $Y(t)$ change as t increases from zero. We may also note a further property of the 'degenerate' processes (1) and (3), that if $X(t)$ is known at times t_0 and t_1, say, the realized values of A and B can be deduced and the entire future behaviour of $X(t)$ predicted exactly. These are examples of *deterministic* processes; our main interest will be in non-deterministic processes whose future behaviour cannot be completely predicted in terms of past observations.

We shall not continue with any complete catalogue of particular types of processes, as it is more convenient to introduce them as they arise (see also Glossary). For example, the historically and practically important processes known as *Markov processes* will be first discussed in Chapter 2. One general point of interpretation should be noted. It was suggested that a process in the real world would usually be thought of in the actual 'process of change', like a growing population or a physical system in motion. We shall find this idea relevant to much of the theoretical development, for example, to the concept of a Markov process (as a process with no memory extending before the previous instant) or to problems of prediction. But a random function $X(t)$ does not in general convey the notion of any such 'dynamic' aspect; it may alternatively refer to a random variable depending on some more 'static' parameter t. We have already cited the space parameters in turbulence. In the model for a population we shall see that it is necessary to specify the number of individuals not only at a particular epoch but also of a given age. Although the idea of a *process* may be more relevant in some contexts than others, it is convenient to retain the general term *stochastic processes* in all cases.

For definiteness, we have so far in this section been considering the mathematical representation of stochastic processes involving one (real) random quantity X changing with one auxiliary parameter t. There is strictly no loss of generality in speaking of *quantitative* variables, since distributions associated with 'states' or 'attributes' can always be defined in relation to random variables taking the values 0 and 1, corresponding to non-occurrence or occurrence of the state in question. We shall, however, sometimes wish to deal with more than one quantity X and (or) more than one parameter. As physical examples, we may wish to deal simultaneously with the position $X(t)$, $Y(t)$, $Z(t)$ and velocity $U(t)$, $V(t)$, $W(t)$ of a given particle, or to consider the velocity $U(x, y, z, t)$, $V(x, y, z, t)$, $W(x, y, z, t)$ of a fluid for given space and time coordinates. There is no essential difficulty in extending definitions and methods to deal with such cases. Processes involving more than one random variable in this sense will be called *multivariate, simultaneous* or *vector* processes, the term *multidimensional* processes or *fields* being reserved for processes depending on more than one parameter. It is even sometimes convenient to introduce complex quantities and to consider $Z(t) \equiv X(t) + i Y(t)$, or $X(z) \equiv X(x + iy)$, but such representations are of course introduced purely for mathematical convenience or generality and not as direct models of 'real' physical processes.

1·31   The characteristic functional.

In §1·21 was noted the theoretical equivalence of a distribution function $F(x)$ and the characteristic function $C(\phi) = E\{e^{i\phi X}\}$, or, for a vector random variable, $F(\mathbf{x})$ and $C(\boldsymbol{\phi})$. It is possible to obtain the characteristic function $C_n(\boldsymbol{\phi})$ corresponding to any vector random variable $\mathbf{R}_n$ extracted from a stochastic process $X(t)$. This is the expectation of e^{iS_n}, where S_n is the scalar product of the random vector $\mathbf{R}_n$ and the corresponding mathematical vector $\boldsymbol{\phi}$. Formally at least S_n may be replaced by the more general 'product sum'

$$S = \int X(t) \, d\Phi(t), \tag{1}$$

the previous quantities S_n being particular cases of S given by particular choices of the arbitrary function $\Phi(t)$. The *functional* $C(\Phi)$ may be considered as another possible method of specifying a stochastic process theoretically. While having a less direct interpretation, it is potentially useful as a concise form of definition which moreover lends itself to mathematical manipulations, especially linear transformations or operations on the random function $X(t)$. For example, for the process

$$X(t) = A \cos \lambda t + B \sin \lambda t$$

defined in the last section, we easily find

$$\log C(\Phi) = -\frac{1}{2} \iint \cos \lambda(t - \tau) \, d\Phi(t) \, d\Phi(\tau). \tag{2}$$

In some cases it is useful to introduce the characteristic functional of a process in relation to parameters other than the time. Examples are those previously referred to such as the space parameters in the vector velocity field of a turbulent fluid, or the age parameter in a population. In the latter case it is convenient to specify the cumulative number $N(x)$ of persons of age less than or equal to x, and introduce an alternative form of generalized 'product sum' to (1), viz.

$$S = \int \phi(x) \, dN(x). \tag{3}$$

This is because the number $N(x)$ is necessarily discontinuous with increment either 0 or a positive integer at each point x. The various methods available for handling this last type of process, sometimes referred to as a *point process*, will be discussed in more detail later (see Chapters 3 and 4).

Chapter 2

RANDOM SEQUENCES

2·1 The random walk

The most elementary examples of stochastic processes are classical enough to be discussed in most text-books on probability. They are random sequences in which the variable X_r at time t_r is independent of the entire previous set of X's. The statistical interest of such sequences lies in the properties of derived variables or sequences, such as the cumulative sums

$$S_r = X_1 + X_2 + \dots + X_r. \tag{1}$$

The process S_r is called a 'random walk', as it represents the position at time t_r of a person taking a random step X_r independently of his previous ones. As a special case we may have

$$P\{X_r = 1\} = p, \quad P\{X_r = -1\} = q \equiv 1 - p.$$

The first example of such a process goes back to the gambling problems of the seventeenth century. We may suppose that two gamblers A and B play a sequence of games, the probability of A winning any particular game being p. If he wins this game, he acquires a unit stake from B and if he loses the game, he loses his own unit stake.

The distribution of the unrestricted sum S_n is the n-fold convolution of the distribution function $F(x)$ of each X_r, or equivalently in terms of cumulant functions

$$K_n = nK, \tag{2}$$

where K_n is the cumulant function of S_n and K that of each X_r. As noted in §1·21, the distribution function of S_n (suitably scaled to zero mean and finite variance) tends to normality as n increases, provided the mean m and variance σ^2 (> 0) of the common distribution of each X_r exist.

In the original gambling problem a point of interest was the nature of the sequence if S_r was limited by the initial capital of each gambler, and, in particular, the 'duration of play'. This problem has acquired further interest in its modern application to sequential sampling (see Chapter 4) or in its relation to physical diffusion processes with boundary conditions. Let us consider therefore boundaries a (< 0) and b (> 0), so that if S_n *first* reaches or goes outside the ends of the interval (a, b) at time t_n, the process is terminated. Denote the modified 'distribution function' of S_n by $F_n(x)$, i.e. $F_n(x) = P\{S_n \leqslant x \text{ and } a < S_r < b \ (r = 1, 2, ..., n - 1)\}$. Then we have the recurrence relation

$$F_n(x) = \int_a^{b-} F(x - y) \, dF_{n-1}(y), \tag{3}$$

since $F_{n-1}(y)$ only contributes to $F_n(x)$ for $a < y < b$. (Here $F_n(x)$ denotes a distribution function in an extended sense convenient in 'absorption' problems, in that owing to previous absorption at the boundary $F_n(\infty)$ is in general no longer unity.) The probability p_n of reaching the boundary b for the first time at n is $F_n(\infty) - F_n(b-)$, and similarly for a we write $p'_n = F_n(a) - F_n(-\infty)$. If we put

$$P(\lambda) = \sum_{r=1}^{\infty} \lambda^{r-1} p_r, \quad P'(\lambda) = \sum_{r=1}^{\infty} \lambda^{r-1} p'_r, \tag{4}$$

then

$$P(\lambda) = G(\infty, \lambda) - G(b-, \lambda), \quad P'(\lambda) = G(a, \lambda) - G(-\infty, \lambda), \tag{5}$$

where

$$G(x, \lambda) = \sum_{r=1}^{\infty} \lambda^{r-1} F_r(x) \tag{6}$$

satisfies the Fredholm integral equation

$$G(x, \lambda) = F(x) + \lambda \int_a^{b-} F(x - y) \, dG(y, \lambda). \tag{7}$$

This equation, due to Samuelson (1948), may in principle be solved (e.g. by going back to (3) and (6)), but it is often more convenient to proceed by other methods. Special cases, such as the classical gambling problem mentioned above, may be considered directly; detailed discussion of these is available in books on probability (see, for example, Uspensky (1937), Chapters 5 and 8; or Feller (1950), Chapter 14), and we shall proceed at once to another general method, due to Wald.

We establish first two lemmas.

LEMMA I. *Denote the random value of* n *at which* S_n *first reaches one of the boundaries by* N. *Then*

$$n_0^s P\{N \geqslant n_0\} \to 0, \quad as \quad n_0 \to \infty, \tag{8}$$

for any finite s. For since $\sigma^2 > 0$, $E\{S_n^2\} \to \infty$ as $n \to \infty$. Hence there is a k such that

$$P\{S_k^2 < (b-a)^2\} < 1, \quad \leqslant 1 - \lambda, \quad \text{say.}$$

Hence for $n_0 = hk$,

$$P\{(S_{k(r+1)} - S_{kr})^2 < (b-a)^2 \text{ for all } r = 0, ..., h-1\} \leqslant (1-\lambda)^{n_0/k},$$

since the increments in X_n are independent. Any of the h inequalities $|S_{k(r+1)} - S_{kr}| \geqslant (b-a)$ ensures that $N < n_0$. Thus finally we have shown that

$$P\{N \geqslant n_0\} \leqslant (1-\lambda)^{n_0/k} = O(e^{-\mu n_0}) = o(n_0^{-s}) \quad \text{as} \quad n_0 \to \infty.$$

LEMMA II. *Denote the m.g.f. of each* X_r *by* $M(\theta)$, *assumed to exist for all real* θ, *and to have the property*

$$M(\theta) \to \infty \quad \text{as} \quad \theta \to \pm\infty \tag{9}$$

(this last property follows if

$$P\{e^X < 1-\delta\} > 0 \quad \text{and} \quad P\{e^X > 1+\delta\} > 0$$

for some $\delta > 0$). It follows also that

$$M'(\theta) = E\{Xe^{\theta X}\}, \quad M''(\theta) = E\{X^2 e^{\theta X}\}, \tag{10}$$

where dashes denote differentiation with respect to θ. *Then if* $m = M'(0) \neq 0$, *there is one and only one real root* $\theta_0 \neq 0$ *such that* $M(\theta_0) = 1$.

For from (10) we have $M''(\theta) > 0$ for all real θ; hence from (9) $M(\theta)$ has only one minimum, θ_1, say. But $M'(0) \neq 0$, hence $\theta_1 \neq 0$, and $M(\theta_1) < M(0) = 1$. The required result then obviously follows.

Wald's identity.

$$E\{[M(\theta)]^{-N} \exp(\theta S_N)\} = 1 \quad (|M(\theta)| \geqslant 1). \tag{11}$$

To prove (11), let j denote a constant integer, and $P_j \equiv P\{N \leqslant j\}$. Then

$$E\{e^{\theta S_j}\} = P_j E_j\{e^{\theta S_j}\} + Q_j E_j'\{e^{\theta S_j}\},$$

where $Q_j = 1 - P_j$, E_j denotes expectation conditional on $N \leqslant j$, E'_j expectation conditional on $N > j$. Now for any fixed $j > N$, $S_j - S_N$ is independent of S_N, and

$$E_j\{e^{\theta S_j}\} = E_j\{e^{\theta S_N}[M(\theta)]^{j-N}\}.$$

Hence as $E\{e^{\theta S_j}\}$ is also $[M(\theta)]^j$, we have

$$P_j E_j\{e^{\theta S_N}[M(\theta)]^{-N}\} + Q_j E'_j\{e^{\theta S_j}[M(\theta)]^{-j}\} = 1.$$

Now $Q_j \to 0$ as $j \to \infty$, and $E'_j\{e^{\theta S_j}\}$ is bounded (as this expectation is conditional on $N > j$, so that $|S_j| < b - a$). Hence finally, for values of θ for which $|M(\theta)| \geqslant 1$, we obtain in the limit the identity (11).

In the above proof we have only made use of Lemma I for $s = 0$; it may be shown further that (11) may be differentiated with respect to θ any number of times under the expectation sign. It should also be noticed that (11) still holds if $a \to -\infty$, provided $P_j \to 1$ and that the real part of θ is non-negative. A special case is when $X \geqslant 0$; here the conditions of Lemma II (which we make use of presently) break down, and it is readily seen that then there is *no* root $\theta_0 \neq 0$ for which $M(\theta) = 1$.

To apply the above result to cases when one or other of the boundaries is *exactly* reached at stage N, we put $S_N = a$ with probability P_a, say, and $S_N = b$ with probability $P_b = 1 - P_a$. Then

$$P_a E_a\{e^{a\theta}[M(\theta)]^{-N}\} + P_b E_b\{e^{b\theta}[M(\theta)]^{-N}\} = 1, \tag{12}$$

where E_a denotes expectation conditional on $S_N = a$, etc. If in (12) we put $\theta = \theta_0$, we obtain

$$P_a = \frac{1 - e^{b\theta_0}}{e^{a\theta_0} - e^{b\theta_0}}. \tag{13}$$

For example, in the gambling problem we have

$$M(\theta) = pe^{\theta} + qe^{-\theta},$$

$$\theta_0 = \log(q/p),$$

whence $\qquad P_a = \dfrac{1 - (q/p)^b}{(q/p)^a - (q/p)^b}. \tag{14}$

Moreover, according to Lemma II (note that $M(\theta)$, as a function of the complex variable θ, is not singular at 0 or θ_0), the equation

$$-\log M(\theta) = i\phi \tag{15}$$

has two roots $\theta_1(\phi)$, $\theta_2(\phi)$ such that $\theta_1(\phi) \to 0$ and $\theta_2(\phi) \to \theta_0$ as $\phi \to 0$. Hence

$$\left. \begin{aligned} P_a e^{a\theta_1(\phi)} C_a(\phi) + P_b e^{b\theta_1(\phi)} C_b(\phi) &= 1, \\ P_a e^{a\theta_2(\phi)} C_a(\phi) + P_b e^{b\theta_2(\phi)} C_b(\phi) &= 1, \end{aligned} \right\} \tag{16}$$

where $C_a(\phi)$ is the characteristic function of N conditional on $S_N = a$, etc., so that

$$E\{e^{i\phi N}\} \equiv C(\phi) = P_a C_a(\phi) + P_b C_b(\phi). \tag{17}$$

Equations (13), (16) and (17) determine the complete distribution of N, the number of stages required to reach the boundaries.

The importance of these formulae lies also in their *approximate* applicability even when the boundaries are passed rather than precisely reached, for we may still put† $S_N \sim a$ or b, leading to the same formulae (13), (16) and (17) as approximations which become asymptotically exact as the individual steps become small in comparison with the distances to be traversed. As a further approximation we might treat X itself as a *normal* variable, since we know that the unrestricted distribution of S_n tends to normality, and this will apply also to the resultant of several small steps. In the limit when the individual steps become infinitesimally small, these approximate formulae will again become exact.

Now, for X normal, equation (15) becomes

$$m\theta + \tfrac{1}{2}\sigma^2\theta^2 = -i\phi,$$

$$\theta_1, \theta_2 = \{-m \pm \sqrt{(m^2 - 2i\sigma^2\phi)}\}/\sigma^2. \tag{18}$$

As $\theta_0 = -2m/\sigma^2$, formula (13) gives

$$P_a = \frac{1 - e^{-2mb/\sigma^2}}{e^{-2ma/\sigma^2} - e^{-2mb/\sigma^2}}. \tag{19}$$

If $a \to -\infty$, we require the real part of θ non-negative. The relevant root in (18) $(m > 0)$ is the one with the $+$ sign. As $P_a \to 0$, (17) becomes

$$C(\phi) = e^{-b\theta_1(\phi)}, \tag{20}$$

---

† The symbol $\sim$ will be used *either* for 'asymptotically equivalent to' *or*, more loosely, for 'approximately equal to'; which meaning is intended should be clear from the context.

whence $\quad \log C(\phi) = -b\theta_1(\phi) = \dfrac{bm}{\sigma^2} - \dfrac{bm}{\sigma^2}\left(1 - \dfrac{2i\sigma^2\phi}{m^2}\right)^{\frac{1}{2}}$ (21)

for the normal case. Expanding (21) in powers of $i\phi$ we obtain the cumulants of N,

$$\kappa_1 = b/m, \quad \kappa_2 = b\sigma^2/m^3, \quad \dots. \tag{22}$$

Returning to the identity (11), we will obtain some moment formulae by differentiation. Differentiating once,

$$E\{S_N e^{\theta S_N}[M(\theta)]^{-N} - N e^{\theta S_N}[M(\theta)]^{-N-1} M'(\theta)\} = 0,$$

or for $\theta = 0$, $\qquad E\{S_N\} = E\{N\}\,E\{X\}.$ (23)

On our approximating assumptions, this gives

$$E\{N\} = (aP_a + bP_b)/m, \tag{24}$$

provided $m \neq 0$. If $m = 0$, we obtain from a second differentiation of (11) (which is still valid, as no condition on m was used in the proof),

$$E\{S_N^2\} = E\{N\}\,E\{X^2\} \quad (m = 0). \tag{25}$$

Now in (13) as $m \to 0$, and $\theta_0 \to 0$, P_b has a definite limit $-a/(b-a)$, $E\{S_N^2\}$ becomes $-ab$, and hence

$$E\{N\} = -ab/\sigma^2 \quad (m = 0). \tag{26}$$

As a second limiting case, we may note that if $-a$ becomes large compared with the individual steps, but $m > 0$ so that $\theta_0 < 0$, then $P_b \to 1$. Even if $m \to 0$, we had $P_b = -a/(b-a)$, and this still tends to unity when $-a$ increases, although in this case from (26) $E\{N\}$ has no finite limit. It is this last case which covers the classical problem of the gambler's ruin, where a player competing even on fair terms ($m = 0$) with a rich enough adversary ($-a$ large) is almost certain to go bankrupt ($P_b \to 1$).

If $m < 0$, so that $\theta_0 > 0$, note that

$$P_b \to e^{-b\theta_0} \quad (a \to -\infty). \tag{27}$$

It may be noticed that Wald's identity (11) is equivalent to showing that the product

$$\Pi_r = e^{\theta S_r}/[M(\theta)]^r$$

with expectation 1 may have an optional stopping-rule when $r = N$, the value when the boundaries are first exceeded, without affecting the expectation. The sequence Π_r has the obvious

property
$$E\{\Pi_r \,|\, \Pi_1, \Pi_2, \ldots, \Pi_{r-1}\} = \Pi_{r-1}, \tag{28}$$

and is an example of what is, from the relation (28), known as a *martingale* (see Doob, 1953).

2·11 Renewals. As a special case of the random variable X, consider a distribution for which $X > 0$ (and hence $m > 0$). This is relevant to the case of an article having an effective random lifetime X before being replaced by a new one, and so on. The total lifetime obtained from n articles is just S_n and tends to normality as n increases. If we consider the number N of renewals or replacements required up to time t, including the initial article, we note that N is the first value of n for which $S_n > t$. The equations (3)–(7) of § 2·1 reduce to

$$F_n(x) = \int_0^t F(x-y)\, dF_{n-1}(y), \tag{1}$$

$$P(\lambda) = \sum_{r=1}^{\infty} \lambda^{r-1} p_r, \tag{2}$$

where p_n is the probability $F_n(\infty) - F_n(t)$ of reaching the boundary t with the number n,

$$G(x, \lambda) = \sum_{r=1}^{\infty} \lambda^{r-1} F_r(x) \tag{3}$$

$$= F(x) + \lambda \int_0^t F(x-y)\, dG(y, \lambda). \tag{4}$$

Putting $\lambda = 1$ in (3) and (4), and noting that $F_r(x)$, *for $x < t$*, denotes the unrestricted r-fold convolution of $F(x)$, we obtain the standard renewal equations for this problem, either the direct Volterra integral equation (4) for the total expectation of renewals of all orders or 'generations' in the interval $(0+, x)$, or (1) and (3) providing an iterative solution in terms of successive 'generations'.

In the case when $F(x)$ and hence $G(x, \lambda)$ have densities $f(x)$, $g(x, \lambda)$, we have by differentiating (4) with respect to x,

$$g(x, \lambda) = f(x) + \lambda \int_0^t f(x-y)\, g(y, \lambda)\, dy. \tag{5}$$

The solution of this equation (when $x = t$) has been discussed by Feller (1941), and we summarize the relevant results.

(i) Let the Laplace transforms of $f(x)$ and $h(x) = \lambda f(x)$ (the solution applies in the case of a more general function $h(x)$ occurring inside the integral) be

$$\alpha(\psi) = \int_0^\infty e^{-\psi x} f(x)\, dx, \quad \gamma(\psi) = \int_0^\infty e^{-\psi x} h(x)\, dx,$$

convergent at least for $\psi > \sigma$. Then the unique non-negative solution $g(x)$ has Laplace transform $\beta(\psi)$ given by

$$\beta(\psi) = \alpha(\psi)/(1 - \gamma(\psi)). \tag{6}$$

(If $\lim_{\psi \to \sigma+} \gamma(\psi) > 1$, let $\sigma' > \sigma$ be the real root of the equation $\gamma(\psi) = 1$; otherwise, if $\lim_{\psi \to \sigma+} \gamma(\psi) \leqslant 1$, write $\sigma' = \sigma$. Then (6) holds for $\psi > \sigma'$.)

(ii) In order that $g(x)$ defined by (6) can be represented by a series solution of the form

$$g(x) = \Sigma_i c_i e^{\psi_i x}, \tag{7}$$

where ψ_i are the roots, assumed simple, of $\gamma(\psi) = 1$ (the series converging absolutely for $x \geqslant 0$), it is necessary and sufficient that

$$\beta(\psi) = \Sigma_i \frac{c_i}{\psi - \psi_i} \tag{8}$$

and that $\Sigma_i c_i$ converges absolutely. The coefficients c_i are then given by

$$c_i = -\alpha(\psi_i) \bigg/ \frac{d\gamma(\psi_i)}{d\psi_i}. \tag{9}$$

In particular, it is necessary that $\beta(\psi)$ is a one-valued function.

Example 1. As an illustration consider the simplest case where $f(x) = \mu e^{-\mu x}$, which is well known to represent the case of a purely random renewal process (see, for example, the next chapter). We easily find that $\beta(\psi) = (1 - \lambda + \psi/\mu)^{-1}$, whence

$$g(x, \lambda) = \mu e^{\mu x(\lambda - 1)}. \tag{10}$$

This result is obviously true for $x < t$ as well as $x = t$, for if $x < t$ the upper limit of integration in (5) is x instead of t, since $f(x) = 0$ for $x < 0$. To obtain $g(x, \lambda)$ for $x > t$ we evaluate it directly from (5), obtaining

$$g(x, \lambda) = \mu e^{\mu t \lambda - \mu x} \quad (x > t). \tag{11}$$

Putting $\lambda = 1$ we obtain $g(x, 1) = \Sigma_r f_r(x) = \mu$ $(x \leqslant t)$, representing a constant renewal rate from all generations. Further,

$$P(\lambda) = \int_t^\infty g(x, \lambda)\, dx = e^{\mu t(\lambda - 1)}, \tag{12}$$

which is the p.g.f. of a Poisson distribution with mean μt, indicating that $(\mu t)^r e^{-\mu t}/r!$ is the probability that the number of further renewals (not including the initial article) in the interval $(0, t)$ is r.

$P(\lambda)$ in (12) may be obtained rather more directly if desired by noting that from (4)

$$G(\infty, \lambda) = 1 + \lambda G(t, \lambda),$$

whence in general

$$P(\lambda) = 1 + (\lambda - 1) G(t, \lambda). \tag{13}$$

The above results are exact. If we make use of the alternative approach via Wald's identity and equation (20) of § 2·1, we derive in this example

$$M(\theta) = (1 - \theta/\mu)^{-1}, \quad \theta_1(\phi) = \mu(1 - e^{i\phi}),$$

$$C(\phi) = \exp\{\mu t(e^{i\phi} - 1)\}.$$

As the last replacement does not occur exactly at t, the last equation is only established as an approximate result valid for large t. Comparison with (12) shows that in this case it also happens to be exact, but this is not of course generally so (see the next example).

Example 2. As a less trivial example, suppose

$$f(x) = \mu^2 x e^{-\mu x}. \tag{14}$$

Then we find $\beta(\psi) = ([1 + \psi/\mu]^2 - \lambda)^{-1}$, whence

$$g(x, \lambda) = (\mu/\sqrt{\lambda}) e^{-\mu x} \sinh(\mu x \sqrt{\lambda}) \quad (x \leqslant t). \tag{15}$$

To obtain $P(\lambda)$ we may use formula (13), giving

$$P(\lambda) = e^{-\mu t}\{\cosh(\mu t \sqrt{\lambda}) + [\sinh(\mu t \sqrt{\lambda})]/\sqrt{\lambda}\}. \tag{16}$$

The asymptotic approach via Wald's identity gives at once

$$C(\phi) \sim \exp\{\mu t(e^{\frac{1}{2}i\phi} - 1)\},$$

whence

$$\kappa_1 \sim \tfrac{1}{2}\mu t, \quad \kappa_2 \sim \tfrac{1}{4}\mu t,$$

compared with the exact results from (16)

$$\left. \begin{array}{l} \kappa_1 = \tfrac{1}{2}\mu t - \tfrac{1}{2}e^{-\mu t}\sinh \mu t = \tfrac{1}{2}\mu t - \tfrac{1}{4} + O(e^{-\mu t}), \\ \kappa_2 = \mu t e^{-\mu t}\sinh \mu t + \tfrac{1}{4}e^{-\mu t}\sinh \mu t - \tfrac{1}{4}\mu t - \tfrac{1}{4}e^{-2\mu t}\sinh^2 \mu t \\ \quad = \tfrac{1}{4}\mu t + \tfrac{1}{16} + O(e^{-\mu t}). \end{array} \right\} \tag{17}$$

An interesting application of renewal theory is to the theory of genetic recombination. The frequency distribution of 'lifetimes' between renewals becomes replaced by the frequency distribution of chromosomal length between points of exchange, and the number of renewals by the number of points of exchange. In this theory one significant question is whether the 'recombination fraction', which is the probability of an *odd* number of points of exchange, rises above 50 %. To examine this question for the last example, which has been used as one model in this theory, we note that this probability $p(t)$ is given by

$$p(t) = \tfrac{1}{2}\{1 - P(-1)\} \tag{18}$$

$$= \tfrac{1}{2} - \tfrac{1}{2}e^{-\mu t}\{\cos \mu t + \sin \mu t\}, \tag{19}$$

which rises to 0·5216 at $\mu t = \pi$ (for technical details of this application, see Owen (1949), where further general formulae— for example, for the number of 'renewals' between t_1 and t_2— will also be found).

Further aspects of these random walk and renewal problems will be taken up later, after the theory of stochastic processes in continuous time has been developed.

2·2 Markov chains

By considering the sums S_n rather than the original random sequence elements X_r we obtained in § 2·1 new sequences whose elements were no longer independent. We shall now consider elementary examples of stochastic processes where the dependence between successive terms of the sequence is a more intrinsic property of the process. We first define generally *Markov processes*, so termed after the famous Russian mathematician who systematically studied particular cases of them. By a Markov process we shall mean a stochastic process for which the values of X_r at any set of times t_r $(r = 1, 2, ..., n)$ depend on the values X_s at any set of *previous* times t_s $(s = 0, -1, ..., -j)$ only through the last available value X_0. For such a process the simultaneous distribution at times t_r $(t_1 < t_2 < ... < t_n)$ is consequently (in terms of the formal notation used in § 1·21, x_r denoting any realized value of X_r)

$$p(x_1, x_2, ..., x_n) = p(x_1)\, p(x_2 \mid x_1)\, p(x_3 \mid x_1, x_2) ...$$

$$= p(x_1)\, p(x_2 \mid x_1)\, p(x_3 \mid x_2) ...$$

$$= p(x_1) \prod_{r=2}^{n} p(x_r \mid x_{r-1}), \tag{1}$$

so that a Markov process is defined by the conditional distribution $p(x_r \mid x_{r-1})$ for any r $(1, ..., n$, and arbitrary choice of the set $t_r)$, together with initial distribution $p(x_1)$.

Theoretically, this appears a severe restriction on the type of process covered, but in practice many processes are found to be so characterized, especially if it is noted that processes which are not at first sight Markovian may become so in a 'phase-space' of higher dimensions, i.e. by a vector characterization of the process. For example, in classical physics the velocity as well as the position of a particle are required at any instant before the future motion of the particle can be predicted. Again, in population studies the number of births in the past year at time t cannot be adequately predicted from the number of births at a previous time t_0, but may be so from a detailed knowledge of the numbers of individuals in all age-groups at that time.

For any Markov process we obtain at once from (1)

$$
\begin{aligned}
p(x_r \mid x_{r-2}) &= \int_{x_{r-1}} p(x_r, x_{r-1} \mid x_{r-2}) \\
&= \int_{x_{r-1}} p(x_r \mid x_{r-1}, x_{r-2}) \, p(x_{r-1} \mid x_{r-2}) \\
&= \int_{x_{r-1}} p(x_r \mid x_{r-1}) \, p(x_{r-1} \mid x_{r-2}), \quad (2)
\end{aligned}
$$

an important relation known as the Chapman-Kolmogorov equation. We may obtain $p(x_r \mid x_1)$ by its repeated use, and hence from the further formula

$$
p(x_r) = \int_{x_1} p(x_r \mid x_1) \, p(x_1), \quad (3)
$$

we may obtain also $p(x_r)$.

In the case of Markov processes with discrete X, to be considered in this section, and called *Markov chains*, we may write equations such as (2) and (3) in matrix and vector notation. Let $p_i(t_r)$ denote the probability of the ith state or possible value of X, at time t_r, and $q_{ij}(t_r, t_{r-1})$ the corresponding conditional probability of the ith state at time t_r, given the jth state at time $t_{r-1}\, (< t_r)$. Writing (p_j) as a column vector $\mathbf{p}$ and (q_{ij}) as the square matrix $\mathbf{Q}$ (with constant i in any row), we have, for example,

$$
\mathbf{Q}(t_r \mid t_{r-2}) = \mathbf{Q}(t_r \mid t_{r-1}) \, \mathbf{Q}(t_{r-1} \mid t_{r-2}), \quad (4)
$$

$$
\mathbf{p}(t_r) = \mathbf{Q}(t_r \mid t_{r-1}) \, \mathbf{p}(t_{r-1}). \quad (5)
$$

(Notice that to preserve both the usual convention for the order of suffices in the matrix Q and the column vectors of Q as conditional distributions, the second suffix and time in Q refer to the *previous* time. To emphasize this, the previous time is put to the right of a vertical stroke, corresponding to the given previous state.)

There are two main classes of Markov chains, according to the nature of t. In this chapter we consider random sequences, i.e. t is discrete.† As a further subdivision we may have the number of states, even if discrete, finite or infinite. We consider first the simplest case of a finite number k of possible states (this case has been discussed at length by Fréchet (1937–8), *second livre*). Under these conditions equation (5) represents a finite set of simultaneous linear difference equations with an immediate solution for $p(t_r)$ in terms of $p(t_0)$, say, given by

$$p(t_r) = \prod_{s=0}^{r-1} Q(t_{r-s} \mid t_{r-s-1})\, p(t_0), \tag{6}$$

where the matrix product is *ordered* from left to right. This direct solution is quite important in practice, as, if Q and $p(t_0)$ are given numerically, the successive vectors $p(t_{r-s})$ $(s = r-1, ..., 0)$ are obtained by routine computational procedure. For constant Q we have simply

$$p(t_r) = Q^r p(t_0). \tag{7}$$

However, to study the structure and ultimate behaviour of the solution (7), it is often convenient to make use of standard expansions or 'spectral resolutions' of Q in terms of its latent or characteristic roots λ. The λ's are the roots of the equation

$$|\lambda I - Q| = 0, \tag{8}$$

where $|\lambda I - Q|$ is the determinant of the characteristic matrix $\Phi(\lambda) \equiv \lambda I - Q$, I denoting the unit matrix. The detailed expansion is simplest in the non-degenerate case when the λ's are all different. In this case we define for each λ_i a column and row vector s_i, t_i' respectively, where

$$Qs_i = \lambda_i s_i, \quad t_i' Q = \lambda_i t_i', \tag{9}$$

or, if $S \equiv (s_1, s_2, ..., s_k), \quad T \equiv (t_1, t_2, ..., t_k),$

$$QS = S\Lambda, \quad T'Q = \Lambda T', \tag{10}$$

† Some writers restrict the use of the word 'chain' to this case.

where $\mathbf{\Lambda}$ is the diagonal matrix of roots λ_i. Hence

$$\mathbf{Q} = \mathbf{S\Lambda S}^{-1} = (\mathbf{T}')^{-1}\mathbf{\Lambda T}'. \tag{11}$$

As
$$\mathbf{t}'_j \mathbf{Q}\mathbf{s}_i = \mathbf{t}'_j(\lambda_i \mathbf{s}_i) = (\mathbf{t}'_j \lambda_j)\,\mathbf{s}_i,$$

$$\mathbf{t}'_j \mathbf{s}_i = 0 \quad \text{for} \quad j \neq i \text{ (and } \lambda_i \neq \lambda_j).$$

We may choose the scale of $\mathbf{t}'_i$ and $\mathbf{s}_i$ so that $\mathbf{t}'_i \mathbf{s}_i = 1$, and then $\mathbf{T}'\mathbf{S} = \mathbf{I}$, and (11) becomes

$$\mathbf{Q} = \mathbf{S\Lambda T}' = \sum_{i=1}^{k} \lambda_i \mathbf{s}_i \mathbf{t}'_i. \tag{12}$$

The *spectral set* of matrices $\mathbf{A}_i \equiv \mathbf{s}_i \mathbf{t}'_i$ is easily seen to have the properties

$$\mathbf{A}_i \mathbf{A}_j = \begin{cases} 0 & (i \neq j), \\ \mathbf{A}_i & (i = j), \end{cases} \quad \text{and} \quad \sum_{i=1}^{k} \mathbf{A}_i = \mathbf{I},$$

the last equation following from the relations

$$\left(\sum_{i=1}^{k} \mathbf{s}_i \mathbf{t}'_i \right) \mathbf{s}_j = \mathbf{s}_j, \quad \mathbf{t}'_j \left(\sum_{i=1}^{k} \mathbf{s}_i \mathbf{t}'_i \right) = \mathbf{t}'_j,$$

whence the equation analogous to (11) reads

$$\sum_{i=1}^{k} \mathbf{A}_i = \mathbf{SIS}^{-1} = \mathbf{I}.$$

We now obtain
$$\mathbf{Q}^r = \sum_{i=1}^{k} \lambda_i^r \mathbf{A}_i. \tag{13}$$

As the $\mathbf{s}_i$ form a linearly independent set, we may express the initial vector $\mathbf{p}(t_0) \equiv \mathbf{p}_0$, say, in terms of them. In fact

$$\mathbf{p}_0 = \left(\sum_{i=1}^{k} \mathbf{s}_i \mathbf{t}'_i \right) \mathbf{p}_0 = \sum_{i=1}^{k} (\mathbf{t}'_i \mathbf{p}_0)\,\mathbf{s}_i.$$

Hence the solution $\mathbf{p}(t_r) \equiv \mathbf{p}_r$ has the structure

$$\mathbf{p}_r = \sum_{i=1}^{k} \alpha_i \lambda_i^r \mathbf{s}_i, \tag{14}$$

where $\alpha_i = \mathbf{t}'_i \mathbf{p}_0$. We may note further that

$$(1, 1, \ldots)\,\mathbf{Q} = (1, 1, \ldots),$$

since the sum of the elements in any column of Q is necessarily unity from its definition. Hence there is always one root $\lambda_1 = 1$, and correspondingly we may choose $t_1' \equiv (1, 1, \ldots)$. Since $t_1' p_0$ is also unity, $\alpha_1 = 1$. Thus (14) is of the form

$$p_r = s_1 + \sum_{i=2}^{k} \alpha_i \lambda_i^r s_i. \tag{15}$$

The root $\lambda_1 = 1$ cannot be exceeded by any other root. For as $|\Phi(\lambda)| = 0$, there is at least one non-trivial solution of the equation $t'(Q - \lambda I) = 0$. Let x_m be the maximum element of $\|x\|$ for any x, where $\|x\|$ denotes the absolute values of the components of x. Then

$$t_m \geqslant (t'Q)_m = \|\lambda\| t_m,$$

whence $\|\lambda\| \leqslant 1$ (cf. Fréchet, 1937–8, p. 105).

From (15) it follows that for the case of simple roots the latent vector s_1 will give a limiting distribution for p_r (independent of p_0) as $r \to \infty$, provided that no other latent roots have modulus equal to one. A more general discussion of the asymptotic behaviour of p_r will be given in § 2·21. Still considering for the moment the case of simple roots, we note the identity

$$\Phi(\lambda_i) \operatorname{adj} \Phi(\lambda_i) = |\Phi(\lambda_i)| I = 0, \tag{16}$$

where $\operatorname{adj} \Phi(\lambda_i)$ denotes the adjoint of $\Phi(\lambda_i)$. But $\Phi(\lambda_i) s_i = 0$, whence a solution s_i can be obtained from any column of $\operatorname{adj} \Phi(\lambda_i)$. Alternatively, as any matrix Q satisfies its own characteristic equation,

$$\prod_{j=1}^{k} (Q - \lambda_j I) = (Q - \lambda_i I) \prod_{j \neq i} (Q - \lambda_j I) = 0, \tag{17}$$

whence s_i is proportional also to any column of $\prod_{j \neq i} (Q - \lambda_j I)$. Similarly t_i' can be taken proportional to any row of either matrix, with the adjustment $t_i' s_i = 1$.

If $Q' = Q$, it is evident that s_1' is proportional to t_1', and in the limiting distribution *all states are equally probable*.

Example 1. As an example consider a very simple model of an individual learning to make a correct response to a certain stimulus. We postulate (cf. Bush and Mosteller, 1951) that two possible responses to the stimulus are governed by the matrix

$$Q = \begin{pmatrix} 1 - b & a \\ b & 1 - a \end{pmatrix},$$

where the *correct* response at any trial corresponds to the first state. The constants a and b depend on the nature of the successive responses and the learning process. Thus if $a = b = \frac{1}{2}$, each response is independent of previous responses; if $\frac{1}{2} < a = b < 1$, there is a tendency to oscillate from one response to the other. If now we take $a > b$, it implies a higher 'transition probability' from an incorrect to a correct response than vice versa, due to some kind of preferential treatment for correct responses. We readily find, consistently with these ideas, that the expansion (12) reads

$$\begin{pmatrix} 1-b & a \\ b & 1-a \end{pmatrix} = \begin{pmatrix} a/(a+b) \\ b/(a+b) \end{pmatrix} (1, \quad 1) + (1-a-b) \begin{pmatrix} 1 \\ -1 \end{pmatrix} (b/(a+b), -a/(a+b)),$$

whence for the probability of a correct response at the rth trial we obtain

$$\frac{a}{a+b} + \frac{b}{a+b} (1-a-b)^r \quad \text{or} \quad \frac{a}{a+b} - \frac{a}{a+b} (1-a-b)^r,$$

depending on whether the response at the zeroth trial was correct or not. Unless a and b are both unity, there is a limiting probability $a/(a+b)$ of a correct response. When $a = b = 1$, we obtain the extreme case of oscillation from one state to the other, and no limiting probability exists. Notice that unless b can be reduced to zero by inhibition of the tendency to revert to the incorrect response, the limiting probability of a correct response does not reach unity.

Example 2. As a slightly more complicated example, consider an application to the theory of inbreeding. The most powerful form of inbreeding possible in the selection for a single pair of allelomorphs A and a (alternative genes at the same chromosome locus) is by self-fertilization, possible with some plants. The progeny of a mating $Aa \times Aa$ is represented by

$$(\tfrac{1}{2}A + \tfrac{1}{2}a)(\tfrac{1}{2}A + \tfrac{1}{2}a) = \tfrac{1}{2}Aa + (\tfrac{1}{4}AA + \tfrac{1}{4}aa),$$

or 50 % heterozygotes Aa, the proportion of heterozygotes being reduced by one-half at each generation. For one of the next most powerful methods, practised with some animals, of brother-sister mating, we may classify the different possible situations in terms of the possible matings. From a mating of the progeny of the mating $Aa \times Aa$ we obtain

$$(\tfrac{1}{4}AA + \tfrac{1}{2}Aa + \tfrac{1}{4}aa) \times (\tfrac{1}{4}AA + \tfrac{1}{2}Aa + \tfrac{1}{4}aa).$$

Similarly, from $Aa \times aa$ we obtain in the next generation matings

$$(\tfrac{1}{2}A + \tfrac{1}{2}a) a \times (\tfrac{1}{2}A + \tfrac{1}{2}a) a.$$

In this way our classification gives the table shown, which we treat as our Q-matrix. (The bracketed mating types are entirely equivalent in relation to the heterozygotes Aa, and thus need not be separated. The pooling of states in a Markov chain is not, however, generally admissible, as will be seen later.)

For the latent roots we have the equation

$$(\lambda - 1)(\lambda - \tfrac{1}{4})(\lambda^2 - \tfrac{1}{2}\lambda - \tfrac{1}{4}) = 0,$$

or

$$\lambda_1 = 1, \quad \lambda_{2,3} = \tfrac{1}{4} \pm \tfrac{1}{4}\sqrt{5}, \quad \lambda_4 = \tfrac{1}{4}.$$

Progeny mating \ Parent mating	$aa \times aa$ $AA \times AA$	$aa \times Aa$ $AA \times Aa$	$Aa \times Aa$	$AA \times aa$
$aa \times aa$ $\}$ $AA \times AA$	1	$\tfrac{1}{4}$	$\tfrac{1}{8}$	0
$aa \times Aa$ $\}$ $AA \times Aa$	0	$\tfrac{1}{2}$	$\tfrac{1}{2}$	0
$Aa \times Aa$	0	$\tfrac{1}{4}$	$\tfrac{1}{4}$	1
$AA \times aa$	0	0	$\tfrac{1}{8}$	0

In this problem we are most interested in the proportion of heterozygotes in any generation. If we call the probabilities for the classes of mating in the table p, q, r and s (in that order), the proportion of heterozygotes will be $h = r + \tfrac{1}{2}q$. Hence, without finding the complete solution for p, q, r and s, we may write for h

$$h_j = A(\tfrac{1}{4} + \tfrac{1}{4}\sqrt{5})^j + B(\tfrac{1}{4} - \tfrac{1}{4}\sqrt{5})^j + C(\tfrac{1}{4})^j,$$

and determine A, B and C from the values h_0, h_1 and h_2, which are readily found to be 1, $\tfrac{1}{2}$, $\tfrac{1}{4}$ when $r_0 = 1$. (In the above formula for h_j advantage has been taken of the obvious fact that there will be no term $D(1)^j$ in h_j, but this of course need not be assumed.) Thus we obtain

$$h_j = \frac{2}{\sqrt{5}} \{(\tfrac{1}{4} + \tfrac{1}{4}\sqrt{5})^{j+1} - (\tfrac{1}{4} - \tfrac{1}{4}\sqrt{5})^{j+1}\}.$$

It should perhaps be pointed out that this type of solution gives the probability distribution of the various genotypes, or in other words, the *expected* proportions when the numbers are large. The sampling *fluctuations* from such expected numbers occurring in any actual inbreeding system is also a stochastic process problem, but one requiring for its investigation a more complete stochastic specification of the system, such as the distribution of numbers of progeny in a family (cf. Bartlett, 1937).

2·21 Classification by asymptotic behaviour.

In the first of the preceding two examples we saw that an oscillating or periodic solution was possible. The extension of this type of solution to k states is provided by a *deterministic* or *permuting* set of transitions from one state to another. We suppose that $\mathbf{Q}$ does not split up into closed groups of permutations, but that each state is occupied in turn. By a suitable numbering of the states the matrix Q is then

$$\begin{pmatrix} 0 & 1 & 0 & \ldots \\ 0 & 0 & 1 & \ldots \\ \vdots & \vdots & \vdots & \\ 1 & 0 & 0 & \ldots \end{pmatrix},$$

with characteristic equation $\lambda^k = 1$, and latent roots

$$\lambda_j \equiv \omega_j = e^{2\pi i j/k} \quad (i = \sqrt{-1},\ j = 1, \ldots, k). \tag{1}$$

Thus

$$\mathbf{p}_r = \sum_{j=1}^{k} \alpha_j \omega_j^r \mathbf{s}_j, \tag{2}$$

permanently representing a process in which each element of the initial probability distribution moves deterministically and cyclically through the k states, the whole cycle being repeated indefinitely.

By evaluating $\mathrm{adj}\,(Q - \omega_j I)$, we find (as is directly obvious) that $\mathbf{s}_j'$ is proportional to

$$(1, \omega_j, \omega_j^2, \ldots, \omega_j^{k-1}),$$

and $\mathbf{t}_j'$ to

$$(1, \omega_j^{-1}, \omega_j^{-2}, \ldots, \omega_j^{-(k-1)}),$$

i.e. to

$$(1, \omega_j^*, \omega_j^{*2}, \ldots, \omega_j^{*(k-1)}),$$

where ω_j^* is the complex conjugate of ω_j. Thus in this case $\mathbf{t}_j$ is the complex conjugate of $\mathbf{s}_j$, provided each is adjusted in scale by the same factor $1/\sqrt{k}$.

From (2) we see that a limiting distribution in the strict sense does not exist, though it does if we average over a large enough number of consecutive times, i.e.

$$\lim_{r \to \infty} \frac{1}{r} \sum_{j=1}^{r} \mathbf{p}_j = \boldsymbol{\pi}_\infty, \tag{3}$$

say, where obviously $\boldsymbol{\pi}_\infty$ satisfies the equation

$$(Q - I)\,\boldsymbol{\pi}_\infty = 0, \tag{4}$$

and represents here a uniform distribution.

Let us consider now the asymptotic behaviour of the general finite chain. From the relations $|\,\mathbf{\Phi}(\mathbf{Q})\,| = 0$ and $\mathbf{Q}\mathbf{p}_r = \mathbf{p}_{r+1}$, we see that $\mathbf{p}_r$ satisfies the kth order difference equation

$$|\,\mathbf{\Phi}(E_r)\,|\,\mathbf{p}_r = 0,$$

where $E_r \equiv 1 + \Delta$ denotes the 'shift' operator in the calculus of finite differences. It follows (see, for example, Milne-Thomson (1933), Chapter XIII) that the solution $\mathbf{p}_r$ has the general form

$$\mathbf{p}_r = \Sigma_i\, \mathbf{q}_i(r)\, \lambda_i^r, \tag{5}$$

where the summation is over all the distinct non-zero latent roots λ_i of $\mathbf{Q}$, and the $\mathbf{q}_i(r)$ are vectors whose components are polynomials in r of degree at most m_i, where m_i is the multiplicity of λ_i. The expression (5) will of course result from the operation $\mathbf{Q}^r$ on $\mathbf{p}_0$, where $\mathbf{Q}^r$ is in general (Frazer, Duncan and Collar, 1946, §3·10)

$$\mathbf{Q}^r = \Sigma_i \frac{1}{(m_i - 1)!} \left[\frac{d^{m_i-1}}{d\lambda^{m_i-1}} \frac{\lambda^r \operatorname{adj} \mathbf{\Phi}(\lambda)}{\psi_i(\lambda)} \right]_{\lambda = \lambda_i}, \tag{6}$$

with $$\psi_i(\lambda) = \prod_{j \neq i} (\lambda - \lambda_j)^{m_j}.$$

Equation (5) only holds for $r \geqslant m_0$, if zero is a latent root with multiplicity m_0, but we are going to make use of it for $r \to \infty$. Let r^m be the highest power of r in the set of polynomials associated with roots of unit modulus. As $\mathbf{p}_r$ always remains a probability distribution (from the definition of $\mathbf{Q}^r$), its elements are bounded. Consequently the coefficient (a trigonometric polynomial) of r^m must vanish identically unless $m = 0$. Thus as $r \to \infty$ each element of $\mathbf{p}_r$ must tend to a trigonometric polynomial; it follows that a limiting distribution in the extended sense of equation (3) always exists. If there are no roots of unit modulus (other than 1 itself), the trigonometric polynomials reduce to constants, and hence the strict limit $\mathbf{p}_\infty$ exists, whether or not the roots λ_i are simple. In general, however, this limiting distribution will not be independent of $\mathbf{p}_0$. (The above argument is due to D. G. Kendall.)

A detailed classification of the various possibilities has been given by Fréchet (1937-8) (cf. also Feller, 1950). We shall conclude this section by enumerating (without proof) some of the more important types and properties.

(i) We call the process *ergodic* if $\boldsymbol{\pi}_\infty$ (defined by equation (3) above) is independent of $\mathbf{p}_0$. This is true if and only if the matrix $\boldsymbol{\Phi}(1)$ is of rank $k-1$ (or equivalently, if the latent root 1 is simple).

(ii) The process is called *regular* if $\mathbf{p}_\infty$ exists and is independent of $\mathbf{p}_0$. In terms of the latent roots λ_i this is true if and only if there is only one simple root $\lambda_1 = 1$ of modulus unity. An alternative necessary and sufficient condition is that $\mathbf{Q}^r$ for some finite r has at least one row with all non-zero elements.

(iii) The *positively regular* case has, moreover, all positive (non-zero) elements in its ultimate distribution. An important *sufficient* condition is that $\mathbf{Q}$ has all non-zero elements. More generally, a necessary and sufficient condition is that $\mathbf{Q}^r$ for some finite r has all non-zero elements.

To see further how these cases arise in relation to the structure of the matrix $\mathbf{Q}$, we classify the possible matrix types:

(a) The matrix $\mathbf{Q}$ we call *reducible* if it can be put in the form

$$\left(\begin{array}{c|c} \mathbf{P} & \mathbf{R} \\ \hline \mathbf{0} & \mathbf{S} \end{array} \right),$$

where $\mathbf{P}$ is a square submatrix and $\mathbf{0}$ denotes a submatrix of zeros. If $\mathbf{P}$ cannot be further so reduced, it forms an irreducible closed set.

(b) The general canonical form for a matrix with a latent root 1 of multiplicity n (>1) is, after a suitable numbering of states,

$$\left(\begin{array}{cccc|c} \mathbf{P}_1 & \mathbf{0} & \cdots & \mathbf{0} & \\ \mathbf{0} & \mathbf{P}_2 & \cdots & \mathbf{0} & \\ \vdots & \vdots & & \vdots & \mathbf{R} \\ \mathbf{0} & \mathbf{0} & \cdots & \mathbf{P}_n & \\ \hline & \mathbf{0} & & & \mathbf{S} \end{array} \right).$$

The submatrices $\mathbf{P}_i$ correspond to *closed sets* or *final groups* of states (an 'absorbing' state, which permits entry but no exit, provides a final group of one state).

(c) We call $\mathbf{Q}$ *primitive* if it has no roots λ_i of modulus unity other than the root 1. The canonical form for irreducible but non-primitive matrices (or submatrices) is

$$
\begin{pmatrix}
0 & \mathbf{Q}_1 & 0 & \ldots & 0 \\
0 & 0 & \mathbf{Q}_2 & \ldots & 0 \\
\vdots & \vdots & \vdots & & \vdots \\
0 & 0 & 0 & \ldots & \mathbf{Q}_{l-1} \\
\mathbf{Q}_l & 0 & 0 & \ldots & 0
\end{pmatrix},
$$

where $\mathbf{Q}_1 \mathbf{Q}_2 \ldots \mathbf{Q}_l$ exists. Such a matrix is called *cyclic*.

We recall that the ultimate transition matrix $\mathbf{Q}^\infty$ exists if $\mathbf{Q}$ is primitive, though the ultimate distribution $\mathbf{Q}^\infty \mathbf{p}_0 \equiv \mathbf{p}_\infty$ is only independent of $\mathbf{p}_0$ in the regular case. In terms of the above matrix classification, the important positively regular case holds if and only if $\mathbf{Q}$ is both irreducible and primitive. A useful necessary and sufficient condition for the positively regular case among primitive matrices (no cyclic groups) is that all 'paths' from any state i to any other state j (including i) are possible (i.e. have non-zero probability for some finite number r of time-intervals).

Infinite number of states. The above discussion has referred to a finite number of states, whereas the set of possible states in a Markov chain may sometimes be enumerably infinite. The limits defined in equation (3) have been shown still to exist under these wider conditions, but a new feature is that the sum of the limits (which in general depend on the initial distribution) will not necessarily be unity. If it always is, the process may be called *non-dissipative*.

An alternative approach, due to Feller, to the asymptotic behaviour of $\mathbf{p}_r$, enabling the enumerable case to be included, makes use of recurrence theory, which will be developed in Chapter 3.

2·22   Nearest neighbour systems. An important application of Markov chain technique is in the statistical mechanics of 'nearest neighbour' systems, used, for example, as models for crystal lattices (see Montroll (1947), and other references given by him). We shall, however, consider here only 'linear' or one-dimensional systems. Physical systems are more often two- and three-dimensional, and their detailed treatment becomes complicated, though they may in principle be regarded as linear systems with a one- or two-dimensional layer of elements (see also § 6·54).

From the probability standpoint we define a linear nearest neighbour system in general as a one-dimensional array of elements with each of which a random variable X is associated, the conditional distribution of any set of X's, say $S \equiv X_1, ..., X_n$, depending only on the two nearest values known, one to the left and one to the right. For definiteness we shall in the numbering of these variables think of a chain evolving from the left to right. In a Markov chain X_r let any set of X's up to t_0 be denoted by $U \equiv (U', X_0)$ and any set after t_n by $V \equiv (X_{n+1}, V')$. Then

$$P\{S \mid U, V\} = \frac{P\{U, S, V\}}{P\{U, V\}} = \frac{P\{V' \mid X_{n+1}, S, U\} P\{X_{n+1}, S \mid U\} P\{U\}}{P\{V' \mid X_{n+1}, U\} P\{X_{n+1} \mid X_0, U'\} P\{U\}},$$

or from the Markov chain property for the ordered set U', X_0, S, X_{n+1}, V',

$$P\{S \mid U, V\} = \frac{P\{X_{n+1}, S \mid X_0\}}{P\{X_{n+1} \mid X_0\}} = P\{S \mid X_0, X_{n+1}\}, \qquad (1)$$

which is the nearest neighbour property. Conversely, it may be shown by similar arguments that in general the *only* linear nearest neighbour systems as above defined are Markov chains.

We may similarly deduce that a Markov chain preserves its Markovian character if reversed in time. For

$$P\{S \mid V\} = \frac{P\{S, V\}}{P\{V\}} = \frac{P\{V' \mid X_{n+1}, S\} P\{X_{n+1}, S\}}{P\{V' \mid X_{n+1}\} P\{X_{n+1}\}}$$

$$= \frac{P\{X_{n+1}, S\}}{P\{X_{n+1}\}} = P\{S \mid X_{n+1}\}. \qquad (2)$$

It should be pointed out that the conditional probabilities for this reversed process do not in general bear a simple relation to the forward conditional probabilities, for they will depend also on the absolute probabilities of the various states at the different instants. However, if we consider only *stationary* Markov chains (for definition of stationarity see § 1·3) the absolute probabilities will be independent of the time, i.e. in the present application to nearest neighbour systems, independent of the element of the system. If $\{q_{ij}\} \equiv \mathbf{Q}$ is the 'forward' conditional probability matrix, and $\{r_{ij}\} \equiv \mathbf{R}$ is the corresponding matrix for the reversed process, the simultaneous probability of $X_n = j$, $X_{n+1} = i$, is $q_{ij} P_j = r_{ji} P_i$. It is reasonable to suppose in the present context that $r_{ji} = q_{ji}$; it then follows that $P_j / P_i = q_{ji} / q_{ij}$. A particular case mentioned in § 2·2 was a symmetric matrix $\mathbf{Q}$, for which $q_{ji} = q_{ij}$. The above relation then gives $P_j = P_i$, consistently with the uniform probability distribution noted for this case.

A feature of importance in the physical applications is the existence of 'long-range order', its absence corresponding to the independence of the probabilities of particular states of the elements of the system far apart. We have seen that this will ultimately follow between the zeroth and rth elements for large r if the root $\lambda_1 = 1$ is simple and dominant. More precisely, the extent of the dependence will be related to the magnitude of the second largest root (in absolute value), λ_2, say.

A further use of the technique in the physical applications is based on the following method of obtaining average values. Suppose a function $G_n(X_0, X_1, ..., X_n)$ associated with a Markov chain $X_0, X_1, ..., X_n$ is of the *product* form

$$g(X_0, X_1) g(X_1, X_2) \dots g(X_{n-1}, X_n).$$

Now by definition

$$E_n\{G_n\} = E_{n-1}\{G_{n-1} E_{n \mid n-1}\{g(X_{n-1}, X_n)\}\}, \tag{3}$$

where E_n denotes averaging with respect to all the variables $X_0, X_1, ..., X_n$, and $E_{n \mid n-1}$ conditional averaging over X_n for given $X_0, X_1, ..., X_{n-1}$. For a Markov chain the right-hand side of (3) has the same structure as the evaluation of $\mathbf{p}_n$ with $p(x_r \mid x_{r-1}) g(x_{r-1}, x_r)$ in place of $p(x_r \mid x_{r-1})$. Hence if we denote the new matrix replacing $\mathbf{Q}$ by $\mathbf{R}$, then after final summation over the $\mathbf{p}_0$ distribution for X_0,

$$E_n\{G_n\} = \mathbf{t}_1' \mathbf{R}^n \mathbf{p}_0, \tag{4}$$

where $\mathbf{t}_1'$ denotes $(1, 1, ...)$ as before. If in particular $\mathbf{R}$ has k simple roots μ_i and hence a spectral resolution

$$\mathbf{R} = \sum_{i=1}^{k} \mu_i \mathbf{u}_i \mathbf{v}_i', \tag{5}$$

and if
$$\mathbf{p}_0 = \sum_{i=1}^{k} \beta_i \mathbf{u}_i,.$$

then
$$E_n\{G_n\} = \mathbf{t}_1' \sum_{i=1}^{k} \beta_i \mu_i^n \mathbf{u}_i. \tag{6}$$

Applying this technique to the function $\exp(\theta F_n)$, where F_n is a function of the *sum* form

$$h(X_0, X_1) + h(X_1, X_2) + \ldots + h(X_{n-1}, X_n),$$

we note that the complete distribution of F_n (through its m.g.f.) may be investigated. In this case, since $G_n \equiv \exp(\theta F_n)$ is a function of θ with $G_n(0) = 1$, we may conveniently write (4) and (6) in the respective forms

$$E_n\{\exp(\theta F_n)\} = \mathbf{t}_1' \mathbf{Q}^n(\theta) \mathbf{p}_0 \tag{7}$$

$$= \mathbf{t}_1' \sum_{i=1}^{k} \alpha_i(\theta) \lambda_i^n(\theta) \mathbf{s}_i(\theta), \tag{8}$$

where $\mathbf{Q}(0) \equiv \mathbf{Q}$, $\lambda_i(0) \equiv \lambda_i$, etc. (It is possible to apply this technique also to the case of simultaneous variables F_{jn} $(j = 1, 2, \ldots)$; see, for example, §8·2.)

We assume that the process is regular, so that the moduli of all roots $\lambda_i(0)$, other than the simple root $\lambda_1(0) = 1$, are less than one. With this condition the cumulant function may for small θ as n increases be written in the asymptotic form

$$K_n(\theta) \sim n \log \lambda_1(\theta) + \log[\alpha_1(\theta) \mathbf{t}_1' \mathbf{s}_1(\theta)] \tag{9}$$

$$\sim n \log \lambda_1(\theta) \quad (\lambda_1(\theta) \neq 1). \tag{10}$$

Even if (7) is not expressible in the simple form (8), this last asymptotic result (10) still follows under the regularity condition (from the general form of Sylvester's theorem; see, for example, Frazer, Duncan and Collar, 1946, §4·15). This result shows that $K_n(\theta)$ may be regarded, asymptotically, as the cumulant function of a sum of equal independent components. Hence the distribution of F_n tends to normality under these conditions, provided that its variance is such that σ_n^2/n is finite and non-zero, and F_n is suitably scaled.

As a simple example of this method, consider a chain of trials with absolute probability P of an event occurring at any trial, but a Markov chain dependence between successive trials with conditional or 'transition' probability matrix

$$\mathbf{Q} = \begin{pmatrix} p_1 & q_2 \\ q_1 & p_2 \end{pmatrix}.$$

The number of 'transitions', treating the juxtaposition of the occurrence and non-occurrence of the event as a transition, can be studied by constructing an F_n with contribution 0 if no transition occurs and 1 otherwise (e.g. if $X_r = 1$ if the event occurs and 0 otherwise, take

$$h(X_r, X_{r+1}) = X_r + X_{r+1} - 2X_r X_{r+1}).$$

The modified matrix becomes

$$\mathbf{Q}(\theta) = \begin{pmatrix} p_1 & q_2 e^\theta \\ q_1 e^\theta & p_2 \end{pmatrix},$$

with latent roots

$$\lambda = \tfrac{1}{2}(p_1 + p_2) \pm \tfrac{1}{2} \sqrt{\{(p_1 - p_2)^2 + 4q_1 q_2 e^{2\theta}\}}.$$

The square root with the $+$ sign gives the root $\lambda_1(\theta)$ with larger modulus, so that, *provided neither q_1 nor q_2 is zero*, the distribution of the total number of transitions tends to normality with mean nm and variance $n\sigma^2$, where m and σ^2 are the coefficients of θ and $\tfrac{1}{2}\theta^2$ in the expansion of $\log \lambda_1(\theta)$; that is,

$$m = \frac{2q_1 q_2}{q_1 + q_2}, \quad \sigma^2 = \frac{4q_1 q_2 (q_1^2 p_2 + q_2^2 p_1)}{(q_1 + q_2)^3}.$$

When, however, either q_1 or q_2, say q_1, is zero, $\lambda_1(\theta) = 1$, and the condition for a limiting normal distribution breaks down. It is directly obvious that when $q_1 = 0$ the event will always occur, once it has occurred at all, so that the quantity F_n has then only two possible values 0 and 1. In this case the exact expression for $E\{\exp(\theta F_n)\}$ is $P + Q\{p_2^n + (1 - p_2^n) e^\theta\}$ (where $Q \equiv 1 - P$). Such a result will of course be obtained by the present technique, for we readily find

$$\lambda_1(\theta) = 1, \qquad \lambda_2(\theta) = p_2,$$
$$\mathbf{s}_1'(\theta) = (1, 0), \qquad \mathbf{s}_2'(\theta) = (e^\theta, -1),$$
$$\mathbf{t}_1'(\theta) = (1, e^\theta), \qquad \mathbf{t}_2'(\theta) = (0, -1),$$
$$(P, Q) = (P + Q e^\theta)\, \mathbf{s}_1'(\theta) - Q \mathbf{s}_2'(\theta),$$

whence the result follows. For large n, it becomes $P + Qe^\theta$, which again is given by the asymptotic formula (9) with

$$\log \lambda_1(\theta) = 0.$$

In this example on the number of transitions, the exceptional case arises from the 'freezing' or 'absorption' of the variable into a particular state once it has reached it; it is excluded if we modify the condition of regularity to that of positive regularity.

In the applications in statistical mechanics the possible simultaneous configurations or states $S_r \equiv (X_0, X_1, ..., X_n)$ of the system have associated energies ϵ_r, and the configuration probabilities are assumed proportional to

$$\exp(\vartheta \epsilon_r), \quad \text{where} \quad \vartheta \equiv -1/(kT)$$

(in this expression k is Boltzmann's constant and T is the absolute temperature). The method of studying the thermodynamic properties of the system is to form the partition function

$$Z(\vartheta) = \Sigma_r \exp(\vartheta \epsilon_r), \tag{11}$$

this being equivalent to obtaining the m.g.f. of the total energy.

For $\qquad\qquad E\{\exp(\theta \epsilon_r)\} = Z(\vartheta + \theta)/Z(\vartheta),$

whence $\qquad\qquad K(\theta) = \log Z(\vartheta + \theta) - \log Z(\vartheta),$

and differentiating $K(\theta)$ with respect to θ and then putting $\theta = 0$ is equivalent to differentiating $\log Z(\vartheta)$ with respect to ϑ. When it is assumed that ϵ_r is a potential energy function of the form

$$\epsilon_r \equiv V(X_0, ..., X_n) = v(X_0, X_1) + v(X_1, X_2) + ... + v(X_{n-1}, X_n),$$

the summation in (11) may be treated like the averaging of the function $\exp(\theta F_n)$ previously considered. The form of the simultaneous configuration probability assumed is in fact equivalent to taking the symmetric form $\exp\{\vartheta v(X_{r-1}, X_r)\}$ for the simultaneous probability $P\{X_{r-1}, X_r\}$ (apart from a constant factor), compatible with a stationary Markov chain in either direction.

2·23   Wald's identity for Markov chains.

We shall have occasion to note later (see § 4·1) that the identity established in § 2·1 for the random walk with absorbing boundaries is not restricted to discrete time. Another useful extension, to Markov chains, is outlined below. Consider the case

$$S_n = X_1 + X_2 + \dots X_n,$$

where X_s is of the form $h(Z_{s-1}, Z_s)$, Z_s being a Markov chain with a transition probability matrix $\mathbf{Q}$ of the type treated in §2·2. As in §2·22, we now extend $\mathbf{Q} \equiv \{q_{ij}\}$, where

$$q_{ij} = P\{Z_s = z_i | Z_{s-1} = z_j\},$$

to $$\mathbf{Q}(\theta) \equiv \{q_{ij}(\theta)\} = \{q_{ij} \exp [\theta h(z_j, z_i)]\}$$

and we assume that $\mathbf{Q}(\theta)$ has a spectral expansion

$$\mathbf{Q}(\theta) = \sum_{r=1}^{k} \lambda_r(\theta)\, \mathbf{s}_r(\theta)\, \mathbf{t}'_r(\theta),$$

where $\lambda_1(0) = \lambda_1 = 1$. Then

$$E\{e^{\theta S_n}\} = \mathbf{t}'_1 \sum_{r=1}^{k} \alpha_r(\theta)\, \lambda_r^n(\theta)\, \mathbf{s}_r(\theta)$$

(equation (8) in §2·22), where $\mathbf{t}_1 = \mathbf{t}_1(0)$ and $\alpha_r(\theta) = \mathbf{t}'_r(\theta)\, \mathbf{p}_0$, $\mathbf{p}_0$ being the initial distribution of Z_0. Then under appropriate conditions the identity becomes

$$E\{[\lambda_1(\theta)]^{-N} e^{\theta S_N} \beta(\theta | X_N)/\alpha_1(\theta)\} = 1, \qquad (1)$$

where $\beta(\theta | X_N)$ is the component of $\mathbf{t}_1(\theta)$ corresponding to the value of X_N when the walk S_N terminated. This result was first indicated by Bellman (1957) in the special case $X_s \equiv Z_s$; but for a detailed derivation of the general result (1) reference may be made to Miller (1962) or Phatarfod (1965). Here we shall merely make one or two relevant comments on the use of (1):

(a) The role of $M(\theta)$ in the independence case is taken over by $\lambda_1(\theta)$, for which we require $|\lambda_1(\theta)| \geqslant 1$.

(b) The identity is exact when the factor $\beta(\theta | X_N)/\alpha_1(\theta)$ is included, depending on the terminating increment X_N. In some problems this will be known, but in most applications the asymptotic form of (1) is used, and $\beta(\theta | X_N)/\alpha_1(\theta) \to 1$ for small θ. This asymptotic form should be valid for more general spectral expansions, provided the process is regular and $\lambda_1(\theta)$ provides the asymptotic substitute for $M(\theta)$.

(c) It is interesting to notice that (1) even in its exact form does not depend on $\lambda_r(\theta)$, $r > 1$. Other identities involving the remaining roots have been obtained by Tweedie (1960), but are less relevant for applications.

As an example we take the gambler's ruin problem (§ 2·1), but now with dependence between successive games governed by the matrix (cf. top of p. 38)

$$Q = \begin{pmatrix} p_1 & q_2 \\ q_1 & p_2 \end{pmatrix},$$

where the first column corresponds to the probabilities after a win by the first gambler.

The modified matrix is

$$Q(\theta) = \begin{pmatrix} p_1 e^{\theta} & q_2 e^{-\theta} \\ q_1 e^{-\theta} & p_2 e^{\theta} \end{pmatrix},$$

and we have

$$\lambda_{1,2}(\theta) = \tfrac{1}{2}(p_1 e^{\theta} + p_2 e^{-\theta}) \pm \tfrac{1}{2}\sqrt{\{(p_1 e^{\theta} - p_2 e^{-\theta})^2 + 4q_1 q_2\}},$$

$$t_1'(\theta) = \{e^{-\theta}, [p_2 e^{-\theta} - \lambda_2(\theta)]/q_1\}, \quad \alpha_1(\theta) = e^{-\theta},$$

given, say, the initial condition that the first gambler wins his first game from his opponent ($X_1 = +1$). Moreover, if we include X_1 in S_N, the identity takes the form

$$E\{[\lambda_1(\theta)]^{-N+1} e^{\theta(S_N-1)}\beta(\theta|X_N)/\alpha_1(\theta)\} = 1. \tag{2}$$

If he ruins his opponent, we know X_N must have been $+1$, whereas if he is first ruined, $X_N = -1$. Hence, putting $\theta = \theta_0 \neq 0$ in (2) such that $\lambda_1(\theta) = 1$, i.e. $e^{\theta} = p_2/p_1$, we obtain (for the gamber's initial capital $-a$, and his opponent, b, as in § 2·1)

$$P_a e^{\theta_0(a-1)}\frac{\beta(\theta_0|-1)}{\alpha_1(\theta_0)} + P_b e^{\theta_0(b-1)}\frac{\beta(\theta_0|+1)}{\alpha_1(\theta_0)} = 1,$$

whence

$$P_b = \frac{1-(p_2/p_1)^a q_2/q_1}{(p_2/p_1)^{b-1} - (p_2/p_1)^a q_2/q_1}. \tag{3}$$

The characteristic function for the distribution of N may also be obtained by methods similar to those indicated in § 2·1.

2·3 Multiplicative chains

A Markov chain of particular importance that is best discussed directly is one where each 'state' of the variable X refers to the number of individuals existing at time t_r; the time t_{r+1}

corresponds to the next generation, which is obtained on the assumption that each individual of the rth generation independently gives rise to an entire probability distribution of possible numbers of offspring in the next generation. If we denote this distribution (for simplicity assumed constant) by its p.g.f. $G(z)$, the p.g.f. from two individuals is $G^2(z)$ (by the probability rules summarized in Chapter 1) from three $G^3(z)$, and so on. The recurrence relation for the resulting distribution of individuals in the rth generation is thus seen to be

$$\Pi_{r+1}(z) = \Pi_r(G(z)). \tag{1}$$

As Π_r is itself a further functional iteration of G (for $\Pi_0(z) \equiv z$), equation (1) can be equivalently written, for one initial individual,

$$\Pi_{r+1}(z) = G(\Pi_r(z)). \tag{2}$$

These equations are not in general explicitly soluble, but imply various important consequences. For the mean and variance of $\Pi_r(z)$ we have

$$m_r = \left[\frac{\partial \log \Pi_r(z)}{\partial \log z}\right]_{z=1}, \quad v_r = \left[\frac{\partial^2 \log \Pi_r(z)}{\partial(\log z)^2}\right]_{z=1}.$$

Hence from (1)

$$m_{r+1} = \left[\frac{\partial \log \Pi_r}{\partial \log G}\frac{\partial \log G}{\partial \log z}\right]_{z=1} = mm_r,$$

$$v_{r+1} = \left[\frac{\partial^2 \log \cdot \Pi_r}{\partial(\log G)^2}\left(\frac{\partial \log G}{\partial \log z}\right)^2 + \frac{\partial \log \Pi_r}{\partial \log G}\frac{\partial^2 \log G}{\partial(\log z)^2}\right]_{z=1}$$

$$= m^2 v_r + v m_r,$$

where m and v are the mean and variance of $G(z)$. This readily gives

$$\left.\begin{aligned} m_r &= m^r m_0, \\ v_r &= m^{2r} v_0 + m^{r-1}(1 - m^r)\, m_0 v/(1 - m). \end{aligned}\right\} \tag{3}$$

It will be noticed that m_r and v_r cannot both be constant except in the trivial case $m = 1$, $v = 0$. In the case $m = 1$, $v \neq 0$, the second term for v_r becomes $r m_0 v$, and increases indefinitely with r.

This theory of population fluctuations is linked with a problem investigated by Francis Galton at the end of the last century in connection with the extinction of family surnames. If each male individual in a population independently has a family containing n sons, where n has a distribution with p.g.f. $G(z)$, and so on for the next generation, what is the chance of any particular male line becoming extinct? A complete solution to this problem was not given until many years later, by Steffensen (1930).[†] The chance required obviously exists as a limit, since $\Pi_r(0)$ must increase with r and is bounded by 1. From the relation (1), since the sequence $0, G(0), G(G(0)), \ldots$ is steadily increasing (from the increasing character of $G(z)$ as z increases from 0 to 1) and has the minimum solution z_m of the equation $G(z) = z$ as its limit, we have for the required probability of extinction $\Pi_0(z_m)$, or if $\Pi_0(z) = z$, simply z_m. If $z_m = 1$, the probability of extinction is unity. As $\partial^2 G(z)/\partial z^2$ is necessarily positive for all z from 0 to 1, there is at most one root of $G(z) = z$ other than $z = 1$, and this differs from $z = 1$ only if $[\partial G(z)/\partial z]_{z=1} > 1$. *Thus the necessary and sufficient condition for certain extinction is* $m \leqslant 1$.

This chance of extinction was investigated by Lotka (1931) for the United States white population in 1920, and found to be nearly 0·9. For example, if for $G(z)$ he substituted the approximate expression $(0·482 - 0·041z)/(1 - 0·559z)$ fitted to the statistics of family sizes, the equation $G(z) = z$ became

$$0·482 - 1·041z + 0·559z^2 = (1-z)(0·482 - 0·559z) = 0,$$

giving $z_m = 0·482/0·559 = 0·86$. The average value $E(n)$ of n was correspondingly of course greater than 1, being 1·175.

Another application of these results occurs in the theory of natural selection. A gene mutation occurring in a finite number N of a species is certain to die out eventually if it confers no selective advantage. On the other hand, if there is a small advantage, so that $m = 1 + \epsilon$, we have an equation for z_m by expanding the equivalent equation $\log G(z) = \log z$ in powers of $\log z$,

$$\log z = \log G(1) + \log z \left[\frac{\partial \log G}{\partial \log z}\right]_{z=1} + \tfrac{1}{2}(\log z)^2 \left[\frac{\partial^2 \log G}{\partial(\log z)^2}\right]_{z=1} + \ldots.$$

This gives the approximate solution

$$\log z_m \sim 2(1 - m)/v = -2\epsilon/v,$$

or as $\Pi_0(z) = z^N$, the probability of extinction is approximately

† See also Bienaymé (1845).

$\exp(-2N\epsilon/v)$. For a Poisson distribution for $G(z)$, so that $v = m = 1 + \epsilon$, and $\epsilon = 0.01$, say, we obtain $(0.9804)^N$ or odds of over 100 to 1 *against* extinction if N is as much as 250 (the slightly more exact result $(0.9803)^N$ given by Fisher (1930) may be obtained by solving the equation $G(z) = z$ more exactly, given the Poisson p.g.f. $G(z) = e^{m(z-1)}$).

The fact that extinction is certain even when $m = 1$ stresses the danger of our assuming that the mean remains representative of an evolutionary stochastic process. The result is reminiscent of the gambler's ruin problem (§ 2·1), and indeed can be identified with it if we think of the contribution to the next generation from *each* individual as an independent addition to a random walk. In terms of successive generations, however, the number of such additional components is equal to the number of individuals in the population at that epoch, so a time to extinction reckoned in terms of random walk components is in general much longer than the time in terms of generations.

When $m < 1$, ultimate extinction may be avoided by an extra immigration component, so that the iteration relation for the cumulant function becomes, say (where $H(\theta) \equiv \log G(e^\theta)$),

$$K_{r+1}(\theta) = K_r(H(\theta)) + J(\theta). \tag{4}$$

The limiting distribution, if it exists, is then given by the functional relation

$$K_\infty(\theta) = K_\infty(H(\theta)) + J(\theta). \tag{5}$$

For example, its mean and variance, obtained by differentiation, are easily found to be

$$m_\infty = \frac{m_J}{1-m}, \quad v_\infty = \frac{v_J}{1-m^2} + \frac{m_J v}{(1-m)(1-m^2)}, \tag{6}$$

where m_J, v_J and m, v correspond to $J(\theta)$ and $H(\theta)$ respectively. These formulae reduce to those given by Haldane (1949) when $J(\theta)$ and $H(\theta)$ both represent Poisson distributions, so that $m_J = v_J$, $m = v$. It may also easily be seen from (6) that $v_\infty/m_\infty > 1$ if $v/m \geqslant 1$, $v_J/m_J \geqslant 1$, and a negative binomial distribution, defined by

$$\Pi(z) = [1 - \alpha(z-1)]^{-\beta} \quad (\alpha, \beta > 0), \tag{7}$$

is sometimes in the absence of an exact solution a useful approximation (with $\alpha = v_\infty/m_\infty - 1$, $\beta = m_\infty/\alpha$).

Returning to the case $m > 1$, with no immigration, we no longer have a strict limiting distribution for the number N_r of individuals, but we can consider the limiting distribution of the variable $U_r \equiv N_r/m^r$. From the m.g.f. equation equivalent to (2), viz.

$$M_{r+1}(\theta) = G(M_r(\theta)), \tag{8}$$

and the relation

$$L_r(\theta) \equiv E\{e^{\theta N_r/m^r}\} = E\{e^{N_r(\theta/m^r)}\} = M_r(\theta/m^r),$$

we obtain
$$L_{r+1}(\theta) = G(L_r(\theta/m)), \tag{9}$$

or a limiting distribution must satisfy (for one initial individual)

$$L(m\theta) = G(L(\theta)) \tag{10}$$

(it has been shown by Harris (1948) that such a limiting distribution exists and is uniquely determined by (10)). From equation (9) we find in particular

$$m_r = 1, \quad m^2 v_{r+1} = m v_r + v,$$

or
$$m_\infty = 1, \quad v_\infty = v/[m(m-1)]. \tag{11}$$

It will be recalled that even for $m > 1$ there is still a non-zero chance of extinction given by z_m, and for the remaining continuous component of the limiting distribution we have

$$m'_\infty = \frac{1}{1-z_m}, \quad v'_\infty = \frac{v}{m(m-1)(1-z_m)} - \frac{z_m}{(1-z_m)^2}, \tag{12}$$

values which could be used to obtain a rapid empirical fit, e.g. by a standard χ^2 distribution of the type $(2x = \chi^2)$

$$f(x)\,dx = x^{\gamma-1} e^{-x}\,dx/\Gamma(\gamma) \quad (\gamma > 0).$$

For example, in the case

$$G(z) = 0.4z + 0.6z^2,$$

for which $m = 1.6$, $v = 0.24$, $z_m = 0$, we find $v_\infty = 0.25$. The distribution of $x \equiv \frac{1}{2}\chi^2$ above has $m(x) = \gamma$, $v(x) = \gamma$, and hence we take $u \equiv \frac{1}{4}x$ as our variable, with $\gamma = 4$. This approximation is

Table 1

u	$P\{U \leqslant u\}$	1st diff.	Approximation	1st diff.
0	0·0000		0·0000	
0·25	0·0475	0·0475	0·0190	0·0190
0·50	0·1728	0·1253	0·1429	0·1239
0·75	0·3455	0·1727	0·3528	0·2099
1·00	0·5312	0·1857	0·5665	0·2137
1·25	0·6993	0·1681	0·7350	0·1685
1·50	0·8304	0·1311	0·8488	0·1138
1·75	0·9186	0·0882	0·9182	0·0694
2·00	0·9678	0·0492	0·9576	0·0394
2·50	0·9975	(0·0297)	0·9897	(0·0321)
3·00	0·9999	(0·0024)	0·9977	(0·0080)

compared in table 1 with the exact distribution given by Harris (1948), who computed this by first expanding the characteristic function as a power series and then inverting it.

More than one type of individual. The above methods are in principle immediately applicable to processes involving more than one type of individual. Thus equation (1) generalizes to

$$\Pi_{r+1}(\mathbf{z}) = \Pi_r(\mathbf{G}(\mathbf{z})), \qquad (13)$$

where the changes from one generation to the next are represented by the *vector* operation $\mathbf{z} \to \mathbf{G}(\mathbf{z})$, and Π denotes the vector or set of simultaneous p.g.f.'s arising from each possible type of initial individual. It is convenient to stress that (13) applies to each component of Π, to indicate its relation with the alternative extension of equation (2),

$$\Pi_{r+1}(\mathbf{z}) = \mathbf{G}(\Pi_r(\mathbf{z})), \qquad (14)$$

in which the use of the vector Π is essential. We shall see later when we consider corresponding equations in continuous time that the two alternative forms (13) and (14) *both* have important analogues.

2·31 Extinction probabilities for variable environments.

In the discussion earlier in this chapter on Markov chains the matrix product (equation (6), § 2·2) for determining the evolution of the chain and the distribution $p(t_r)$ at time t_r was in general not necessarily time-homogeneous (the transition matrix might, for example, vary with the season), though the subsequent theory concentrated on the time-homogeneous case. Similarly, the p.g.f. $G(z)$ (reverting for simplicity to one type of individual) in § 2·3 may more generally vary from one generation to another, representing for example the interaction with an environment which may vary with time.

Equation (1) of § 2·3 is then replaced by

$$\Pi_{r+1}(z) = \Pi_r(G_r(z)) \tag{1}$$

and in particular the chance of extinction by

$$\Pi_{r+1}(0) = \Pi_r(G_r(0)). \tag{2}$$

In one class of models of this kind $G_r(z)$ may be determined for each r as an independent random case from a countable set $\{G_r(z)\}$ with prescribed probabilities, ρ_r, say. Some relevant theorems for this class are noted (without proof) from Wilkinson (1969):

(A) Let q_1 be the probability of ultimate extinction if, initially, $N_0 = 1$, and suppose $\quad \Sigma_r p_r |\log m_r| < \infty,$

where $m_r = [\partial G_r(z)/\partial z]_{z=1} < \infty$. Then a N. & S.C. that $q_1 < 1$ is

$$\Sigma_r p_r \log m_r > 0 \quad \text{and} \quad \Sigma_r p_r |\log(1 - G_r(0))| < \infty. \tag{3}$$

(B) For $N_0 = s$ let the probability of ultimate extinction be q_s. Denote $\Pi_r(z|N_0 = s)$ by $\Pi_r^{(s)}(z)$. Then

$$\Pi_r^{(s)}(z) \to q_s \leqslant 1 \quad \text{as} \quad r \to \infty \quad (z \text{ in } 0, 1). \tag{4}$$

(C) Define the auxiliary stochastic process $\{X_n\}$ by

$$X_0 = z_0 \text{ (in } 0, 1), \quad X_{r+1} = G_r(X_r).$$

Then
$$E\{X_n^s | X_0 = z_0\} \to q_s. \tag{5}$$

Because $0 \leqslant X_n \leqslant 1$, this result ensures that $X_n \to X$ in distribution (independent of $z_0 < 1$), such that $E\{X^s\} = q_s$. Denote

$$P\{X \leqslant x\} \text{ by } F(x).$$

(D) If the conditions (3) do not hold, then $q_s = 1$ for all s. If they do hold, then

$$q_s = \int_0^1 x^s dF(x) \to 0 \quad \text{as} \quad s \to \infty. \tag{6}$$

Notice that while no longer is $q_s = q_1^s$ in general, the sequence $\{q_s\}$ forms a *moment* sequence.

Unfortunately these theorems do not provide a simple solution for the sequence $\{q_s\}$. The q_s may in fact be seen to satisfy the set of equations

$$\left.\begin{aligned} q_1 &= p_0^{(1)} + p_1^{(1)} q_1 + p_2^{(1)} q_2 + \dots \\ q_2 &= p_0^{(2)} + p_1^{(2)} q_1 + p_2^{(2)} q_2 + \dots \\ &\vdots \end{aligned}\right\} \tag{7}$$

where $\qquad p_i^{(s)} = P\{N_1 = i \,|\, N_0 = s\}.$

It may, however, be possible, under the conditions for (6) to hold, to curtail these equations and hence solve them approximately. If we denote the curtailed equations (7) up to q_n by

$$(\mathbf{I}_n - \mathbf{P}_n)\,\mathbf{q}_n = \mathbf{p}_n,$$

where $\mathbf{p}_n' \equiv (p_1^{(1)}, p_0^{(2)}, \dots, p_0^{(n)})$, $\mathbf{I}_n$ is the $n \times n$ unit matrix and $\mathbf{P}_n = (p_j^{(i)})$, then

$$\mathbf{q}_n \sim (\mathbf{I}_n - \mathbf{P}_n)^{-1}\,\mathbf{p}_n. \tag{8}$$

More precisely, it is shown by Wilkinson (1969) that

$$(\mathbf{I}_n - \mathbf{P}_n)^{-1}\mathbf{p}_n \leqslant \mathbf{q}_n \leqslant (\mathbf{I}_n - \mathbf{P}_n)^{-1}\mathbf{p}_n + \gamma_{n+1}\mathbf{1}, \tag{9}$$

where $\{\gamma_i\}$ is a sequence of real numbers tending to zero, such that $q_j \leqslant \gamma_j$, and $\mathbf{1}$ is the vector with components unity. If

$$G_r(z) \leqslant z \quad \text{for} \quad z \geqslant \gamma, \tag{10}$$

which implies that $m_r > 1$ for all r, then we may use in (9) the sequence

$$\gamma_{n+1} = \gamma^{n+1}. \tag{11}$$

For the example

$$G_1(z) = \tfrac{1}{4} + \tfrac{3}{4}z^2, \quad G_2(z) = \tfrac{1}{3}z + \tfrac{2}{3}z^2, \quad p_1 = p_2 = \tfrac{1}{2},$$

Wilkinson gives for $n = 8$ in (9)

$$q_1 \sim 0{\cdot}19040, \quad q_2 \sim 0{\cdot}04753, \quad q_3 \sim 0{\cdot}01294, \quad q_4 \sim 0{\cdot}00368, \dots$$

with an error of order γ^9, where $\gamma = \tfrac{1}{3}$.

Notice that this extinction probability problem is a particular case of an analogous probability problem for appropriate Markov chain sequences with *random* transition matrices, for which a set of equations like (7) would still apply, with the conditional probabilities $p_i^{(s)}$ simply the averaged transition probabilities.

Chapter 3

PROCESSES IN CONTINUOUS TIME

3·1 The additive process

We saw at the beginning of the last chapter that the distribution of the cumulative sum S_n of a number of completely independent and homogeneous components X_r was determined by its cumulant function equation $K_n = nK$. When we consider the precise formulation of the corresponding problem in continuous time, it is natural to postulate that, for any intervals t_1 and t_2,

$$K(t_1 + t_2) = K(t_1) + K(t_2), \tag{1}$$

or equivalently, for continuous functions,

$$K(t) = tK(1) \equiv tK, \tag{2}$$

say. We obtain one particular solution of this equation by recalling the limiting case, referred to in § 2·1, of an indefinitely large number of independent components in any interval Δt, with finite total mean and variance, for which

$$K(t) = mti\phi - \tfrac{1}{2}\sigma^2 t\phi^2. \tag{3}$$

As another case, suppose that events are occurring independently in time, such that the chance of an event in any small interval Δt is $m\Delta t + o(\Delta t)$ (and the chance of none, $1 - m\Delta t + o(\Delta t)$). Then for the total number of events, equation (1) gives the relation†

$$\Delta K \equiv K(t + \Delta t) - K(t) = \log \left\{ 1 + m(e^{i\phi} - 1)\,\Delta t + o(\Delta t) \right\}$$
$$= m(e^{i\phi} - 1)\,\Delta t + o(\Delta t), \tag{4}$$

or
$$\frac{\partial K(t)}{\partial t} = m(e^{i\phi} - 1),$$

† For any function $f(t)$ we shall denote the difference $f(t + \Delta t) - f(t)$ by Δf: δf may alternatively be used if Δt is a 'small' interval δt. The strict differential is denoted by df.

whence $$K(t) = mt(e^{i\phi} - 1), \tag{5}$$

which is the cumulant function of a Poisson distribution with mean mt.

The results (3) and (5) are of great importance in understanding the possible structure of additive processes, and indeed of Markov processes in general. The normal component (3) may be referred to as the *diffusion* component, in contrast with any non-zero contributions to the cumulative sum $S(t)$ associated with the sudden occurrences of events at random times, as indicated by Poisson components of the type (5). A detailed study of the general equation (1) by various writers (see, for example, Cramér, 1937) has in fact shown that the general solution is a linear superposition of the normal or diffusion component and discontinuous or *transition* components of this second type. The solution is given by Cramér in the form

$$K(t) = -\tfrac{1}{2}\sigma^2 t\phi^2 + t \int_{-\infty}^{\infty} \frac{e^{i\phi x} - 1 - i\phi x}{x^2} \, d\Omega(x), \tag{6}$$

where $\Omega(x)$ is bounded, never decreasing and continuous at $x = 0$, and where in both terms contributions to the mean (which we shall assume below to be finite) have been removed. (The interpretation of $\Omega(x)$ in the second component is that it represents the total variance contribution from discontinuities of different amounts $\leqslant x$, the non-zero contribution from $x = 0$ given by σ^2 having been removed; the total variance increment per unit time is thus $\sigma^2 + \Omega(\infty)$, assumed finite.)

As a particular case of the second component, we may have a single process of the type (5) with an associated random variable X with distribution $F(x)$. Then we obtain

$$K(t) = mt \int_{-\infty}^{\infty} (e^{i\phi x} - 1) \, dF(x)$$
$$= mt[C_X(\phi) - 1], \tag{7}$$

where $C_X(\phi)$ is the characteristic function for the random variable X.

The equations (3), (5) and (6) refer to homogeneous additive processes, in which the cumulant function in (1) depends only on the interval and not otherwise on the time. $K(t)$ is, however, the sum of increments $\Delta K \to 0$ with the interval Δt, and more general increments of this type depending on t may be considered if required. Homogeneous stochastic processes are, however, of especial practical significance, as their evolution depends only on the initial conditions and the invariant structure of the process. A further interesting point is that even if the increment ΔK depends on the value of the variable $S(t)$ at time t, this assumption modifying the process from an additive one to a Markov one, the form of the increment ΔK indicated by (6) will still be correct for a small enough interval Δt.

In §2·1 we considered the random walk problem in the presence of absorbing boundaries, and, in particular, noted that the method making use of Wald's identity could be applied to the limiting case of a large number of small steps, implying in effect its application to the normal or diffusion component (3). A re-examination of this method reveals that it may at once be extended to deal with the general 'random walk' or additive process in continuous time given by (6). In particular, when the boundaries are far away from the starting point, the asymptotic approximations will be applicable. As for large t the general process (6) tends to the normal additive process (when suitably scaled), the normal theory results in §2·1 are still asymptotically relevant.

It is, however, more direct for the normal diffusion process (3) (and hence asymptotically for the additive process (6)) to consider the differential equation obtained from (3), namely,

$$\frac{\partial K(t)}{\partial t} = mi\phi - \tfrac{1}{2}\sigma^2\phi^2, \tag{8}$$

an equation which will be more familiar if inverted back to the corresponding equation for the density function $f(x)$, which depends of course also on t,

$$\frac{\partial f}{\partial t} + m\frac{\partial f}{\partial x} = \tfrac{1}{2}\sigma^2\frac{\partial^2 f}{\partial x^2}. \tag{9}$$

An unrestricted solution of (9) is obviously, from the equivalence to (8),

$$f_1(x) = \frac{1}{\sqrt{(2\pi\sigma^2 t)}}\exp\left\{-\frac{1}{2}\frac{(x-mt)^2}{\sigma^2 t}\right\}, \tag{10}$$

this being for the initial condition that the random walk starts from the origin $x = 0$ at $t = 0$. The general solution of (9) is a linear superposition of such solutions starting from arbitrary points x, and may be used to construct the appropriate solution in the presence of absorbing boundaries, at which $f(x)$ must vanish. We quote the result for $m = 0$ and boundaries $a \, (< 0) = -b$ and $b \, (> 0)$. Such results may be conveniently obtained by a method of images, as in some physical problems (see Chandrasekhar, 1943; Bartlett, 1945),

$$f_2(x) = \sum_{s=-\infty}^{\infty} (-1)^s (2\pi\sigma^2 t)^{-\frac{1}{2}} \exp\left\{-\tfrac{1}{2}(x - 2bs)^2/(\sigma^2 t)\right\}. \quad (11)$$

An alternative case when $a \to -\infty$, obtained more easily as a single absorbing boundary problem, is (for $m = 0$)

$$f_3(x) = (2\pi\sigma^2 t)^{-\frac{1}{2}} \left[\exp\left\{-\tfrac{1}{2}x^2/(\sigma^2 t)\right\} - \exp\left\{-\tfrac{1}{2}(x - 2b)^2/(\sigma^2 t)\right\}\right]. \quad (12)$$

The corresponding solutions for $m \neq 0$ may be obtained by writing $f \exp\left\{(mx - \tfrac{1}{2}m^2 t)/\sigma^2\right\}$ for f.

A result which will be used later (§4·1) follows from the multiplication rule of probability and a comparison of (11) or (12) with (10), namely, that the probability of previously remaining within the prescribed boundaries, *given* that the position of the diffusing point at time t is the same value $x = 0$ as its value at $t = 0$, is $f_2(0)/f_1(0)$ or $f_3(0)/f_1(0)$ respectively. For (11) and (12) respectively, these probabilities become

$$\sum_{s=-\infty}^{\infty} (-1)^s \exp\left(-\frac{2b^2 s^2}{\sigma^2 t}\right) \quad (13)$$

and

$$1 - \exp\left(-\frac{2b^2}{\sigma^2 t}\right). \quad (14)$$

To obtain the time taken to reach the boundaries, we note that the modified solutions (11) and (12) are probability densities in the extended sense used in § 2·1, part of the probability being lost at the boundaries as time proceeds. Thus if the probability density for the time taken to reach the boundaries is $g(t)$, then

$$g(t) = -\frac{\partial}{\partial t} \int_a^b f(x) \, dx. \quad (15)$$

This gives, for example, in the case $a \to -\infty$,

$$g(t) = \frac{b}{\sqrt{(2\pi\sigma^2 t^3)}} \exp\left\{ -\frac{1}{2} \frac{(b-mt)^2}{\sigma^2 t} \right\}. \tag{16}$$

This distribution $g(t)$ is, in the terminology of §3·3, a *passage-time* distribution, and the methods to be developed in that section provide yet another way of deriving it. The distribution for the double-boundary problem is compounded of the conditional distributions for each boundary separately, for each of which the relevant passage time is that *conditional on not first reaching the other boundary*.

Monte Carlo method. It will be noticed that just as the above normal theory results provide an asymptotic solution to the problems for random walk sequences discussed in Chapter 2, so conversely an approximate solution to problems connected with the diffusion equation (9) above could be obtained from actual trials, for example, by simulating the games between two gamblers with the aid of random numbers. For example, in a classical problem propounded by C. Huyghens, the players A and B would win on a throw of fourteen and eleven respectively with three dice, any other score being disregarded. It is easily found that $p = \frac{15}{42}$. The players started with 12 counters each, so that $-a = b = 12$, the stake per game being one counter.

For this problem, equation (14) of §2·1 gives the probability of A winning (absorption at boundary b) as

$$244,140,625/282,673,677,106 = 0.000864,$$

the solution given by Huyghens. It is thus nearly certain that 'absorption' will take place at boundary a, and the 'duration of play' will have a distribution approximately given by (16) (with $b = |a|$ and the sign of $m = p - q$ reversed; $\sigma^2 = 4pq$), the effect of the other boundary, and of the discrete steps in the gambling problem, being neglected.

The results from 100 artificial 'game sequences' (all of which resulted in A losing) are compared in fig. 1 with the distribution so obtained; the agreement, with as few as 12 steps to reach the boundary in the artificial games, is surprisingly good. Conversely, the observed histogram gives an approximate solution for the distribution (16) with these values for its parameters. Of course we should not employ the method in such a case, but the method may be useful in more complicated problems, for example, in solving partial differential equations representing random-walk problems in two or three dimensions, with more complicated boundary conditions. The reader interested in this topic should consult the literature on this subject (see, for example, the National Bureau of Standards (U.S.A.) publication on the Monte Carlo Method, 1951). It should perhaps be added that the use of random numbers and artificial samples for the empirical study of theoretical problems has long been familiar to statisticians, and will be illustrated again in later sections.

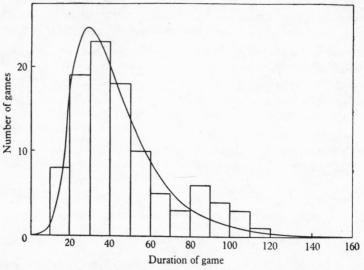

Fig. 1. Frequency distribution of the duration of play in 100 'games', yielding an approximate solution of a diffusion problem whose correct solution is given by the continuous curve. (Reprinted from *Applied Statistics*, **2** (1953), 49.)

3·2   Markov chains

We shall next consider how the basic equations and properties of Markov chains, developed in § 2·2, are modified if these chains are specified for continuous time. We first consider finite chains and suppose (i) that the probability of *any* change in the state in the interval t, $t + \Lambda t$ is specified by the diagonal matrix

$$\mathbf{P}(t) \Delta t + o(\Delta t),$$

and further (ii) that the asymptotic conditional distribution, *given* that a change occurs, is $S(t)$, so that the relation holds

$$\begin{aligned}
Q(t_2 + \Delta t \mid t_1) &= Q(t_2 + \Delta t \mid t_2)\, Q(t_2 \mid t_1) \\
&= [I - P(t_2)\, \Delta t]\, Q(t_2 \mid t_1) + [S(t_2)\, P(t_2)\, \Delta t]\, Q(t_2 \mid t_1) \\
&\quad + o(\Delta t).
\end{aligned}$$

Hence

$$\begin{aligned}
\frac{\partial Q(t_2 \mid t_1)}{\partial t_2} &= \lim_{\Delta t \to 0} \frac{Q(t_2 + \Delta t \mid t_1) - Q(t_2 \mid t_1)}{\Delta t} \\
&= [S(t_2) - I]\, P(t_2)\, Q(t_2 \mid t_1) = R(t_2)\, Q(t_2 \mid t_1), \qquad (1)
\end{aligned}$$

say, where $\qquad\qquad R(t_2) \equiv [S(t_2) - I]\, P(t_2).$

Equation (1) is the continuous analogue of equation (4) of § 2·2, and is to be regarded as a set of differential equations for each column of Q, starting with the corresponding column vector of the unit matrix at time t_1. Equivalently, for an arbitrary initial distribution $p(t_1)$, we obtain the equation for $p(t_2)$, by multiplying (1) to the right by $p(t_1)$,

$$\frac{\partial p(t_2)}{\partial t_2} = R(t_2)\, p(t_2). \qquad (2)$$

If, alternatively, we consider a small change in t_1, we obtain similarly, from the relation

$$Q(t_2 \mid t_1 - \Delta t) = Q(t_2 \mid t_1)\, Q(t_1 \mid t_1 - \Delta t),$$

the equation (we assume $P(t)$ and $S(t)$ are continuous with respect to t)

$$\frac{\partial Q(t_2 \mid t_1)}{\partial t_1} = - Q(t_2 \mid t_1)\, R(t_1), \qquad (3)$$

which is to be interpreted as a set of differential equations for each *row* of Q, the solution to satisfy the condition that it is the corresponding row of the unit matrix at time t_2.

The equations (1) and (3) will be termed the *forward* and *backward* differential equations, and were first derived by Kolmogorov (1931). The second or backward equation seems less natural for evaluating $p(t_2)$ from $p(t_1)$, but, as we shall see later, is related to a second powerful method of studying the solution of Markov processes in general, the first being related to the 'forward' equation, in which the last time change from t_2 to $t_2 + \Delta t$ is considered.

For homogeneous chains, $Q(t_2 \mid t_1)$, will depend merely on the difference $t_2 - t_1$, and R will be independent of t. Moreover R will commute with Q, for

$$[I + R\Delta t + o(\Delta t)]\, Q(t_2 - t_1) = Q([t_2 + \Delta t] - t_1)$$
$$= Q(t_2 - [t_1 - \Delta t])$$
$$= Q(t_2 - t_1)\,[I + R\Delta t + o(\Delta t)].$$

Thus the *matrix* equations (1) and (3) become identical, though their interpretations as *vector* equations with different initial conditions still remain distinct.

The solution of equation (2) for $R(t)$ constant is of course closely related to the solution of equation (7), § 2·2, and must be equivalent to it if we regard the Markov chain in discrete time as a sequence extracted from the complete chain in continuous time (this we can always do if the latter is validly specified, though not conversely). In fact we obviously have the formal equivalence $Q^t = \exp(Rt)$ or $Q = \exp R$, where Q represents the set of conditional probabilities for a unit interval. The latent vectors of R and the matrix Q are consequently the same, and their corresponding latent roots connected by the equation $\exp \lambda_R = \lambda_Q$, so that the latent roots of R are such that there is at least one zero root, and the others have real parts equal to or less than zero. It is, however, possible to say more about limiting distributions than was possible for the discrete time case, for the assumptions made above imply continual probability diffusion between states and the strictly cyclic or deterministic type of chain is no longer admissible. To see this more precisely, we consider the latent roots directly. As $\mid R - \lambda I \mid = 0$, the equation $t'(R - \lambda I) = 0$ has at least one solution t', the maximum component (in absolute value) of t' being t_m, say. Then for the corresponding element in the last equation

$$t_m \parallel r_{mm} - \lambda \parallel = \parallel \sum_{j \neq m} r_{jm} t_j \parallel$$

$$\leqslant t_m \parallel r_{mm} \parallel \quad (r_{mm} = -\sum_{j \neq m} r_{jm}),$$

or $\qquad \parallel r_{mm} - \lambda \parallel \leqslant \parallel r_{mm} \parallel,$

(where, as before, $\| x \|$ denotes the absolute value of any quantity x). This implies that λ is inside or on the circle, centre r_{mm} (< 0) and radius $\| r_{mm} \|$, i.e. either $\lambda = 0$ or the real part of λ is less than zero. The possibility of multiple zero roots is not excluded, and corresponds to the limiting distribution, which now always exists, depending on the initial conditions.

The discussion of limiting properties of the formal solutions

$$Q(t \mid 0) = e^{Rt}, \quad p(t) = e^{Rt} p_0,$$

of (1) and (2) in the homogeneous case also proceeds similarly to the discussion in § 2·2, but with the possibility of cyclic motion now excluded. It will be sufficient to note that the limiting properties for continuous time must be compatible with those for discrete time (but not conversely). Since only primitive matrices Q are now admissible, we have the result

$$\lim_{t \to \infty} p_t = p_\infty$$

always exists. The canonical form for R corresponding to classification by final groups is similar to that given for Q in § 2·21, as may be seen from the relation

$$Q(\epsilon \mid 0) = I + \epsilon R + o(\epsilon)$$

for small ϵ (for a more direct derivation, see Ledermann, 1950). The condition for regularity (p_∞ independent of p_0) in terms of the latent roots is that R is of rank $k - 1$ (or the root $\lambda_1 = 0$ is simple); the condition for positive regularity is that R is irreducible.

Infinite number of states. As in the case of discrete time, we may encounter chains with an infinite number of states. Our notation still formally applies in this case, but we have of course no longer any guarantee that the extension permits mathematically valid and unique solutions. This question has been studied in great detail by several writers, though it is not usually an acute one in practical applications. It will be convenient to defer its discussion at present, and we conclude this section on Markov chains in continuous time with three examples.

Example 1. *Finite chain with simple roots.* In the case of a finite chain, considered for simplicity with simple roots, we have a spectral resolution for R,

$$R = \sum_{i=1}^{k} \lambda_i s_i t_i', \tag{4}$$

say, where $\lambda_1 = 0$ but is included to complete the set of latent row and column vectors, so that $\Sigma_i \mathbf{s}_i \mathbf{t}_i' = \mathbf{I}$. Then

$$\mathbf{p}_t = \sum_{i=1}^{k} e^{\lambda_i t} \mathbf{s}_i (\mathbf{t}_i' \mathbf{p}_0)$$
$$= \mathbf{s}_1 + \sum_{i=2}^{k} e^{\lambda_i t} \mathbf{s}_i \alpha_i. \tag{5}$$

If we suppose that N individuals or particles independently move from one state to another according to the above process, and we label the jth state by a p.g.f. variable z_j, writing $\mathbf{z}' \equiv (z_1, z_2, ..., z_k)$, then the p.g.f. for all N individuals is

$$(\mathbf{z}' \mathbf{p}_t)^N = \left(\mathbf{z}' \mathbf{s}_1 + \sum_{i=2}^{k} e^{\lambda_i t} \alpha_i \mathbf{z}' \mathbf{s}_i \right)^N$$
$$\rightarrow (\mathbf{z}' \mathbf{s}_1)^N, \tag{6}$$

as $t \rightarrow \infty$.

As a particular case, suppose

$$\mathbf{R} = \begin{pmatrix} 0 & \lambda \\ 0 & -\lambda \end{pmatrix} = 0 \begin{pmatrix} 1 \\ 0 \end{pmatrix} (1, \quad 1) - \lambda \begin{pmatrix} 1 \\ -1 \end{pmatrix} (0, \quad -1),$$

corresponding to a rate λ from the second state to the first, but no return. Then all individuals eventually move from the second state. For the number N_t remaining in the second state at time t, for $N_0 = N$, we have the p.g.f. ($z_1 \equiv 1, z_2 \equiv z$)

$$\Pi(z) = [1 + e^{-\lambda t}(z-1)]^N. \tag{7}$$

Example 2. *The general homogeneous birth process.* We consider next a simple case of an enumerable chain, where only one-way transitions can occur from a 'ground' or 'zero' state to the one immediately above it; for example, the total number of occurrences of some event can increase but not decrease. This is termed a 'birth' or 'escalator' process, and was first discussed in a remarkable early paper by McKendrick (1914), who showed that the negative binomial distribution can arise as a special case of it.

Let the transition probability from the ith to $(i+1)$th state be $\lambda_i \Delta t + o(\Delta t)$, so that

$$\left. \begin{aligned} \frac{\partial p_i}{\partial t} &= \lambda_{i-1} p_{i-1} - \lambda_i p_i \quad (i > 0), \\ \frac{\partial p_0}{\partial t} &= -\lambda_0 p_0, \end{aligned} \right\} \tag{8}$$

or by direct solution

$$p_i(t) = e^{-\lambda_i t} \int_0^t \lambda_{i-1} e^{\lambda_i u} p_{i-1}(u) \, du,$$

if $p_i = 0 \ (i > 0)$ at $t = 0$. Hence

$$p_0 = e^{-\lambda_0 t}, \quad p_1 = \frac{\lambda_0 (e^{-\lambda_0 t} - e^{-\lambda_1 t})}{\lambda_1 - \lambda_0}, \ ...,$$

and by induction (the results follow somewhat more simply by the use of Laplace transforms with respect to the time)

$$p_i = (-1)^i \lambda_0 \lambda_1 \dots \lambda_{i-1}$$
$$\times \sum_{j=0}^{i} \frac{e^{-\lambda_j t}}{(\lambda_j - \lambda_0)(\lambda_j - \lambda_1) \dots (\lambda_j - \lambda_{j-1})(\lambda_j - \lambda_{j+1}) \dots (\lambda_j - \lambda_i)}. \tag{9}$$

(The case of some λ_i equal may easily be inferred from (9).) The negative binomial distribution follows from the particular case $\lambda_i = \lambda + i\mu$; it may, alternatively, be derived by the generating-function methods used in the further discussion in § 3·4 of multiplicative chains.

The necessary and sufficient condition for the birth process equations (8) to give a unique and valid solution for finite t, such that $\Sigma_i p_i = 1$, was first shown by Feller to be the divergence of the sum $\Sigma_i 1/\lambda_i$. This condition will appear later as a special case of a more general condition when 'deaths', i.e. transitions downwards by one, are also possible.

Example 3. *'Contagion' processes.* The particular case $\lambda_i = \lambda + i\mu$ is sometimes contrasted with the simple Poisson additive process $\lambda_i = \lambda$ and called a *contagion* effect, as the chance of an additional event (e.g. an accident) now depends on the number that has already occurred. Now it is well known in statistics that alternatively the negative binomial can be obtained from the Poisson by a *heterogeneity* effect, the observed distribution being a superposition of individual Poisson distributions with variable mean Λ per unit time interval. For the resultant p.g.f. will then be

$$E\{e^{\Lambda t(z-1)}\} = M_\Lambda\{t(z-1)\}, \tag{10}$$

where if (and only if) we choose

$$M_\Lambda(\theta) = (1 - \beta\theta)^{-\alpha}, \tag{11}$$

corresponding to the frequency function for Λ,

$$f(\lambda)\,d\lambda = \beta^{-\alpha}\lambda^{\alpha-1}e^{-\lambda/\beta}\,d\lambda/(\alpha-1)! \quad (\lambda \geqslant 0),$$

then

$$M_\Lambda\{t(z-1)\} = (1 + \beta t - \beta tz)^{-\alpha}, \tag{12}$$

a negative binomial distribution.

This dual interpretation of the negative binomial, indicating that the observed frequency distribution is not sufficient to discriminate between true contagion and heterogeneity, and that a more detailed study of the numbers of events for different time-intervals is necessary, has been generalized by O. Lundberg for the more general compound distribution of the type (10) above, which may also alternatively be generated by a contagion process.

A Taylor expansion of (10) about $z = 0$ gives

$$M\{t(z-1)\} = M(-t) + tzM'(-t) + \dots + (tz)^n M^{(n)}(-t)/n! + \dots, \tag{13}$$

where $M'(\theta) = \partial M(\theta)/\partial\theta$, etc. Define

$$\lambda_n(t) = M^{(n+1)}(-t)/M^{(n)}(-t), \tag{14}$$

and the conditional probability connecting the numbers of events n and m at times t and s ($\leqslant t$),

$$q_{nm}(t, s) = \frac{(t-s)^{n-m} M^{(n)}(-t)}{(n-m)! \, M^{(m)}(-s)} \quad (n \geqslant m). \tag{15}$$

The differential equation

$$\frac{\partial q_{nm}(t, s)}{\partial t} = -\lambda_n(t) \, q_{nm}(t, s) + \lambda_{n-1}(t) \, q_{n-1, m}(t, s) \tag{16}$$

is satisfied by (15) when $\lambda_n(t)$ is given by (14), and also $q_{nm}(t, s) \to \delta_{n, m}$ as $t \to s$ ($\delta_{n, m} = 1$ if $n = m$, 0 otherwise). The distribution at time t, given $m = 0$ at $s = 0$, is

$$p_n(t) \equiv q_{n0}(t, 0) = t^n M^{(n)}(-t)/n!, \tag{17}$$

which agrees with the coefficient of z^n in (13).

In spite of this generalization, the negative binomial has a distinctive place in that it is only for this compound Poisson distribution that the corresponding contagion process can be made homogeneous by suitable choice of the time scale. For if

$$\lambda_n(t) = \mu_t \nu_n, \tag{18}$$

then the relation

$$\lambda_{n+1}(t) = \lambda_n(t) - \lambda'_n(t)/\lambda_n(t) \quad (n = 0, 1, \ldots), \tag{19}$$

which readily follows from (14), becomes

$$\nu_{n+1} - \nu_n = -\mu'_t/\mu_t^2.$$

This equation must hold for any $n \geqslant 0$ and any $t > 0$, so that both sides must be constant; this gives

$$\nu_n = \lambda + \mu n; \quad \mu_t = 1/(\nu + \mu t). \tag{20}$$

This conclusion does not mean that this contagion process is the only homogeneous Markov chain (on transformed time-scale) leading to the negative binomial; there may be Markov chains of more complicated nature leading to this distribution, and in fact the birth-death-and-immigration process of § 3·41 (equation (3) with the initial number n zero) is such an example.

3·3 Recurrence and passage times for renewal processes

The problem of the probable time of recurrence of a given state of a physical system which changes according to a specified stochastic process is an important one in practice, with fundamental implications in the interpretation of statistical mechanics. It is also of relevance in purely theoretical discussions on the ergodic properties of Markov processes, as has been pointed out by Feller. In view of this general relevance for some of the neighbouring sections on Markov processes, it seems desirable to consider the theory at this stage. We shall include the appropriate results for sequences, the formalism being sufficiently similar for the various typical cases all to be discussed together.

Suppose we wish to study the recurrence of a particular state S for a random sequence, given that S occurred at time zero. We denote the probability of S at any subsequent time r, regardless of the states occurring between the times 0 and r, by

$$P\{S_r \mid S_0\} \equiv \{e_r\},$$

this being assumed non-zero for at least some $r > 0$; all such probabilities are conditional on S having occurred at time zero. We use the further symbolic notation

$$P\{\text{not } S_r \mid S_0\} \equiv \{\bar{e}_r\} = 1 - \{e_r\}, \quad P\{S_r, S_s \mid S_0\} \equiv \{e_r e_s\},$$

$$P\{\text{not } S_r, S_s \mid S_0\} \equiv \{\bar{e}_r e_s\} = \{(1 - e_r) e_s\} = \{e_s\} - \{e_r e_s\},$$

etc. Then for the probability of a first recurrence at time n we have

$$\{\bar{e}_1 \bar{e}_2 \dots \bar{e}_{n-1} e_n\} = \{(1 - e_1)(1 - e_2) \dots (1 - e_{n-1}) e_n\}$$

$$= \{e_n\} - \sum_{r=1}^{n-1} \{e_r e_n\} + \sum_{r=1}^{n-2} \sum_{s=r+1}^{n-1} \{e_r e_s e_n\} - \dots. \quad (1)$$

This formula in theory determines the recurrence distribution, for the probabilities on the right are theoretically available for any validly specified random sequence. For stationary processes they will be independent of the choice of the time origin 0.

If the discrete time r is replaced by a continuous time t, and we assume that the probability of S occurring in the *time* interval t, $t + \Delta t$ is $[e_t] \Delta t + o(\Delta t)$ (and $o(\Delta t)$ of occurring more than once), the formula (1) becomes replaced by the (formal) relation for the recurrence probability *density* at time t,

$$[e_t] - \int_0^t [e_u e_t] \, du + \int_0^t \int_u^t [e_u e_v e_t] \, du \, dv - \dots$$

$$= [e_t] - \int_0^t [e_u e_t] \, du + \frac{1}{2!} \int_0^t \int_0^t [e_u e_v e_t] \, du \, dv - \dots. \quad (2)$$

The formal solution (2) may be useful when the set of different states is non-enumerable, for example, if S refers to the value of a continuous spatial coordinate; it is of no interest without modification if S is one of a finite or enumerable group of states, as in a Markov chain, for $\{e_t\}$ would in general not be of the form $[e_t]\,dt$, and recurrence on the above definition would almost certainly be instantaneous. In such cases we exclude instantaneous recurrence or continuation in state S, as we shall see presently.

We define a process to be of the *renewal* or *regenerative* type with respect to the state S if the occurrence of S is sufficient to regenerate the process, probabilities being no longer dependent on other past history. This definition is wider than that of a Markov process, for which *every* state has the regenerative property[†]. It is consistent with the 'renewals' treated in § 2·11, the state S being then taken to be the wearing out of an article and its replacement by a new one. In the renewal case, if we also assume homogeneity, there is a great simplification in formulae (1) and (2). For we then have

$$\{e_r e_n\} = \{e_r\}\{e_{n-r}\}, \quad \{e_r e_s e_n\} = \{e_r\}\{e_{s-r}\}\{e_{n-s}\},$$

etc., and forming generating functions on both sides of (1), i.e. writing

$$\Pi_R(z) = \sum_{n=1}^{\infty} z^n\{\bar{e}_1 \bar{e}_2 \dots \bar{e}_{n-1} e_n\}, \quad \Pi(z) = \sum_{n=1}^{\infty} z^n\{e_n\},$$

we obtain (for $0 \leqslant z < 1$)

$$\Pi_R = \Pi - \Pi^2 + \Pi^3 \dots = \Pi/(1+\Pi). \tag{3}$$

This formula can be obtained more simply (see Feller, 1950, Chapter 12), but it is instructive to make use of the above procedure, which will be used again in the case of passage times. It should be emphasized that (3) determines Π_R in terms of Π or conversely; the converse situation arises in the renewal or replacement theory of an article with the 'lifetime' distribution specified by Π_R.

For continuous time, in problems for which a recurrence density in the sense of equation (2) exists, we replace generating functions by Laplace transforms in t, so that z^n is replaced by $e^{-\psi t}$. Thus for

$$M \equiv \int_0^{\infty} e^{-\psi u}[e_u]\,du, \tag{4}$$

[†] Throughout this book only 'strong' Markov processes are considered; otherwise this regenerative property would not necessarily hold (see, for example, Chung or Dynkin, 1960).

and similarly for M_R, (3) is replaced by

$$M_R = M/(1+M), \tag{5}$$

with the inverse relation of renewal theory (see equation (6) of § 2·11)

$$M = M_R/(1-M_R). \tag{6}$$

The purely random occurrence of an event or state S at average rate μ (considered in § 2·11 and again in § 3·1), so that

$$M = \int_0^\infty e^{-\psi u}\mu\,du = \frac{\mu}{\psi}, \tag{7}$$

gives from (5) $M_R = 1/[1+\psi/\mu]$, which is the transform of the exponential distribution $\mu e^{-\mu x}$.

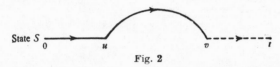

State S $\quad 0 \qquad\qquad u \qquad\qquad\qquad v \qquad\qquad t$

Fig. 2

In the case of continuous time, but a finite or enumerable set of states as in a Markov chain, we could proceed to modify our recurrence time in (3) by removing the term $\{e_1\} = p$, say, representing continuation in the state S, and measure the recurrence time from $t = 1$ for all processes for which we know the state at $t = 1$ is *not* S. The modified generating function is

$$\Pi'_R = \frac{\Pi_R - zp}{z(1-p)}, \tag{8}$$

and for continuous time this will usually give a definite limiting function M'_R if we write $z = e^{-\psi\tau}$ and let $\tau \to 0$. However, it is easier to use other arguments. The probability $P\{S_t|S_0\} \equiv \{e_t\} \equiv p_t$ is assumed non-zero, and we apply a renewal type of argument to the sequence of events (see fig. 2): S from $t = 0$ to u, then departure from S and first return at v, then occurrence of S at t (without conditions on the state during the interval v, t). The chance of leaving S in time dt is assumed to be $\lambda\,dt$, and the recurrence distribution for the time from departure to return is given by the *density* function $f(t)$. Then we must have

$$p_t = \int_{v=0}^t \int_{u=0}^v \lambda e^{-\lambda u} f(v-u)\,p_{t-v}\,dv\,du + e^{-\lambda t},$$

whence, taking Laplace transforms, we obtain

$$G(\psi) = \frac{M'_R G}{1 + \psi/\lambda} + \frac{1}{\lambda} \frac{1}{1 + \psi/\lambda},$$

whence finally $\qquad M'_R = 1 + \psi/\lambda - 1/(\lambda G).$ (9)

Passage times. These formulae may be extended to passage times from one state S to a different state T. For sequences let

$$P\{S_r \mid S_0\} \equiv \{e_r\}, \qquad P\{T_r \mid T_0\} \equiv \{f_r\},$$
$$P\{S_r \mid T_0\} \equiv \{g_r\}, \qquad P\{T_r \mid S_0\} \equiv \{h_r\},$$

where it is now important to specify explicitly whether S or T is given as the initial state. Consider

(a) the first passage from S to T, *irrespective of the recurrence of S.* We have

$$\{(1-h_1)(1-h_2)\dots(1-h_{n-1})h_n\} = \{h_n\} - \sum_{r=1}^{n-1} \{h_r h_n\}$$
$$+ \sum_{r=1}^{n-2} \sum_{s=r+1}^{n-1} \{h_r h_s h_n\} - \dots. \quad (10)$$

For homogeneous renewal or regenerative processes (the renewal property now of course being assumed with respect both to S and T), (10) becomes

$$\{h_n\} - \sum_{r=1}^{n-1} \{h_r\}\{f_{n-r}\} + \sum_{r=1}^{n-2} \sum_{s=r+1}^{n-1} \{h_r\}\{f_{s-r}\}\{f_{n-s}\} - \dots,$$

or $\qquad \Pi_P^{ST} = \Pi^{ST}/(1 + \Pi^{TT}),$ (11)

where Π_P^{ST} is the (unconditional) first passage time generating function, and

$$\Pi^{ST} \equiv \sum_{r=1}^{\infty} \{h_r\} z^r,$$

etc. If we alternatively require

(b) first passage to T, *without recurrence of S,* i.e. the passage time to T measured from the *last* occurrence of S, then the formulae become somewhat more complicated. We must now consider

$$\{(1-h_1-e_1)(1-h_2-e_2)\dots(1-h_{n-1}-e_{n-1})h_n\}$$
$$= \{h_n\} - \sum_{r=1}^{n-1} \{(h_r+e_r)h_n\}$$
$$+ \sum_{r=1}^{n-2} \sum_{s=r+1}^{n-1} \{(h_r+e_r)(h_s+e_s)h_n\} - \dots. \quad (12)$$

Using such relations for homogeneous renewal processes as

$$\{h_r e_n\} = \{h_r\}\{g_{n-r}\}, \quad \{h_r e_s e_n\} = \{h_r\}\{g_{s-r}\}\{e_{n-s}\},$$

we find that

$$\Pi_Q^{ST} = \frac{\Pi^{ST}}{(1 + \Pi^{TT})(1 + \Pi^{SS}) - \Pi^{ST}\Pi^{TS}}, \tag{13}$$

where Π_Q^{ST} is the first passage generating function *without intermediate recurrence of S*.

These formulae also extend immediately to the case of continuous time with densities $[e_t]$. Thus the formula analogous to (11) is

$$M_P^{ST}(\psi) = M^{ST}/(1 + M^{TT}). \tag{14}$$

It is perhaps useful in the case of (13) to check the completely analogous formula by deriving it by an alternative argument. The recurrence of S from 0 to t may be split up into two contingencies: (i) recurrence of S *conditional on no passage through* T, and (ii) first passage to T *conditional on no recurrence of S* followed by first *unconditional* passage from T to S. This readily gives, on taking transforms,

$$M_R^S = C_R^S + M_Q^{ST} M_P^{TS}, \tag{15}$$

where C_R^S is the conditional recurrence function for S as above defined.

Similarly, the passage from S to T may consist either of passage conditional on no recurrence of S, or recurrence of S conditional on no passage through T followed by unconditional passage. Hence also

$$M_P^{ST} = M_Q^{ST} + C_R^S M_P^{ST}. \tag{16}$$

From these two equations, we find the analogue of (13),

$$M_Q^{ST} = \frac{M^{ST}}{(1 + M^{TT})(1 + M^{SS}) - M^{ST}M^{TS}}, \tag{17}$$

and also the further formula

$$C_R^S = \frac{M^{SS}(1 + M^{TT}) - M^{ST}M^{TS}}{(1 + M^{TT})(1 + M^{SS}) - M^{ST}M^{TS}}. \tag{18}$$

Formula (18) also of course has its analogue for discrete time,

$$\frac{\Pi^{SS}(1 + \Pi^{TT}) - \Pi^{ST}\Pi^{TS}}{(1 + \Pi^{TT})(1 + \Pi^{SS}) - \Pi^{ST}\Pi^{TS}}, \tag{19}$$

this being alternatively derivable by the method used for (13). It should be noticed that even if $M_R^S(0)$, $M_P^{ST}(0)$ and $M_P^{TS}(0)$ are all unity (see § 3·31) the equations (15) and (16) imply that $C_R^S(0)$ and $M_Q^{ST}(0)$ are less than unity. This merely means that *any* recurrence or passage may not satisfy the additional requirement. To convert formulae like (13) or (18) into valid conditional distributions, we may always divide by the total probability sum $\Pi_Q^{ST}(1)$ or $C_R^S(0)$ respectively.

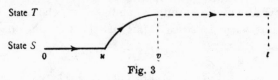

Fig. 3

Corresponding formulae may also be obtained if the time is continuous but the number of states enumerable. For example, for first passage regardless of intermediate recurrence we obtain the relation for $p_{ST}(t) \equiv \{h_t\}$,

$$p_{ST}(t) = \int_{v=0}^{t} \int_{u=0}^{v} \lambda_S e^{-\lambda_S u} f_{ST}(v-u) \, p_{TT}(t-v) \, dv \, du,$$

from the sequence of events (fig. 3): S from 0 to u, then (unconditional) first passage to T at v, and also occurrence of T at t (without conditions on the state during the interval v, t). In this equation $p_{TT}(t) \equiv \{f_t\}$, $f_{ST}(t)$ is the first passage density† from S to T, and we now write for clarity $\lambda_S \equiv \lambda$. For the transforms we obtain

$$G^{ST}(\psi) = \frac{1}{(1+\psi/\lambda_S)} M_P^{ST}(\psi) \, G^{TT}(\psi), \tag{20}$$

or $$M_P^{ST}(\psi) = (1+\psi/\lambda_S) G^{ST}(\psi) / G^{TT}(\psi). \tag{21}$$

For conditional recurrence and passage times we may employ a similar argument to that used for the relations (15) and (16). The only change is the insertion of a lifetime in S after the conditional recurrence of S and a lifetime in T after the conditional passage to T. Hence (15) and (16) are replaced by the relations

$$M_R' = C_R' + M_Q^{ST}[1+\psi/\lambda_T]^{-1} M_P^{TS}, \tag{22}$$

$$M_P^{ST} = C_R'[1+\psi/\lambda_S]^{-1} M_P^{ST} + M_Q^{ST}, \tag{23}$$

† For formal simplicity it is written as a density, though in general the distribution will have a discrete component at zero, corresponding to a direct transition to T after leaving S.

where C'_R denotes the transform of the conditional recurrence distribution. Solving these equations, we find

$$M_Q^{ST} = \frac{G^{ST}}{\lambda_S[G^{SS}G^{TT} - G^{ST}G^{TS}]}, \tag{24}$$

$$C'_R = 1 + \frac{\psi}{\lambda_S} - \frac{G^{TT}}{\lambda_S[G^{SS}G^{TT} - G^{ST}G^{TS}]}. \tag{25}$$

The consistency of formulae (9), (21), (24) and (25) is checked by the following elementary example of a Markov chain with the transition rates between its only two states S and T given by the matrix

$$\begin{pmatrix} -1 & 2 \\ 1 & -2 \end{pmatrix}.$$

Here

$$\{e_t\} = \tfrac{2}{3} + \tfrac{1}{3}e^{-3t}, \quad \{g_t\} = 2\{h_t\} = \tfrac{2}{3}(1 - e^{-3t}), \quad \{f_t\} = \tfrac{1}{3} + \tfrac{2}{3}e^{-3t},$$

$$G^{SS} = \frac{2 + \psi}{\psi(3 + \psi)}, \quad G^{TS} = 2G^{ST} = \frac{2}{\psi(3 + \psi)}, \quad G^{TT} = \frac{\psi + 1}{\psi(3 + \psi)}.$$

We find on applying the above formulae

$$M'_R = (1 + \tfrac{1}{2}\psi)^{-1}, \quad C'_R = 0, \quad M_P^{ST} = M_Q^{ST} = 1,$$

which are obviously correct, as the recurrence time for S is identical in this example with lifetime in T, no recurrence is possible without traversing T, and conditional and unconditional passages from S to T are identical and take zero time as no other state is possible.

Finally, we may require conditional first passages of other types. Thus for three different states S, T, U, say, let us consider first passage from S to T, *conditional on no intermediate passage to U*. This is merely an extension of the previous conditional passage time (for which $U \equiv S$), and an obvious extension of the previous arguments may be made. Thus formulae (15) and (16) generalize to

$$\left.\begin{aligned} M_P^{SU} &= M_Q^{SU} + M_Q^{ST}M_P^{TU} \\ M_P^{ST} &= M_Q^{ST} + M_Q^{SU}M_P^{UT} \end{aligned}\right\}, \tag{26}$$

whence now

$$M_Q^{ST} = \frac{M_P^{ST} - M_P^{UT}M_P^{SU}}{1 - M_P^{UT}M_P^{TU}}. \tag{27}$$

Application to diffusion processes. It should perhaps be pointed out that these formulae cannot immediately be applied to a normal diffusion process, as, for example, represented by equation (10) of § 3·1. This is because of the absence of a finite 'velocity' for such processes (see Chapter 5), and correspondingly of a finite density of occurrence of a state $S \equiv x$ *in time*, even although the probability density with respect to the continuum of states x is finite. However, it is not difficult to make the appropriate modifications. Thus as one possibility we may re-interpret M^{ST}, say, in formula (14) as the transform of the *space* density function, and consider the first passage time not to the 'point state' x, but to the interval $(x, x + dx)$. An integral equation argument then leads to the result $M_P^{ST} = M^{ST}/M^{TT}$ in place of (14). Formula (27) correspondingly becomes

$$M_Q^{ST} = (M^{ST}M^{UU} - M^{UT}M^{SU})/(M^{TT}M^{UU} - M^{UT}M^{TU}).$$

From equation (10) of § 3·1, we find (for $S \equiv 0$, $T \equiv b$)

$$M^{ST} = \frac{1}{\sqrt{(2\sigma^2\psi')}} \exp\left\{\frac{bm}{\sigma^2} - \frac{b\sqrt{(2\psi')}}{\sigma}\right\} \quad \left(\psi' = \psi + \frac{1}{2}\frac{m^2}{\sigma^2}; \; b > 0\right), \quad (28)$$

whence also

$$M^{TT} = 1/\sqrt{(2\sigma^2\psi')}. \quad (29)$$

These formulae give, for example,

$$\frac{M^{ST}}{M^{TT}} = \exp\left\{\frac{bm}{\sigma^2} - \frac{b\sqrt{(2\psi')}}{\sigma}\right\} \quad (b > 0),$$

in agreement with equation (21) of § 2·1 ($\psi \equiv -i\phi$).

Diffusion processes in more than one dimension. The theory of unrestricted random walks in more than one dimension is of course very analogous to the one-dimensional case. The theory of random walks in the presence of absorbing boundaries, and the associated passage-time problems, obviously can become much more complicated, even in the pure diffusion case. We shall not attempt to develop a general theory here, pertaining also to more general additive processes and even to Markov processes, but merely discuss a particular example (for other more general methods, see, for example, Keilson 1963; Whittle, 1964: for the mathematical foundations, see Dynkin, 1960).

Consider an unrestricted continuous and isotropic walk in three dimensions starting at the origin at time zero, and with position co-ordinates $r \equiv (x, y, z)$ at time t with distribution represented by its characteristic function

$$C_t(\boldsymbol{\phi}) = \exp\left(-\tfrac{1}{2}t\sigma^2\phi^2\right), \tag{30}$$

where $\sigma^2 = \sigma_x^2 = \sigma_y^2 = \sigma_z^2$ and $\phi^2 = \phi_x^2 + \phi_y^2 + \phi_z^2$. To obtain the passage-time distribution to the spherical boundary of radius a, it is convenient to use an extension of Wald's identity to the three-dimensional case (the proof of which is straightforward) viz.

$$E\{e^{\boldsymbol{\theta}'\cdot\mathbf{R}}[M(\boldsymbol{\theta})]^{-T}\} = 1 \quad (|M(\boldsymbol{\theta})| \geqslant 1), \tag{31}$$

where $\boldsymbol{\theta}'.\mathbf{R} \equiv \theta_x X + \theta_y Y + \theta_z Z$, $M(i\boldsymbol{\theta}) \equiv C_1(\boldsymbol{\theta})$, and $\mathbf{R}$ and T are the random position and time of hitting the boundary. The result (31) is in general not immediately useful[†] as $\mathbf{R}$ and T would not be independent for a starting point other than the origin. However, in this particular case they are obviously independent by symmetry, and we have

$$E\{[M(\boldsymbol{\theta})]^{-T}\} E\{e^{\boldsymbol{\theta}'\cdot\mathbf{R}}\} = 1. \tag{32}$$

Moreover, by straightforward integration (cf. equation (19), §6·51)

$$E\{e^{\boldsymbol{\theta}'\cdot\mathbf{R}}\} = \sinh(\theta a)/(\theta a), \tag{33}$$

where $\theta^2 = \theta_x^2 + \theta_y^2 + \theta_z^2$. Hence, with the substitution

$$\log M(\boldsymbol{\theta}) = \psi$$

so that $\theta = \sqrt{(2\psi)}/\sigma$, we obtain for the Laplace transform of the passage-time distribution to the spherical boundary,

$$\frac{a\sqrt{(2\psi)}/\sigma}{\sinh\{a\sqrt{(2\psi)}/\sigma\}}. \tag{34}$$

Expanding in powers of $-\psi$ will give the moments of the passage-time distribution, whence we easily find for the mean and variance respectively

$$\kappa_1 = \frac{a^2}{3\sigma^2}, \quad \kappa_2 = \frac{2a^4}{45\sigma^4}, \tag{35}$$

† Cf. the discussion by Barnett (1965).

compared with the symmetric one-dimensional case (cf. equation (26), §2·1)

$$\kappa_1 = \frac{a^2}{\sigma^2}, \quad \kappa_2 = \frac{2a^4}{3\sigma^4}. \tag{36}$$

In fact, the general formulae in n dimensions are similarly derivable as

$$\kappa_1 = a^2/(n\sigma^2), \quad \kappa_2 = 2a^4/\{n^2(n+2)\,\sigma^4\}. \tag{37}$$

If we consider further the distribution in the presence of an absorbing boundary at the (three-dimensional) sphere, and let the Laplace transform of the unrestricted characteristic function $C_t(\boldsymbol{\phi})$ in (30), say $L_\psi(\boldsymbol{\phi})$, be modified to $M_\psi(\boldsymbol{\phi})$, then the type of argument made use of in §3·3 gives immediately the relation

$$L_\psi(\boldsymbol{\phi}) = M_\psi(\boldsymbol{\phi}) + Q_\psi(\boldsymbol{\phi})\, L_\psi(\boldsymbol{\phi}), \tag{38}$$

where $Q_\psi(\boldsymbol{\phi})$ is the Laplace transform in (34).

Hence

$$M_\psi(\boldsymbol{\phi}) = \frac{1 - Q_\psi(\boldsymbol{\phi})}{\psi + \tfrac{1}{2}\sigma^2\phi^2}. \tag{39}$$

3·31　Ergodic properties.

In the relation $\Pi_R = \Pi/(1+\Pi)$ for renewal sequences obtained in the previous section (equation (3)), valid for $0 \leqslant z < 1$, it is evident that Π_R also exists for $z = 1$, but may not necessarily be unity; in fact, the event S is only *certain* to recur if $\Pi_R(1) = 1$, the recurrence of S being otherwise uncertain. $\Pi_R(1) < 1$ implies (and conversely) that $\Pi(1) < \infty$, which is thus an equivalent condition for S being uncertain (transient).

When S is certain, we assume for convenience that S is not periodic, i.e. does not recur only at times t, $2t$, $3t$, ... $(t > 1)$ (in this latter case we can always regard z^t as the argument of the generating functions instead of z). Then a theorem given by Erdös, Feller and Pollard (1949) (cf. also Kolmogorov, 1936) establishes the ergodic formula

$$\lim_{r \to \infty} \{e_r\} = 1/\mu, \tag{1}$$

where μ is the mean $[\partial \Pi_R/\partial z]_{z=1}$. This formula remains true if μ is infinite, when $\{e_r\} \to 0$ as $r \to \infty$.

In the case of continuous time with a time density of occurrence $[e_t]$, we may deduce from equation (5), §3·3, that $M_R(\psi) \to 1$ as $\psi \to 0$ if $M(\psi) \to \infty$, and S is then certain to recur. The relevant theorem† for such a process (for which no periodic case arises) is now one given by Blackwell (1948), namely, that if the expected number of recurrences or renewals in the interval $(0, t)$ is $U(t)$, and μ is the mean recurrence or renewal interval, then

$$\lim_{t \to \infty} [U(t+h) - U(t)] = h/\mu \qquad (2)$$

for every $h > 0$. This formula also remains true if μ is infinite. In the present case, $[e_t] = dU(t)/dt$.

In the case of continuous time but an enumerable set of states, so that equation (9) of §3·3 holds, we note that $G(0) = \infty$ still implies $M'_R(0) = 1$ and alternatively, $G(0) < \infty$ implies $M'_R(0) < 1$ (and conversely). To obtain a formula corresponding to (1) or (2), apply Blackwell's theorem to a 'renewal' defined for the immediate purpose as the complete lifetime plus recurrence of S. As contributions to $\{e_t\}$ can arise from renewals of S in this sense occurring at a previous time u and surviving in state S to time t, we have the relation

$$\{e_t\} = \int_0^t e^{-\lambda_S(t-u)} dU(u). \qquad (3)$$

$U(u)$ is now the 'convolution' of the lifetime and recurrence distributions, each with a density, so that $U(t)$ is still differentiable. It follows that $\{e_t\} \to 1/(\lambda_S \mu)$ as $t \to \infty$, where μ is the sum of the lifetime mean, $1/\lambda_S$, and the true recurrence mean μ_R, say; that is, $\mu = \mu_R + 1/\lambda_S$ and

$$\lim_{t \to \infty} \{e_t\} = 1/(1 + \lambda_S \mu_R). \qquad (4)$$

We may now indicate briefly the implications of the above results for Markov processes, for which every state has the renewal or regenerative property. We consider only irreducible processes, for which any state can be reached with non-zero probability from any other state after some interval; processes with 'absorbing' states are thus excluded. Then for Markov chain sequences‡ with a state X_i certainly recurrent, we may,

---

† A more recent theorem in this class, covering both results (1) and (2) (and also the limit of $\{e_t\}$ in (3)), has been given by Smith (1954).

‡ For a more detailed and exhaustive discussion of this case, reference should be made to Feller (1950, Chapter 15).

making use of an argument due to Feller, choose intervals of minimum sufficient length to allow passage from state X_i to another state X_j and back to occur, and write

$$\left.\begin{aligned} q_{ii}(t+\tau_1+\tau_2) &\geqslant q_{ji}(\tau_1)\,q_{jj}(t)\,q_{ij}(\tau_2) = \alpha\beta q_{jj}(t), \\ q_{jj}(t+\tau_1+\tau_2) &\geqslant q_{ij}(\tau_2)\,q_{ii}(t)\,q_{ji}(\tau_1) = \alpha\beta q_{ii}(t), \end{aligned}\right\} \tag{5}$$

where $\alpha, \beta > 0$ (here, as in § 2·2, $q_{ji}(t)$ denotes $P\{S_j \text{ at } t \mid S_i \text{ at } 0\}$). The relations (5) imply that $q_{ii}(t)$ and $q_{jj}(t)$ have the same type of limiting behaviour, corresponding to the three alternatives: X_j also certainly recurrent, with (i) $q_{ii}(t)$ and $q_{jj}(t)$ both tending to zero, (ii) $q_{ii}(t)$ and $q_{jj}(t)$ both tending to non-zero limits, (iii) X_i and X_j both periodic, with the same period. (To show that X_j is also certainly recurrent, sum equations (5) over t, and use the result that X_i certainly recurrent implies the divergence of $\Sigma_t q_{ii}(t)$.)

Moreover, from the relation $\Pi_P^{ST} = \Pi^{ST}/(1 + \Pi^{TT})$ (equation (11), § 3·3), with $S \equiv X_i$, $T \equiv X_j$, as $\Pi_P^{ST}(1) \leqslant 1$, we must have $\Pi^{ST}(1) < \infty$ if $\Pi^{TT}(1) < \infty$ (the case of uncertain recurrence). Alternatively, for certain recurrence we must have $\Pi_P^{ST}(1) = 1$, hence $\Pi^{ST}(1)/\Pi^{TT}(1) = 1$. In this last case (if also non-periodic, so that $q_{jj}(t) \to 1/\mu_j$) it readily follows from the probability relation equivalent to equation (11), § 3·3, that as t increases $q_{ji}(t)$ also tends to $1/\mu_j$. This is still true if $\mu_j = \infty$.

For a finite or enumerable set of states but continuous time, a similar argument makes use of equations (9) and (21) of § 3·3, together with the relation (4) above. The periodic case does not now arise; and the intervals τ_1 and τ_2 in (5) may now be any non-zero intervals. It has of course been implicitly assumed in this recurrence theory that a valid Markov process in continuous time has been specified; the question of whether this is achieved by means of differential equations of the type derived in § 3·2 is returned to in the last section of this chapter.

For continuous time and a continuous infinity of states with finite time densities of occurrence $f_{ii}(t) \equiv [e_t]$, $f_{ji}(t) \equiv [h_t]$, etc. (excluding periodic processes), we may for certainly recurrent X_i replace (5) by the relations

$$
\left.
\begin{aligned}
&\int_t^{t+h} f_{ii}(\tau + \tau_1 + \tau_2 + h)\, d\tau \\
&\qquad \geqslant \iiint f_{ji}(\tau_1 + u)\, f_{jj}(\tau + h - u - v)\, f_{ij}(\tau_2 + v)\, du\, dv\, d\tau, \\
&\int_t^{t+h} f_{jj}(\tau + \tau_1 + \tau_2 + h)\, d\tau \\
&\qquad \geqslant \iiint f_{ij}(\tau_2 + v)\, f_{ii}(\tau + h - u - v)\, f_{ji}(\tau_1 + u)\, du\, dv\, d\tau,
\end{aligned}
\right\} \tag{6}
$$

where the integration is over the interval $(0, h)$ for u and v. These relations ensure the same type of limiting behaviour for $\int_t^{t+h} f_{ii}(v)\, dv$ and $\int_t^{t+h} f_{jj}(v)\, dv$. The formula $M_P^{ST} = M^{ST}/(1 + M^{TT})$ (equation (14) of § 3·3) ensures that since $M_P^{ST}(0) \leqslant 1$, $M^{TT}(0) < \infty$ implies $M^{ST}(0) < \infty$. Alternatively, if $M^{TT}(0) = \infty$, so that recurrence is certain, we must have $M_P^{ST}(0) = 1$ and $\int_t^{t+h} f_{ji}(u)\, du$ and $\int_t^{t+h} f_{jj}(u)\, du$ have the same limit h/μ_j (which is zero if μ_j is infinite).

3·32　Alternative method for Markov chains.

An alternative to the use of 'renewal' arguments for obtaining recurrence distributions in the case of Markov chains with a finite, or effectively finite, number of states is sometimes convenient for direct calculations. It consists of modifying the transition or conditional probability matrix by prohibiting all transitions *out from* the specified state S. This automatically 'freezes' the state S once it occurs, and the probability of any lifetime in the composite state 'not S' (which determines the recurrence time back to S) is identified with the probability of still being in 'not S' for the modified process. Thus for discrete time and a matrix $\mathbf{Q}$ of transition probabilities, the modified matrix is, say,

$$
\mathbf{Q}_m = \begin{pmatrix} 1 & \mathbf{x}_1' \\ \hline 0 & \mathbf{B} \end{pmatrix},
$$

where for convenience the top left-hand corner corresponds to the specified state S. The recurrence time θ_1, say, will be defined analogously to the definition for continuous time, as the time to first recurrence after having left S (i.e. continuation in state S is reckoned as the 'lifetime' of S, and no longer counted as recurrence; this corresponds to the modified recurrence distribution Π'_R in equation (8) of § 3·3). The probability column vector $\mathbf{p}_0$, given 'not S' at time zero (and S at time -1), will have the form $\mathbf{p}'_0 = (0 \mid \mathbf{y}')$, say. Hence

$$Q_m^n \mathbf{p}_0 = \begin{pmatrix} 1 & \mathbf{x}'_n \\ \hline 0 & \mathbf{B}^n \end{pmatrix} \begin{pmatrix} 0 \\ \mathbf{y} \end{pmatrix} = \begin{pmatrix} \mathbf{x}'_n \mathbf{y} \\ \hline \mathbf{B}^n \mathbf{y} \end{pmatrix},$$

and the cumulative distribution is $1 - [\mathbf{B}^n \mathbf{y}]$, where $[\mathbf{y}]$ denotes (in the present context) the *sum* of the elements of the vector $\mathbf{y}$. The generating function is thus given by

$$\Pi'_R = \sum_{n=1}^{\infty} [(\mathbf{B}^{n-1} - \mathbf{B}^n)\,\mathbf{y}]\,z^n = \left[\frac{(\mathbf{I} - \mathbf{B})\,z}{\mathbf{I} - \mathbf{B}\,z}\,\mathbf{y} \right]. \tag{1}$$

If $\mathbf{p}_0$ is not conditional on S at time -1, the above procedure determines the distribution Π''_R, say, of another recurrence time θ_2, say, which is recurrence to S, measured from a time for which the state is not S. For example, in the case of only two states, let $\mathbf{B} = 1 - \lambda$, and with $\mathbf{y} = 1$ for either θ_1 or θ_2, we confirm the trivial solution

$$\Pi'_R = \Pi''_R = \lambda z / [1 - (1 - \lambda)\,z]. \tag{2}$$

Similarly for continuous time, let the transition matrix be of the form $\mathbf{Q} = \mathbf{I} + \mathbf{R}\delta t \mathbin{\vcenter{\hbox{$\cdot$}}} o(\delta t)$. Let

$$\mathbf{R}_m = \begin{pmatrix} 0 & \mathbf{x}' \\ \hline 0 & \mathbf{C} \end{pmatrix}$$

be the modification of $\mathbf{R}$ in which the first column has all zeros. The cumulative recurrence distribution for θ_1 becomes $1 - [e^{\mathbf{C}t}\mathbf{y}]$, and the Laplace transform

$$M'_R(\psi) \equiv \int_0^{\infty} e^{-\psi u} [-\mathbf{C}\,e^{\mathbf{C}u}\,\mathbf{y}]\,du$$

$$= \left[\frac{\mathbf{C}}{\mathbf{C} - \mathbf{I}\psi}\,\mathbf{y} \right]. \tag{3}$$

Again for the case of only two states, and $C = -\lambda$,

$$M'_R(\psi) = M''_R(\psi) = 1/(1 + \psi/\lambda), \tag{4}$$

where $M''_R(\psi)$ is the similar distribution for θ_2.

The general type of recurrence-time distribution for finite Markov chains in continuous time will obviously be a sum of exponential distributions if C has simple roots; if multiple roots exist, the expression $[e^{Ct} \mathbf{y}]$ will involve components of the form $P(t) e^{-\nu t}$, where $P(t)$ is a polynomial in t (the distribution is thus expressible as a sum of χ^2 distributions).

3·4   Multiplicative chains

We return now to the discussion of important types of Markov chain begun at the end of § 3·2. It is very illuminating to consider directly the equations for multiplicative chains when the time becomes continuous. It will be recalled that our general equation for multiplicative chains in the case of more than one type of individual was expressible in the dual form (end of § 2·3)

$$\mathbf{\Pi}_n(\mathbf{z}) = \mathbf{\Pi}_{n-1}(\mathbf{G}(\mathbf{z})) = \mathbf{G}(\mathbf{\Pi}_{n-1}(\mathbf{z})).$$

Let us now postulate that as the time interval δt becomes small

$$\mathbf{G}(\mathbf{z}) - \mathbf{z} = \mathbf{\dot{g}}(\mathbf{z})\, \delta t + o(\delta t),$$

then we see that in the limit we must have

$$\frac{\partial \mathbf{\Pi}_t(\mathbf{z})}{\partial t} = \left(\sum_{r=1}^{k} g^{(r)} \frac{\partial}{\partial z_r} \right) \mathbf{\Pi}_t(\mathbf{z}) \tag{1}$$

for k types of individual, where $g^{(r)}$ denote the components of $\mathbf{\dot{g}}$. Alternatively, from the second form of the recurrence relation, we obtain

$$\frac{\partial \mathbf{\Pi}_t(\mathbf{z})}{\partial t} = \mathbf{\dot{g}}(\mathbf{\Pi}_t(\mathbf{z})). \tag{2}$$

In the particular case of one type of individual, these reduce to

$$\frac{\partial \Pi_t(z)}{\partial t} = g(z) \frac{\partial}{\partial z} \Pi_t(z) \tag{3}$$

and
$$\frac{\partial \Pi_t(z)}{\partial t} = g(\Pi_t(z)). \tag{4}$$

The equations (1) and (3) correspond to the forward equations of § 3·2, equations (2) and (4) to the backward equations. From (4) we obtain immediately

$$t = \int_s^{\Pi_t} \frac{du}{g(u)} \tag{5}$$

as $\Pi_0(z) = z$. For example, if

$$g(z) = \lambda(z^2 - z) + \mu(1 - z), \tag{6}$$

representing a 'birth' rate λ per individual and a 'death' rate μ, we obtain

$$\int \frac{du}{g(u)} = \int \frac{1}{\lambda - \mu} \left(\frac{du}{u-1} - \frac{\lambda \, du}{\lambda u - \mu} \right) = \frac{1}{\lambda - \mu} \log \frac{u-1}{\lambda u - \mu},$$

whence we find from (5)

$$\Pi_t(z) = \left[\frac{\mu e^{(\lambda - \mu)t}(z-1)}{\lambda z - \mu} - 1 \right] \Big/ \left[\frac{\lambda e^{(\lambda - \mu)t}(z-1)}{\lambda z - \mu} - 1 \right], \tag{7}$$

starting from one individual, or $\Pi_t^n(z)$ from n. In the case $\lambda = \mu$, this becomes

$$\Pi_t(z) = \frac{1 - (\lambda t - 1)(z-1)}{1 - \lambda t(z-1)}. \tag{8}$$

If we expand $\log \Pi_t(z)$ in (7) or (8) in powers of $\theta = \log z$, we obtain the cumulants of the number N_t of individuals at time t. In particular, it is readily found that the mean and variance are given by

$$\left. \begin{aligned} m_t &= e^{(\lambda - \mu)t}, \\ v_t &= \begin{cases} \dfrac{\lambda + \mu}{\lambda - \mu} e^{(\lambda - \mu)t}[e^{(\lambda - \mu)t} - 1] & (\lambda \neq \mu), \\ 2\lambda t & (\lambda = \mu). \end{cases} \end{aligned} \right\} \tag{9}$$

In the case $\lambda = \mu$, we have an example of a process with constant mean but with ever-increasing variance. For the probability of extinction, we have

$$\Pi_t(0) = \begin{cases} \mu(e^{(\lambda - \mu)t} - 1)/(\lambda e^{(\lambda - \mu)t} - \mu) & (\lambda \neq \mu), \\ \lambda t/(1 + \lambda t) & (\lambda = \mu). \end{cases} \tag{10}$$

It will be seen that as $t \to \infty$, the chance of extinction tends to one for $\lambda \leqslant \mu$, and μ/λ for $\lambda \geqslant \mu$. The corresponding limiting chances for n initial individuals are one and $(\mu/\lambda)^n$.

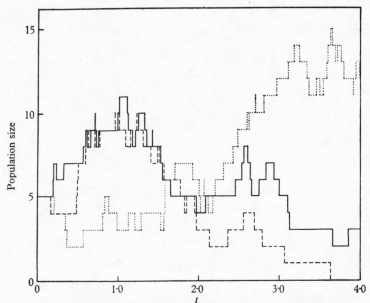

Fig. 4. Three independent realizations of a simple birth-and-death multiplicative process with $\lambda = \mu = 1$, each beginning with five individuals at $t = 0$.

To illustrate the variability between different realizations of this process, especially when $\lambda = \mu$, a number of artificial realizations were constructed for the case $\lambda = \mu = 1$, each beginning with five individuals at $t = 0$. Three are shown in fig. 4, of which one achieved extinction before $t = 4$. The method of construction used the fact that after any event leaving N_t individuals, the chance of the next event (transition) in time dt is $N_t(\lambda + \mu)\,dt$, or the time until the next event is a random variable T following the exponential distribution $e^{-T/m}\,dT/m$, where $1/m = N_t(\lambda + \mu)$. When the event occurs, the chance that it is a birth is $\lambda/(\lambda + \mu)$. (For the estimation of λ and μ from such a realization, see Chapter 8.)

The more general result (5) (for one initial individual) enables the limiting distribution as $t \to \infty$ of $N_t e^{-\nu t}$, where $E\{N_t\} = e^{\nu t}$, to be investigated. We must have for $L(\theta) \equiv \lim L_t(\theta)$, where $L_t(\theta)$ is the m.g.f. of $N_t e^{-\nu t}$ ($\nu > 0$),

$$L(\theta(1+\nu\Delta t)) = L(\theta) + g(L(\theta))\Delta t + o(\Delta t),$$

or
$$\nu\theta\,\partial L/\partial\theta = g(L(\theta)), \qquad (11)$$

whence
$$\int^L \frac{du}{g(u)} = \frac{\log\theta}{\nu} + \text{constant},$$

where the constant may be determined from the condition $(\partial L/\partial\theta)_{\theta=0} = 1$. For example, when

$$g(z) = (\lambda z - \mu)(z-1), \quad \nu = \lambda - \mu,$$
$$L = (1 - C\mu\theta)/(1 - C\lambda\theta),$$

where $C = 1/\nu$, so that finally

$$L = (1 - \mu\theta/\nu)/(1 - \lambda\theta/\nu), \qquad (12)$$

as could of course have been obtained from equation (7). By writing (12) in the form

$$L = \frac{\mu}{\lambda} + \left(1 - \frac{\mu}{\lambda}\right)\frac{1}{(1 - \lambda\theta/\nu)},$$

we see that it has a discrete component μ/λ at the origin, corresponding to the chance of extinction, and a remaining continuous component with exponential distribution.

The non-homogeneous birth-and-death process. The solution (7) for the homogeneous birth-and-death process could alternatively have been obtained from equation (3), which in this case is

$$\frac{\partial\Pi_t(z)}{\partial t} = (z-1)(\lambda z - \mu)\frac{\partial\Pi_t(z)}{\partial z}. \qquad (13)$$

It will be clear from the structure of (3) that equation (13) still holds if λ and μ are functions of t. The solution of (13) is $\Pi_t(z) = \Psi(Z)$, say, where $Z(t)$ is a solution of the equation

$$dz/dt + (z-1)(\lambda z - \mu) = 0.$$

This equation becomes, with $z - 1 = 1/u$,

$$du/dt + (\mu - \lambda)u = \lambda,$$

with solution
$$ue^\rho = \int_0^t \lambda e^\rho\,dt + \text{constant},$$

where
$$\rho = \int_0^t (\mu - \lambda)\,dt.$$

Hence
$$\Pi_t(z) = \Psi\left(\frac{e^\rho}{z-1} - \int_0^t \lambda e^\rho\, dt\right),$$

where, from $\Pi_0(z) = z$, we find $\Psi(x) = 1 + 1/x$, whence

$$\Pi_t(z) = 1 + \cfrac{1}{e^\rho/(z-1) - \int_0^t \lambda e^\rho\, dt}. \tag{14}$$

This solution is due to D. G. Kendall (1948), who deduced from it that $\Pi_t(0)$, the probability of no individuals at time t, tends to unity as t increases, if and only if

$$e^\rho + \int_0^t \lambda e^\rho\, dt = 1 + \int_0^t \mu e^\rho\, dt \to \infty$$

as $t \to \infty$.

A particular case of the above process was considered by Arley (1943) in his discussion of physical cascade showers, where cosmic-ray particles give rise on impact with matter to secondary particles, which in turn may give rise to further particles, before such particles lose their energy and are absorbed. In a simplified model for this process he assumed that λ was constant, but that the chance of the 'extinction' of a particle increased linearly with the time. (This is not an exact treatment, which would have to be based on a probability 'balance-sheet' of the previous random impacts of the particle; cf. § 3·41.) We thus assume that $\mu(t)$ is of the form μt, where μ is constant. Then $\rho = \frac{1}{2}\mu t^2 - \lambda t$, and

$$\Pi_t(z) = 1 + \left[\frac{e^{\frac{1}{2}\mu t^2 - \lambda t}}{z-1} - \int_0^t \lambda e^{\frac{1}{2}\mu\tau^2 - \lambda\tau}\, d\tau\right]^{-1}. \tag{15}$$

The mean growth m_t is found to follow the Gaussian law

$$m_t = e^{\lambda t - \frac{1}{2}\mu t^2}, \tag{16}$$

while for the variance v_t we find

$$v_t = e^{\lambda t - \frac{1}{2}\mu t^2}(1 - e^{\lambda t - \frac{1}{2}\mu t^2}) + 2\lambda e^{2(\lambda t - \frac{1}{2}\mu t^2)}\int_0^t e^{\frac{1}{2}\mu\tau^2 - \lambda\tau}\, d\tau. \tag{17}$$

Two or more types of individual. In the application to cosmic-ray showers there are actually two types of particle, photons and electrons, to be considered (positive and negative electrons not being distinguished). These give rise respectively to pairs of electrons and photon-electron pairs, and (writing for convenience of notation w, z for z_1, z_2) we have as a somewhat more realistic model

$$g_1(w, z; t) = \lambda_1(z^2 - w) + \mu_1(1 - w),$$
$$g_2(w, z; t) = \lambda_2(zw - z) + \mu_2(1 - z),$$

where w corresponds to photons, z to electrons, and the death-rates μ_1 and μ_2 (λ and μ being in general dependent on t) are, as before, an approximate representation of the loss of energy of the particles in time due to previous collisions. The resulting equation

$$\frac{\partial \Pi_t}{\partial t} = g_1 \frac{\partial \Pi_t}{\partial w} + g_2 \frac{\partial \Pi_t}{\partial z} \qquad (18)$$

has been solved for small t (i.e. small distances) when the 'death-rates' μ_1 and μ_2 can be neglected, it being further assumed that $\lambda_1 = \lambda_2 = \lambda$ (constant). The auxiliary equations to (18) then become

$$\frac{dz}{dt} + \lambda(zw - z) = 0, \quad \frac{dw}{dt} + \lambda(z^2 - w) = 0. \qquad (19)$$

It is now possible to obtain a solution, making use of the transformation

$$ze^{-\lambda t} = u, \quad we^{-\lambda t} = v, \quad T = e^{\lambda t},$$

whence $\quad u = A/\sinh[A(T - B)], \quad v = A \coth[A(T - B)]. \quad (20)$

Solving equations (20) for A and B to obtain two independent integrals of (19) and making use of the initial condition

$$\Pi_0(w, z) = z \quad \text{(for one initial electron)},$$

we find that

$$\Pi_t(w, z) = \frac{e^{-\lambda t} \sqrt{(w^2 - z^2)}}{\sinh\left\{(e^{-\lambda t} - 1) \sqrt{(w^2 - z^2)} + \coth^{-1}[w/\sqrt{(w^2 - z^2)}]\right\}}; \qquad (21)$$

and similarly if $\Pi_0 = w$ (one initial photon)

$$\Pi_t(w, z) = e^{-\lambda t} \sqrt{(w^2 - z^2)}$$
$$\times \coth\left\{(e^{-\lambda t} - 1) \sqrt{(w^2 - z^2)} + \coth^{-1}[w/\sqrt{(w^2 - z^2)}]\right\}. \qquad (22)$$

The moments may be obtained by differentiating (21) or (22) with respect to w and z, but it is usually simpler to differentiate the initial equation and solve directly. In fact, for homogeneous chains, we derive directly from (1) and (2) the first and second *factorial moment* equations

$$\left.\begin{aligned}
\frac{\partial m_j^i}{\partial t} &= m_h^i \left[\frac{\partial g^h}{\partial z_j}\right], \\
\frac{\partial m_{jk}^i}{\partial t} &= m_{hk}^i \left[\frac{\partial g^h}{\partial z_j}\right] + m_{hj}^i \left[\frac{\partial g^h}{\partial z_k}\right] + m_h^i \left[\frac{\partial^2 g^h}{\partial z_j \partial z_k}\right],
\end{aligned}\right\} \quad (23)$$

and
$$\left.\begin{aligned}
\frac{\partial m_j^i}{\partial t} &= m_j^h \left[\frac{\partial g^i}{\partial z_h}\right], \\
\frac{\partial m_{jk}^i}{\partial t} &= m_{jk}^h \left[\frac{\partial g^i}{\partial z_h}\right] + m_j^h m_k^n \left[\frac{\partial^2 g^i}{\partial z_h \partial z_n}\right].
\end{aligned}\right\} \quad (24)$$

In these alternative sets of equations, m_j^i and m_{jk}^i refer to the first- and second-order factorial moments

$$E\{N_j(t)\}, \qquad E\{N_j(t) N_k(t)\} \quad (j \neq k),$$

$$E\{N_j(t) [N_j(t) - 1]\} \quad (j = k),$$

for an initial individual of type i, the square bracket denotes that the expression contained in it is to be evaluated for $z = 1$, and the summation convention is to be understood.

To illustrate these general equations in the case of the above example, we put $\lambda = 1$ for convenience and, more consistently with the above notation,

$$g^{(1)}(z_1, z_2) \equiv g_2(w, z) = z_1 z_2 - z_1,$$

$$g^{(2)}(z_1, z_2) \equiv g_1(w, z) = z_1^2 - z_2.$$

Then equations (23) give

$$\left.\begin{aligned}
\frac{\partial m_1^i}{\partial t} &= 2m_2^i, \quad \frac{\partial m_2^i}{\partial t} = m_1^i - m_2^i,
\end{aligned}\right.$$

and
$$\left.\begin{pmatrix} \dfrac{\partial m_{11}^i}{\partial t} \\[2ex] \dfrac{\partial m_{12}^i}{\partial t} \\[2ex] \dfrac{\partial m_{22}^i}{\partial t} \end{pmatrix} = \begin{pmatrix} 0 & 4 & 0 \\ 1 & -1 & 2 \\ 0 & 2 & -2 \end{pmatrix} \begin{pmatrix} m_{11}^i \\[1ex] m_{12}^i \\[1ex] m_{22}^i \end{pmatrix} + \begin{pmatrix} 2m_2^i \\[1ex] m_1^i \\[1ex] 0 \end{pmatrix}, \right\} \quad (25)$$

while (24) give

$$\frac{\partial m_j^{(1)}}{\partial t} = m_j^{(2)}, \quad \frac{\partial m_j^{(2)}}{\partial t} = 2m_j^{(1)} - m_j^{(2)}$$

and

$$\frac{\partial m_{jk}^{(1)}}{\partial t} = m_{jk}^{(2)} + (m_j^{(1)} m_k^{(2)} + m_j^{(2)} m_k^{(1)}), \qquad (26)$$

$$\frac{\partial m_{jk}^{(2)}}{\partial t} = (2m_{jk}^{(1)} - m_{jk}^{(2)}) + 2m_j^{(1)} m_k^{(1)}.$$

The solution of either set, remembering that $m_{jk}^i = 0$ initially for all i, j, k, we find to be

$$m_1^{(1)} = \tfrac{1}{3}(2e^t + e^{-2t}), \quad m_1^{(2)} = \tfrac{1}{3}(2e^t - 2e^{-2t}),$$

$$m_2^{(1)} = \tfrac{1}{3}(e^t - e^{-2t}), \quad m_2^{(2)} = \tfrac{1}{3}(e^t + 2e^{-2t}),$$

$$m_{11}^{(1)} = \tfrac{8}{9}e^{2t} - \tfrac{4}{9}e^{-t} + \tfrac{7}{45}e^{-4t} - \tfrac{14}{15}e^t + \tfrac{1}{3}e^{-2t},$$

$$m_{12}^{(1)} = \tfrac{4}{9}e^{2t} + \tfrac{1}{9}e^{-t} - \tfrac{7}{45}e^{-4t} - \tfrac{2}{5}e^t,$$

$$m_{22}^{(1)} = \tfrac{2}{9}e^{2t} + \tfrac{2}{9}e^{-t} + \tfrac{7}{45}e^{-4t} - \tfrac{4}{15}e^t - \tfrac{1}{3}e^{-2t}, \qquad (27)$$

$$m_{11}^{(2)} = \tfrac{8}{9}e^{2t} + \tfrac{8}{9}e^{-t} - \tfrac{8}{45}e^{-4t} - \tfrac{14}{15}e^t - \tfrac{2}{3}e^{-2t},$$

$$m_{12}^{(2)} = \tfrac{4}{9}e^{2t} - \tfrac{2}{9}e^{-t} + \tfrac{8}{45}e^{-4t} - \tfrac{2}{5}e^t,$$

$$m_{22}^{(2)} = \tfrac{2}{9}e^{2t} - \tfrac{4}{9}e^{-t} - \tfrac{8}{45}e^{-4t} - \tfrac{4}{15}e^t + \tfrac{2}{3}e^{-2t}.$$

3·41   The effect of immigration. The multiplicative chains treated above strictly do not allow 'immigration' into the system from outside ('emigration' may be treated synonymously with 'deaths'). However, as in the case of multiplicative sequences, the effect of immigration is readily included; for simplicity we consider problems with only one type of individual. We assume that immigrants enter with rate coefficient $\nu(t)$. Let the p.g.f. solution at time t ($\geqslant \tau$) with one initial individual at time τ in the absence of immigration be $G(z, t, \tau)$. Then, since the chance of entry of one new individual in the interval $\tau_r, \tau_r + \Delta\tau$ is $\nu(\tau_r)\Delta\tau + o(\Delta\tau)$, the complete solution is, from the assumption of independence of the individuals,

$$G^n(z, t, 0) \lim_{\Delta\tau \to 0} \prod_{(r)} \{1 + \nu(\tau_r)\Delta\tau (G(z, t, \tau_r) - 1)\},$$

where the intervals cover the total period $(0, t)$ and $\Pi_0(z) = z^n$. Hence

$$\Pi_t(z) = G^n(z, t, 0) \exp\left\{ \int_0^t \nu(\tau) \left(G(z, t, \tau) - 1 \right) d\tau \right\}. \tag{1}$$

For $n = 0$ and $G(z, t, \tau) = z$, this gives the familiar Poisson distribution

$$\Pi_t(z) = \exp\left\{ (z - 1) \int_0^t \nu(\tau) \, d\tau \right\}. \tag{2}$$

If we put for $G(z, t, \tau)$ the solution

$$G(z, t, \tau) = 1 + \left[\frac{e^{\rho'}}{(z - 1)} - \int_\tau^t \lambda e^{\rho'} \, dt \right]^{-1}$$

of the birth-and-death process, where

$$\rho' = \int_\tau^t (\mu - \lambda) \, dt,$$

we obtain a general solution for a birth-death-and-immigration process. In the homogeneous case when λ, μ and ν are constant, we find (for $\lambda \neq 0$)

$$\Pi_t(z) = \frac{(\lambda - \mu)^{\nu/\lambda} \left[\mu(T - 1) - (\mu T - \lambda) z \right]^n}{\left[\lambda T - \mu - \lambda(T - 1) z \right]^{n + \nu/\lambda}}, \tag{3}$$

where $T \equiv e^{(\lambda - \mu)t}$. When $n = 0$, this is a negative binomial distribution, and in particular it includes the negative binomial distribution obtained at the end of § 3·2 (put $\mu = 0$, and note that the chance of an increase by one in Δt when n individuals are present is $(\lambda n + \nu) \Delta t + o(\Delta t)$, the form of λ_n assumed for 'birth processes' in the more general sense there defined).

The effect of immigration may be, as in the case of sequences, to produce a stable limiting distribution for N_t. Thus suppose $G(z, t, \tau) \to 1$ as t increases; an equilibrium distribution then exists if

$$\lim_{t \to \infty} \int_0^t \nu(\tau) \left(G(z, t, \tau) - 1 \right) d\tau$$

exists. For a homogeneous process, i.e. ν constant and

$$G(z, t, \tau) \equiv H(z, t - \tau),$$

we then have

$$\Pi_\infty = \exp\left\{ \nu \int_0^\infty (H(z, \tau) - 1) \, d\tau \right\}. \tag{4}$$

For example, for $\mu > \lambda$, equation (3) gives

$$\Pi_\infty = \left(\frac{\mu - \lambda z}{\mu - \lambda}\right)^{-\nu/\lambda}.$$

On the other hand, if $\lambda = 0$, it is easily shown that equation (3) is replaced by

$$\Pi_t(z) = [1 + T(z-1)]^n \exp\{\nu(z-1)(1-T)/\mu\}, \tag{5}$$

where $T \equiv e^{-\mu t}$, this tending to $\exp\{\nu(z-1)/\mu\}$ as $t \to \infty$. This shows how a limiting Poisson distribution arises when deaths (or emigrations) are counterbalanced by immigration; alternatively, a negative binomial distribution can arise if births also occur. The process (5) is an important one in practice, for it may approximately represent a situation where particles move independently in and out of a certain small volume under observation. If we assume that immigrations take place at rate ν, and emigrations are proportional to the number N_t inside the volume, the solution (5) follows. It has been used as a model for colloidal particles in suspension and also for the movement of spermatozoa (see Chandrasekhar (1943) and Rothschild (1953) respectively; see also §§ 5·21, 6·31, 8·3 and 9·13).

3·42 Point processes.

Many stochastic processes, and in particular multiplicative processes, arise in physics and biology where we have to deal with particles or individuals distributed in a continuous infinity of states. The elementary birth-and-death processes discussed in § 3·4 are unrealistic whether applied to population growth or to cosmic-ray showers, for the behaviour of individuals depends on their *age* x and of particles on their *energy* ϵ. We may regard this parameter as a new continuous parameter of a stochastic process, in addition to the time t, but from the present development the natural viewpoint is of an evolving continuous set of random variables.

The immediate difficulty is that only probability densities can be attached to particular values of the age x or energy ϵ, and not non-zero probabilities (stochastic processes of this type will be called *point processes*). Integration of the probability density can only yield the first-order expectation of the numbers of particles or individuals in prescribed ranges of ϵ or of x, whereas we require also to study fluctuations in these numbers. To overcome this obstacle in the cosmic-ray problem, Bhabha and Ramakrishnan introduced higher-order density functions, called by Ramakrishnan product-densities. A somewhat similar procedure was introduced independently in the population problem by D. G. Kendall, who noted its connection with the use of the characteristic functional.

Let $N(x, t)$ represent the number of individuals at time t with the parametric value $X \leqslant x$. We assume that the density relation

$$f_1(x, t)\, dx = E\{dN(x, t)\} \tag{1}$$

exists, such that the probability of one individual in $x, x + \delta x$ is $f_1 \delta x + o(\delta x)$, and the total probability of more than one, $o(\delta x)$. As already noted, the integral of $f_1(x)$ over x yields only the mean number of individuals in the range of integration (the addition law of probability does not apply in its simple form, for the events are not in general mutually exclusive).

However, if we also form the product-density of order two,

$$f_2(x_1, x_2)\, dx_1 dx_2 = E\{dN(x_1)\, dN(x_2)\} \quad (x_1 \neq x_2), \tag{2}$$

this may be interpreted as the simultaneous probability that an individual is in $x_1, x_1 + dx_1$, and another in $x_2, x_2 + dx_2$, provided the two differential elements do not overlap (regardless of the numbers of individuals in other ranges of x). When $x_1 = x_2$,

$$E\{[dN(x_1)]^2\} = E\{dN(x_1)\} = f_1(x_1)\, dx_1. \tag{3}$$

In view of this degeneracy, we obtain the formula

$$E\{[\Delta N(x)]^2\} = \int_x^{x+h} \int_x^{x+h} E\{dN(x_1)\, dN(x_2)\}$$
$$= \int_x^{x+h} f_1(x_1)\, dx_1 + \int_x^{x+h} \int_x^{x+h} f_2(x_1, x_2)\, dx_1 dx_2. \tag{4}$$

The density-functions f_1 and f_2 may consistently be thought of as factorial-moment densities, for (4) is equivalent to

$$E\{\Delta N(x)\, [\Delta N(x) - 1]\} = \int_x^{x+h} \int_x^{x+h} f_2(x_1, x_2)\, dx_1 dx_2. \tag{5}$$

We may similarly define higher-order densities, degeneracies occurring whenever at least two of the differential elements coincide. To perceive the relation between f_r and the rth factorial moment of N in a finite range of x, it is sufficient to note that the contributions ($\geqslant 0$) to f_r and the rth factorial moment of N must vanish together when the number of individuals in the given range of x is $\leqslant r-1$. The general relation between $E\{[\Delta N(x)]^r\}$ and f_r is thus

$$E\{[\Delta N(x)]^r\} = \sum_{s=1}^{r} {}^r c_s \int \dots \int f_s \, dx_1 \dots dx_s, \qquad (6)$$

where the ${}^r c_s$ are obtained from the relations

$$n^r = \sum_{s=1}^{r} {}^r c_s n(n-1) \dots (n-s+1) \quad (n=1, 2, \dots). \qquad (7)$$

The ${}^r c_s$ in terms of the 'finite differences of zero' are $\Delta^s 0^r / s!$ and have been tabulated for $r, s = 2\text{--}25$ by Stevens (1937–8).

In the case of individuals with no interdependence of any kind between different x intervals (a fixed number n of independent individuals does not permit this property, for if a particular individual is not in one range it is more likely to be in another), we have

$$f_r(x_1, \dots, x_r) = f_1(x_1) f_1(x_2) \dots f_1(x_r),$$

whence the rth factorial moment of ΔN is given by $[E\{\Delta N\}]^r$, a property of the Poisson distribution.

To see how these densities are related to the use of the characteristic functional, let us define the complete characteristic functional with respect to $N(x)$

$$C(\phi(x), t) = E\left\{\exp\left[i \int \phi(u) \, dN(u)\right]\right\}. \qquad (8)$$

This functional contains automatically all the above moments; for example, putting $\phi(u) = \phi$ for u between x and $x+h$, and 0 otherwise, and expanding in powers of ϕ, we have

$$C(\phi) = 1 + i\phi E\{\Delta N\} - \tfrac{1}{2}\phi^2 E\{(\Delta N)^2\} \dots.$$

It is convenient to relate this formalism to the previous equations for multiplicative chains by writing $i\phi(u) \equiv \log z(u)$. Consider, for example, a cascade shower problem where one type of particle of energy ϵ gives rise after 'collision' to two new particles of the same type, with energies ϵ_1 and ϵ_2, the probability for an interval dt being

$$w(\epsilon_1, \epsilon_2 \mid \epsilon)\, d\epsilon_1 d\epsilon_2 dt \quad (\epsilon_1 \leqslant \epsilon_2, \epsilon_1 + \epsilon_2 \leqslant \epsilon).$$

(This is the 'nucleon cascade' problem, and as it involves only one type of particle is somewhat simpler than the electron-photon cascade shower already referred to, which may, however, be treated by similar methods; see, for example, Ramakrishnan (1952), Bartlett and Kendall (1951), and further literature referred to.) Referring back to the equations (1) and (2) of § 3·4 for more than one type of individual, the function $\mathfrak{g}(\mathbf{z})$ now generalizes to

$$g(z(u) \mid \epsilon) = \iint w(\epsilon_1, \epsilon_2 \mid \epsilon)\, [z(\epsilon_1)\, z(\epsilon_2) - z(\epsilon)]\, d\epsilon_1 d\epsilon_2. \qquad (9)$$

This immediately yields the equation

$$\frac{\partial \Pi_t(z(u) \mid \epsilon)}{\partial t} = \iint w(\epsilon_1, \epsilon_2 \mid \epsilon)\, [\Pi_t(z(u) \mid \epsilon_1)\, \Pi_t(z(u) \mid \epsilon_2) \\ - \Pi_t(z(u) \mid \epsilon)]\, d\epsilon_1 d\epsilon_2. \qquad (10)$$

The 'forward' equation is not quite so neatly generalized, but could be expressed in 'differential' form

$$d\Pi_t(z(u)) = \Pi_t(z(v) + g(z(u) \mid v)\, dt) - \Pi_t(z(v)), \qquad (11)$$

where $g(z(u) \mid v)$ is given by (9) with v replacing ϵ. This gives the formal extension of equation (1) of § 3·4

$$\frac{\partial \Pi_t(z(u) \mid \epsilon)}{\partial t} = \int \left\{ \iint w(\epsilon_1, \epsilon_2 \mid v)\, [z(\epsilon_1)\, z(\epsilon_2) - z(v)]\, d\epsilon_1 d\epsilon_2 \right\} \\ \times \left\{ \frac{\partial \Pi_t(z(v) \mid \epsilon)}{\partial z(v)\, dv} \right\} dv, \qquad (12)$$

where, as the probability of a particle in the range $v, v + dv$ is $O(dv)$, the 'functional derivative' is formally defined and written as $\partial \Pi_t(z(v) \mid \epsilon)/[\partial z(v)\, dv]$ (cf. Hopf, 1952).

The first and second product-density equations may similarly be written down. It is convenient to define the symmetrical function

$$\omega(\epsilon_1, \epsilon_2 \mid \epsilon) = \tfrac{1}{2}[w(\epsilon_1, \epsilon_2 \mid \epsilon) + w(\epsilon_2, \epsilon_1 \mid \epsilon)],$$

for which all values of ϵ_1, ϵ_2 (such that $\epsilon_1 + \epsilon_2 \leqslant \epsilon$) are relevant. Then if

$$\int \omega(\epsilon_1, \epsilon_2 \mid \epsilon) \, d\epsilon_1 = \tfrac{1}{2}\Omega(\epsilon_2 \mid \epsilon),$$

$$\iint \omega(\epsilon_1, \epsilon_2 \mid \epsilon) \, d\epsilon_1 d\epsilon_2 = W(\epsilon),$$

it is readily found that

$$\left.\begin{aligned}
\frac{\partial f_1(x \mid \epsilon)}{\partial t} &= \int f_1(u \mid \epsilon) \, \Omega(x \mid u) \, du - W(x) f_1(x \mid \epsilon), \\
\frac{\partial f_2(x, y \mid \epsilon)}{\partial t} &= \int f_2(u, y \mid \epsilon) \, \Omega(x \mid u) \, du + \int f_2(x, u \mid \epsilon) \, \Omega(y \mid u) \, du \\
&\quad + 2\int f_1(u \mid \epsilon) \, \omega(x, y \mid u) \, du - [W(x) + W(y)] f_2(x, y \mid \epsilon),
\end{aligned}\right\}$$

$$(13)$$

corresponding to (12), or, alternatively,

$$\left.\begin{aligned}
\frac{\partial f_1(x \mid \epsilon)}{\partial t} &= \int f_1(x \mid u) \, \Omega(u \mid \epsilon) \, du - W(\epsilon) f_1(x \mid \epsilon), \\
\frac{\partial f_2(x, y \mid \epsilon)}{\partial t} &= \int f_2(x, y \mid u) \, \Omega(u \mid \epsilon) \, du - W(\epsilon) f_2(x, y \mid \epsilon) \\
&\quad + 2\iint f_1(x \mid u) f_1(y \mid v) \omega(u, v \mid \epsilon) \, du \, dv.
\end{aligned}\right\}$$

$$(14)$$

These equations may be solved if it is assumed that

$$\omega(\epsilon_1, \epsilon_2 \mid \epsilon) \, d\epsilon_1 d\epsilon_2 \equiv \omega'(\eta_1, \eta_2) \, d\eta_1 d\eta_2,$$

where $\epsilon_1 = \epsilon\eta_1$, $\epsilon_2 = \epsilon\eta_2$. For example, taking the Mellin transforms of f_1 and f_2 in (13) with respect to x and y for given ϵ, so that

$$\gamma_1(s_1) = \int f_1(x \mid \epsilon) \, x^{s_1-1} \, dx, \quad \gamma_2(s_1, s_2) = \iint f_2(x, y \mid \epsilon) \, x^{s_1-1} y^{s_2-1} \, dx \, dy.$$

and choosing for convenience the scale of t so that $\alpha(1, 1) = 1$, where

$$\alpha(s_1, s_2) = \iint \omega'(\eta_1, \eta_2) \, \eta_1^{s_1-1} \eta_2^{s_2-1} \, d\eta_1 d\eta_2,$$

we have

$$\left.\begin{aligned}
\frac{\partial \gamma_1}{\partial t} &= (2\alpha(1, s_1) - 1)\gamma_1, \\
\frac{\partial \gamma_2}{\partial t} &= (2\alpha(1, s_1) + 2\alpha(1, s_2) - 2)\gamma_2 + 2\alpha(s_1, s_2)\gamma_1(s_1 + s_2 - 1).
\end{aligned}\right\}$$

$$(15)$$

This leads to the solution

$$
\left.\begin{aligned}
\gamma_1 &= \epsilon^{s_1-1}\, e^{-[1-2\alpha(1,\,s_1)]t}, \\
\gamma_2 &= \frac{2\alpha(s_1, s_2)\, \epsilon^{s_1+s_2-2}\{e^{-[1-2\alpha(1,\,s_1+s_2-1)]t} - e^{-[2-2\alpha(1,\,s_1)-2\alpha(1,\,s_2)]t}\}}{1 - 2\alpha(1, s_1) - 2\alpha(1, s_2) + 2\alpha(1, s_1+s_2-1)}.
\end{aligned}\right\} \quad (16)
$$

The higher density transforms γ_r may be obtained in succession by an extension of these methods.

3·5 General equations for Markov processes

In the previous section on multiplicative chains in continuous time, the primitive iterative relations for simple multiplicative chains in discrete time led formally but naturally to differential and integro-differential equations of various types. Multiplicative chains are examples of Markov processes, and the classification of these various equations will become even clearer if we examine the character of the 'forward' and 'backward' equations for Markov processes in general.

Let us denote the 'state' of the process by $X(t)$, and form the characteristic function $C_t(\phi)$ of $X(t)$ at the time t. Then by definition

$$
\frac{\Delta C_t(\phi)}{\Delta t} = E\left\{\left[\frac{e^{i\phi \Delta X(t)} - 1}{\Delta t}\right] e^{i\phi X(t)}\right\},
$$

where Δ is the usual symbol for a first difference. Hence if the conditional average for given X at time t of the square-bracketed quantity has a limit (almost always) as $\Delta t \to 0$, say $\Psi(i\phi, t, X)$, then

$$
\frac{\partial C_t(\phi)}{\partial t} = E_X\{\Psi(i\phi, t, X)\, e^{i\phi X}\}. \quad (1)
$$

In operational form, if we may commute the expectation and differentiation symbols,

$$
\frac{\partial C_t(\phi)}{\partial t} = \Psi\left(i\phi, t, \frac{\partial}{\partial i\phi}\right) C_t(\phi), \quad (2)
$$

where the operator $\partial/\partial i\phi$ acts only on $C_t(\phi)$.

This equation will hold if the limiting functions exist even if the process is not Markovian, but the function Ψ may not then be independent of further past history as it must for a Markov process, and so (2) can no longer be used to obtain a solution for given initial conditions. Equation (2) includes the partial differential equation obtained for multiplicative chains, and in fact includes a wider class of partial differential equations available for Markov chains of the population type, where $X(t)$ is a positive integer $N(t)$. In terms of the p.g.f. $\Pi_t(z)$, for such chains it becomes

$$\frac{\partial \Pi_t(z)}{\partial t} = \Psi\left(\log z, t, z\frac{\partial}{\partial z}\right) \Pi_t(z). \tag{3}$$

The use of an equation of this last type has been introduced independently by more than one writer, but appears first to have been made by Palm (1943). The extension of equations (1), (2) and (3) to cover explicitly more than one variable is evident. It will be found that the appropriate equation for many Markov chains can be written down at once in terms of the possible transitions. For example, if we return to the simplified electron-photon cascade model discussed in § 3·4 (equation (18)), we have the following schematic picture of the possible transitions:

Type of transition	Rate	Operator
$w \to z^2$	λ_1	$\partial/\partial w$
$w \to 1$	μ_1	$\partial/\partial w$
$z \to zw$	λ_2	$\partial/\partial z$
$z \to 1$	μ_2	$\partial/\partial z$

This gives rise to the equation

$$\frac{\partial \Pi_t(z, w)}{\partial t} = [\lambda_1(z^2 - w) + \mu_1(1-w)]\frac{\partial \Pi_t(z, w)}{\partial w}$$
$$+ [\lambda_2(zw - z) + \mu_2(1-z)]\frac{\partial \Pi_t(z, w)}{\partial z}.$$

If, as in immigration, the chance of transition were independent of the number of individuals, the transition would be of the type $1 \to z$, and correspondingly, the operator would be 1. If it were proportional to the product of the numbers of two types of individual, as in molecular association in problems of statistical mechanics or in epidemiological problems where the chance of infection depends on contact between infected and uninfected persons (see § 4·4), the transition would be of the type $zw \to \ldots$, and, correspondingly, the operator $\partial^2/\partial z\,\partial w$ would appear (in the epidemiological application if z relates to uninfected persons and w to infected persons, the transition is $zw \to w^2$). Unfortunately higher-order terms in the variables $z, w, \ldots$ or in the operators are liable to lead to intractability in the solution, even if the equation itself is simple to write down. Sometimes when the equilibrium or limiting distribution is of interest, as in statistical-mechanical applications, this is obtainable, even when the complete distributional solution at any time t is not (see, for example, the discussion of the equilibrium distributions in quantum kinetics by Moyal (1949)).

Returning to the general equation (1), we note that if the limit $E_{\Delta X}\{[e^{i\phi \Delta X(t)} - 1]/\Delta t\}$ exists for given X as $\Delta t \to 0$, then we have also the same limit

$$\lim_{\Delta t \to 0} \log E_{\Delta X}\{e^{i\phi \Delta X(t)}\}/\Delta t = \Psi(i\phi, t, X).$$

This indicates, as noted in § 3·1, that Ψ has the local structure of an additive process, and contains two possible components, a normal or *diffusion* term, and a discontinuous or *transition* term. The Markov chains we have been discussing obviously belong to the latter type. For pure diffusion processes containing merely the first component, we have

$$\Psi(i\phi, t, X) = m(t, X)\, i\phi - \tfrac{1}{2}\sigma^2(t, X)\, \phi^2,$$

whence (1) becomes

$$\frac{\partial C_t(\phi)}{\partial t} = E_X\{[m(t, X)\, i\phi - \tfrac{1}{2}\sigma^2(t, X)\, \phi^2]\, e^{i\phi X}\}.$$

Inverting this equation to obtain the corresponding diffusion equation for the probability density, we have in general

$$\frac{\partial f_t(x)}{\partial t} = -\frac{\partial}{\partial x}[f_t(x)\, m(t, x)] + \frac{1}{2}\frac{\partial^2}{\partial x^2}[f_t(x)\, \sigma^2(t, x)]. \tag{4}$$

Consider now the corresponding formalism for the backward equation. We must now insert the initial condition $X = x_0$ at t_0 explicitly in the distribution function $F(x \mid x_0, t_0)$, and may write

$$F(x \mid x_0, t_0) = \int F(x \mid x_0 + \Delta x_0, t_0 + \Delta t_0)\, dG(\Delta x_0 \mid x_0, t_0),$$

or in terms of $C_t(\phi)$,

$$C_t(\phi \mid x_0, t_0) = \int C_t(\phi \mid x_0 + \Delta x_0, t_0 + \Delta t_0) \, dG(\Delta x_0 \mid x_0, t_0),$$

where the integral is for the conditional variable Δx_0 with distribution function G. This leads to the differential equation

$$\frac{\partial C_t(\phi)}{\partial t_0} = - \lim_{\Delta t_0 \to 0} \int [C_t(\phi \mid x_0 + \Delta x_0, t_0 + \Delta t_0) - C_t(\phi \mid x_0, t_0 + \Delta t_0)] / \Delta t_0$$
$$\times \, dG(\Delta x_0 \mid x_0, t_0), \quad (5)$$

considered for variable x_0. By taking some appropriate transform, with respect to x_0, of $F(x)$ (or $C(\phi)$), an equation resembling (2) (or its transform with respect to x_0) may be obtained. However, some precautions on the admissible joint range of integration of x_0 and $x_0 + \Delta x_0$ are necessary, and equation (5) seems often more useful as it stands. For example, if $G(\Delta x_0 \mid x_0, t_0)$ represents a pure diffusion distribution, we easily obtain the analogue of (4)

$$-\frac{\partial f_t(x \mid x_0, t_0)}{\partial t_0} = m(t_0, x_0) \frac{\partial f_t(x \mid x_0, t_0)}{\partial x_0} + \tfrac{1}{2}\sigma^2(t_0, x_0) \frac{\partial^2 f_t(x \mid x_0, t_0)}{\partial x_0^2}. \quad (6)$$

Further, in the important case of multiplicative chains, $C_t(\phi \mid x_0)$ can always be expressed as $[C_t(\phi \mid 1)]^{x_0}$, enabling an equation for variable x_0 to be converted to one for $C_t(\phi \mid 1)$. Even without this simplification equation (5) is valuable for studying the admissible solutions of Markov processes. We shall examine it further in the case of homogeneous transition processes (i.e. with no diffusion component).

Let the chance of a non-zero transition $\Delta X \leqslant y$ from x in time dt be $g(y \mid x) \, dt$ and write

$$\int_{-\infty}^{\infty} dg(y \mid x) = h(x).$$

Then (5) is equivalent to the integral equation, alternatively written down from the renewal or regenerative property of x and $x + y$,

$$C_t(\phi \mid x) = \int_0^t \int_{-\infty}^{\infty} C_u(\phi \mid x + y) \, e^{-(t-u)h(x)} \, du \, dg(y \mid x) + e^{i\phi x} e^{-th(x)}. \quad (7)$$

In this equation one transition has occurred in the first right-hand term at time $t - u$, and if we define $C_t^{(n)}$ as the solution compounded of zero, one, ..., n transitions only (in time t), we have the iterative solution of (7)

$$C_t^{(n)}(\phi \mid x)$$
$$= \int_0^t \int_{-\infty}^\infty C_u^{(n-1)}(\phi \mid x+y)\, e^{-(t-u)h(x)}\, du\, dg(y \mid x) + e^{i\phi x}\, e^{-th(x)}. \quad (8)$$

We could, alternatively, have written down equivalent equations for $F(x_t \mid x)$ and $F^{(n)}(x_t \mid x)$, where $F^{(n)}(x_t \mid x)$ is necessarily increasing with n and bounded above by unity, and hence has a limit $F(x_t \mid x)$. If $F(\infty \mid x) = 1$, this limit is a valid distribution representing the solution to our equation. If $F(\infty \mid x) < 1$, it implies that part of the total probability has disappeared, and $F(x_t \mid x)$ would only represent the solution in a somewhat wider sense. We shall not consider such cases further here (for a detailed discussion see Moyal, 1957).

Taking the Laplace transform of (7) gives us, say,

$$L_\psi(\phi \mid x) = \frac{\displaystyle\int_{-\infty}^\infty L_\psi(\phi \mid x+y)\, dg(y \mid x) + e^{i\phi x}}{\psi + h(x)}. \quad (9)$$

If $F(\infty \mid x) = 1$, $C(0 \mid x) = 1$ and $L_\psi(0 \mid x) = 1/\psi$. This enables the conditions for a valid solution (in the sense just defined) to be examined. Thus for a homogeneous 'birth-and-death' process extending the 'birth' process of §3·2 to include transitions up or down by one, we have

$$dg(y \mid x) = [\lambda_x \delta(y-1) + \mu_x \delta(y+1)]\, dy,$$

where $\delta(y)$ is the Dirac δ-function. Thus (9) becomes

$$L_\psi(\phi \mid x) = \frac{\lambda_x L_\psi(\phi \mid x+1) + \mu_x L_\psi(\phi \mid x-1) + e^{i\phi x}}{\lambda_x + \mu_x + \psi}. \quad (10)$$

For $\phi = 0$, $\psi > 0$, this gives, for the quantity

$$J_\psi(x) \equiv 1 - \psi L_\psi(0 \mid x) \geqslant 0,$$

$$J_\psi(x) = \frac{\lambda_x J_\psi(x+1) + \mu_x J_\psi(x-1)}{\lambda_x + \mu_x + \psi}. \quad (11)$$

This recurrence relation in x should be satisfied only by $J_\psi(x) \equiv 0$. For convenience we assume $\lambda_x \neq 0$ $(x = 0, 1, \ldots)$, for otherwise it is obvious that $J_\psi(x) = 0$ for all initial values not greater than the maximum x for which $\lambda_x = 0$, and we may re-number the states upwards from this x as zero (in the frequent case where $\lambda_0 = 0$ this would imply re-numbering from $x = 1$). In the particular case $\mu_x = 0$ for all x, (11) gives

$$J_\psi(x) = J_\psi(n) \bigg/ \prod_{i=x}^{n-1} (1 + \psi/\lambda_i);$$

$J_\psi(n)$ is from its definition necessarily bounded for all n, so that $J_\psi(x) \equiv 0$ if and only if the infinite product diverges, i.e. if and only if $\Sigma 1/\lambda_i$ diverges, the condition quoted in § 3·2. In the case λ_x and μ_x both non-zero, D. G. Kendall has shown (unpublished; cf. also Dobrušin, 1952) that (11) permits no solution other than $J_\psi(x) = 0$ if and only if ΣW_i diverges, where

$$W_i = \frac{1}{\lambda_i} + \frac{\mu_i}{\lambda_i \lambda_{i-1}} + \ldots + \frac{\mu_i \mu_{i-1} \ldots \mu_1}{\lambda_i \lambda_{i-1} \ldots \lambda_1 \lambda_0}. \tag{12}$$

For example, for the multiplicative birth-and-death process for which λ_n and μ_n are proportional to n, $\Sigma 1/\lambda_i$ and hence also ΣW_i diverge, and thus a valid solution for which $F(\infty) = 1$ exists for all finite t. It may be noticed from the solution of the simple multiplicative birth-and-death process with $\lambda_n = n\lambda$, $\mu_n = n\mu$ given in § 3·4 that $F(\infty)$ may not remain unity as $t \to \infty$; thus, for $\lambda > \mu$, $F(\infty) = \Pi_\infty(1) = (\mu/\lambda)^x$ for x individuals at $t = 0$, showing that the probability not absorbed at zero all moves eventually to infinity, so that the process is still dissipative.

Spectral expansions. It will be recalled that in the comparatively elementary case of finite Markov chains (§ 3·2), the probability solution was expressible in terms of the latent roots and vectors of the matrix e^{Rt}. We would therefore expect such solutions if appropriately generalized to be available in principle for more general homogeneous Markov processes, as was noted in § 3·2 in the case of Markov chains with an enumerably infinite number of states. Particular solutions of this kind are to be found in the literature; perhaps the two most general classes of process which have been discussed in detail in such terms are (i) the homogeneous 'birth-and-death' process as defined above (see, for example, Karlin and McGregor, 1957), (ii) the diffusion equation (4) above (in the time-homogeneous case). In this last case the properties of the spectral resolution (following the development given by J. Keilson, 1963) are briefly indicated below.

We write $f_t(x) = w(x) v(x, t)$, where

$$w(x) \propto \frac{1}{\sigma^2(x)} \exp\left\{ \int^x \frac{2m(u)\,du}{\sigma^2(u)} \right\}. \tag{13}$$

Then the equation for $v(x, t)$ may be written

$$w \frac{\partial v}{\partial t} = \frac{1}{2} \frac{\partial}{\partial x}\left(\sigma^2 w \frac{\partial v}{\partial x} \right) = Lv, \tag{14}$$

say, where L is the linear operator defined by (14). If the solution is sought within a finite range of x, say a to b, with appropriate (e.g. absorbing or reflecting) boundary conditions, then it may be shown that the latent root (or 'eigenvalue') equation

$$Lv = \beta wv$$

has an infinite but simple set of roots $0 \geqslant \beta_0 > \beta_1 \ldots$ and the spectral resolution exists

$$v(x, t) = \sum_{r=0}^{\infty} e^{\beta_r t} v_r(0) \psi_r(x), \tag{15}$$

where $\psi_r(x)$ is the 'eigenfunction' corresponding to β_r, i.e.

$$L\psi_r(x) = \beta_r w(x) \psi_r(x)$$

and $\qquad v_r(0) = \int_a^b \psi_r(u) f_0(u)\,du.$

In terms of the roots β_r and corresponding (normalized) functions $\psi_r(x)$ the complete solution $f_t(x)$ becomes from (15)

$$f_t(x) = w(x) \sum_{r=0}^{\infty} \left\{ \int_a^b \psi_r(u) f_0(u)\,du \right\} \psi_r(x)\, e^{\beta_r t}. \tag{16}$$

For the initial condition $x = x_0$, i.e. $f_0(x) = \delta(x - x_0)$, this gives

$$f_t(x \,|\, x_0) = w(x) \sum_{r=0}^{\infty} \psi_r(x_0)\, \psi_r(x)\, e^{\beta_r t}, \tag{17}$$

and for the particular initial distribution $w(x)$,

$$f_t(x) = w(x)\,w(x_0) \sum_{r=0}^{\infty} \psi_r(x_0)\,\psi_r(x)\,e^{\beta_r t}, \tag{18}$$

a solution which is symmetric in x and x_0. It will be seen that the substitution $f_t(x) = w(x)\,v(x, t)$ has enabled us in this class of cases to express the solution in terms of a *single* orthogonal set of functions $\psi_r(x)$, in contrast with the *bi-orthogonal* set of vectors in § 3·2 (a similar 'symmetrization' is possible with the birth-and-death process). In the case of reflecting boundaries and an 'equilibrium' distribution $f_\infty(x)$, $\psi_0(x) = \psi_0$, $\beta_0 = 0$, and $f_\infty(x) \propto w(x)$. In the absorbing boundary case, the first root β_0 is negative, corresponding to a continual 'loss' of probability at the boundary. We have asymptotically in this case

$$f_t(x) \sim w(x)\,\psi_0(x)\,e^{\beta_0 t}. \tag{19}$$

The difficulty remains of determining $\psi_0(x)$, but Keilson has pointed out that if we define

$$\beta\{\xi\} = [\xi, L\xi]/[\xi, w\xi],$$

where

$$[\xi, \eta] = \int_a^b \xi(u)\,\eta(u)\,du,$$

then $\beta\{\xi\} \leqslant \beta_0$. and for $\xi \sim \psi_0(x)$, any small variation $\delta\beta\{\xi\} \sim 0$, a result which may facilitate the numerical determination of $\psi_0(x)$ in otherwise intractable problems.

The reader should work through the special case $m(x) = 0$, $\sigma^2(x) = \sigma^2$, under the alternative conditions $v|_{a,b} = 0$ (absorbing), $\partial v/\partial x|_{a,b} = 0$ (reflecting), with, say, $a = -b$, $x_0 = 0$. The solution in the absorbing case corresponding to equation (17) is:

$$f_t(x) = \sum_{r=0}^{\infty} \frac{(-1)^r}{b} \cos\left[(r+\tfrac{1}{2})\frac{\pi x}{b}\right] e^{-\frac{1}{2}(r+\frac{1}{2})^2 \pi^2 \sigma^2 t/b^2}, \tag{20}$$

which should be compared with the solution obtained by other methods in this case (cf. for example, equation (11), § 3·1).

First passage times. We will at this stage note some useful formulae for first passage probabilities and times for the diffusion equation (4) or (6). In the backward equation (6) (in the time-homogeneous case) write

$$F_t(x) = \int_{-\infty}^{x} f_t(x)\, dx, \quad P_0(t) = F_t(a), \quad P_1 = 1 - F_t(b),$$

$$P(t) = P_0(t) + P_1(t),$$

$$L(s) = L_0(s) + L_1(s) = \int_0^{\infty} e^{-su} \frac{\partial P(u)}{\partial u}\, du.$$

Then we find that the distribution of first passage time to a or b is determined by the equation (cf. Cox and Miller, 1965, §5·10)

$$\tfrac{1}{2}\sigma^2(x_0) \frac{\partial^2 L_i}{\partial x_0^2} + m(x_0) \frac{\partial L_i}{\partial x_0} = sL_i$$

$$\left(L_i(x_0 = a \text{ or } b) = \begin{cases} 0 & \text{or} \quad 1 \quad (i = 1), \\ 1 & \text{or} \quad 0 \quad (i = 0)). \end{cases}\right) \quad (21)$$

One method of solution of (21) when moments exist is to write

$$L_i = \sum_{r=0}^{\infty} (-s)^r L_i^{(r)}(x_0)/r!$$

then

$$L_i^{(0)} = P_i, \quad P_1 = \int_a^{x_0} \phi(u)\, du \bigg/ \int_a^b \phi(u)\, du, \quad P_0 = 1 - P_1, \quad (22)$$

and
$$L_i^{(r)} = \left[\int_a^b \phi(u)\, du \right] \left[P_1 \int_{x_0}^b 2r P_0 w(u)\, L_i^{(r-1)}\, du \right.$$

$$\left. + P_0 \int_a^{x_0} 2r P_1 w(u)\, L_i^{(r-1)}\, du \right], \quad (23)$$

where
$$\phi(u) = \exp\left\{ -2 \int_a^u m(u)\, du/\sigma^2(u) \right\},$$

and
$$w(u) = 1/[\sigma^2(u)\,\phi(u)]$$

(cf. equation (13)). Alternatively, we may solve for $P_i(t)$ directly. Thus with the spectral expansion solution $f_t(x|x_0)$ in (17) we have immediately

$$P_0(t) = \int_{-\infty}^a f_t(x|x_0)\, dx, \quad P_1(t) = \int_b^{\infty} f_t(x|x_0)\, dx. \quad (24)$$

3·51 Multidimensional diffusion equations with radial symmetry. Finally in this chapter it is of some interest in both

physics and biology to derive the form of two- and three-dimensional diffusion equations with radial symmetry. In particular, in two dimensions D. G. Kendall (1974) has used a Brownian motion with a mean drift towards the origin as a model for bird navigation. In three dimensions a similar drift to the origin produces models analogous to the quantum-mechanical model for the hydrogen atom (see, for example, Cane, 1967).

Consider first the two-dimensional case $\mathbf{r} \equiv (x, y)$. The forward equation (4) of § 3·5 becomes, in the time-homogeneous case,

$$\frac{\partial f_t(\mathbf{r})}{\partial t} = -\frac{\partial}{\partial x}\left[f_t(\mathbf{r})\,m_x(\mathbf{r})\right] - \frac{\partial}{\partial y}\left[f_t(\mathbf{r})\,m_y(\mathbf{r})\right]$$

$$+\frac{1}{2}\left\{\frac{\partial^2}{\partial x^2}\left[f_t(\mathbf{r})\,\sigma_x^2(\mathbf{r})\right] + 2\,\frac{\partial^2}{\partial x\,\partial y}\left[f_t(\mathbf{r})\,w_{xy}(\mathbf{r})\right]\right.$$

$$\left.+\frac{\partial^2}{\partial y^2}\left[f_t(\mathbf{r})\,\sigma_y^2(\mathbf{r})\right]\right\}. \quad (1)$$

In the case of isotropy and homogeneity apart from a drift to the origin we put

$$\sigma_x^2 = \sigma_y^2 = \sigma^2, \quad w_{xy} = 0, \quad m_x(\mathbf{r}) = m(r)\cos\theta,$$

$$m_y(\mathbf{r}) = m(r)\sin\theta,$$

where $x = r\cos\theta$, $y = r\sin\theta$. Changing to polar coordinates gives us

$$\frac{\partial f_t(\mathbf{r})}{\partial t} + \frac{1}{r}\frac{\partial}{\partial r}\left[rm(r)f_t(\mathbf{r})\right] = \frac{\sigma^2}{2r}\frac{\partial}{\partial r}\left[r\,\frac{\partial f_t(\mathbf{r})}{\partial r}\right] + \frac{\sigma^2}{2r^2}\frac{\partial^2 f_t(\mathbf{r})}{\partial^2\theta^2}. \quad (2)$$

Suppose in particular $m(r)$ is constant and negative, and we consider stationary solutions $f(\mathbf{r})$. Then

$$-m\frac{\partial}{\partial r}\left[rf(\mathbf{r})\right] = \frac{\sigma^2}{2}\left\{\frac{\partial}{\partial r}\left[r\,\frac{\partial f(\mathbf{r})}{\partial r}\right] + \frac{1}{r}\frac{\partial^2 f(\mathbf{r})}{\partial^2\theta^2}\right\} \quad (3)$$

and, if $f(\mathbf{r})$ is isotropic and a function of r only,

$$\frac{-2m}{\sigma^2}\frac{\partial}{\partial r}\left[rf(r)\right] = \frac{\partial}{\partial r}\left[r\,\frac{\partial f(r)}{\partial r}\right],$$

whence $$\frac{-2m}{\sigma^2}\,rf(r) = r\,\frac{\partial f(r)}{\partial r} + C.$$

For large r, $f(r)$ must tend to zero, whence $C = 0$, and

$$\frac{1}{f(r)} \frac{\partial f(r)}{\partial r} = \frac{-2m}{\sigma^2},$$

whence

$$f(r) \propto e^{-2mr/\sigma^2}. \tag{4}$$

In three dimensions with a drift $m(r)$ and otherwise isotropy and homogeneity, we have similarly in polar coordinates

$$\frac{\partial}{\partial t} f_t(\mathbf{r}) + \frac{1}{r^2} \frac{\partial}{\partial r} \left[r^2 m(r) f_t(\mathbf{r}) \right]$$

$$= \frac{\sigma^2}{2} \left\{ \frac{1}{r^2} \frac{\partial}{\partial r} \left[r^2 \frac{\partial f_t(\mathbf{r})}{\partial r} \right] + \frac{1}{r^2 \sin\theta} \frac{\partial}{\partial \theta} \left[\sin\theta \frac{\partial f_t(\mathbf{r})}{\partial \theta} \right] \right.$$

$$\left. + \frac{1}{r^2 \sin^2\theta} \frac{\partial^2 f_t(\mathbf{r})}{\partial^2 \phi^2} \right\}, \tag{5}$$

where $x = r\cos\theta\cos\phi$, $y = r\sin\theta\cos\phi$, $z = r\sin\phi$. Again let us consider the case $m(r) = -m$, and stationary solutions $f(r)$. Then

$$\frac{-2m}{\sigma^2} \frac{\partial}{\partial r} [r^2 f(r)] = \frac{\partial}{\partial r} \left[r^2 \frac{\partial f(r)}{\partial r} \right],$$

and we obtain as before

$$f(r) \propto e^{-2mr/\sigma^2}. \tag{6}$$

The probability of being in the range r, $r + dr$ is $2\pi r f(r) \, dr$ in the two-dimensional case, and $4\pi^2 r^2 f(r) \, dr$ in the three-dimensional case.

Notice more generally that for radial symmetry with

$$m(\mathbf{r}) = m(r), \quad \sigma^2(\mathbf{r}) = \sigma^2(r),$$

the equilibrium solution $f(r)$ in either two or three dimensions satisfies

$$m(r)f(r) = \frac{1}{2} \frac{\partial}{\partial r} [\sigma^2(r) f(r)],$$

whence

$$f(r) \propto \frac{1}{\sigma^2(r)} \exp \int \frac{2m(r)}{\sigma^2(r)} \, dr, \tag{7}$$

(cf. equation (13), §3·5). If we write $f_t(\mathbf{r}) = f(r) v(\mathbf{r}, t)$, then the equation for v becomes in the above case with $\sigma^2(r) = \sigma^2$, $m(r) = -m$,

$$\frac{\partial v}{\partial t} + \frac{m \, \partial v}{\partial r} = \frac{\sigma^2}{2} \nabla^2 v, \tag{8}$$

where ∇^2 denotes the Laplacian operator as in equations (2) or (5). Alternatively, with $f_t(\mathbf{r}) = \sqrt{(f(r))}\,\psi(\mathbf{r}, t)$, then we find

$$\frac{\partial \psi}{\partial t} + \psi \left[\frac{Dm}{2r} + \frac{m^2}{2\sigma^2} \right] = \frac{\sigma^2}{2}\,\nabla^2 \psi, \qquad (9)$$

where $D = 1$ in two dimensions, and 2 in three. Note that the above equations represent Brownian-type motion for $\mathbf{r}$; for Brownian motion for the velocity, see § 5·21; for further critical discussion of the analogy with quantum mechanics, see Bartlett (1975).

Chapter 4

MISCELLANEOUS STATISTICAL APPLICATIONS

4·1 Some applications of the random walk or additive process

In this chapter we select some more problems in mathematical statistics (excluding applications in physics), both for their intrinsic interest and in order to demonstrate further the application of the methods so far developed. The topics discussed, especially in this first section, may appear somewhat miscellaneous, but are nevertheless linked by means of the appropriate stochastic process technique. Thus as further examples of 'random walks' with absorbing boundaries we discuss in this section (i) a problem of collective insurance risk, (ii) the sampling theory of distribution functions, and (iii) sequential analysis.

A problem in insurance risk. Consider the insurance risk of a firm with initial capital b, increasing premiums μ per unit time, and claims occurring at random at an average rate ν. The amount of a claim follows the distribution with m.g.f. $M(\theta)$. It is required to know the chance P_b of the firm going bankrupt.

This problem is merely a generalization of the gambler's ruin problem. Although here we have a problem in continuous time, it has been noted that Wald's identity (§ 2·1) is still applicable. We take the other boundary $a \to -\infty$ (measuring for convenience claims positively, bankruptcy corresponding to $X = b$). We assume $\nu M'(0) - \mu < 0$ (otherwise $P_b = 1$), and the relation for P_b is then

$$P_a E_a\{e^{X(T)\theta_0}\} + P_b E_b\{e^{X(T)\theta_0}\} = 1, \tag{1}$$

where θ_0 (> 0) is the root of the equation

$$\exp\{\nu(M(\theta) - 1) - \mu\theta\} = 1 \tag{2}$$

(cf. formula (7) of § 3·1, the extra term $\exp\{-\mu\theta\}$ arising from the negative constant drift μ). In (1) note that for $X(T)$ in the first term, $X(T) \leqslant a$, and as $a \to -\infty$, we obtain

$$P_b = 1/E_b\{e^{X(T)\theta_0}\} \leqslant e^{-b\theta_0}. \tag{3}$$

In the asymptotic case when b is large compared with individual claims, we have the approximate equality $P_b \sim e^{-b\theta_0}$.

This result has been given by various writers using more direct methods (see, for example, Segerdahl, 1939), and it is of interest to demonstrate the equivalence of (3) to the usual solution.

The amount of capital at time t is given by

$$Z(t) = b - X(t) = b + \mu t - Y(t),$$

where the total amount $Y(t)$ of all claims has m.g.f. given by $\exp\{\nu t (M(\theta) - 1)\}$. Let the probability of bankruptcy at some time subsequent to a stage at which the capital is Z be $P(Z)$. Then from the situation at time $t + \Delta t$ we deduce that

$$P(Z) = (1 - \nu\Delta t)\, P(Z + \mu\Delta t) + \nu\Delta t \int_0^Z P(Z - u)\, dF(u)$$
$$+ \nu\Delta t [1 - F(Z)] + o(\Delta t),$$

where $F(x)$ is the distribution function corresponding to $M(\theta)$. This gives as $\Delta t \to 0$ (it is assumed that the derivative $P'(Z)$ exists)

$$-\mu P'(Z) + \nu P(Z) = \nu [1 - F(Z)] + \nu \int_0^Z P(Z - u)\, dF(u),$$

or, for $Z = b$ at $t = 0$, a probability $P(b)$ satisfying the equation

$$-\frac{\mu}{\nu} P'(b) + P(b) = 1 - F(b) + \int_0^b P(b - u)\, dF(u).$$

Integrating the last term by parts, we obtain

$$-\frac{\mu}{\nu} P'(b) + P(b) = 1 - F(b) - \left\{ P(b - u)[1 - F(u)] \right\}_0^b$$
$$- \int_0^b P'(b - u)\,[1 - F(u)]\, du,$$

or $\quad \dfrac{\mu}{\nu} P'(b) = -[1 - F(b)] + \dfrac{\partial}{\partial b} \displaystyle\int_0^b P(b - u)\,[1 - F(u)]\, du.$

Integrating with respect to b from b to ∞, we have finally an integral equation of Volterra type for $P(b)$,

$$P(b) = \int_b^\infty \frac{\nu}{\mu} [1 - F(u)]\, du + \int_0^b \frac{\nu}{\mu} P(b - u)\,[1 - F(u)]\, du.$$

Methods of solution of an integral equation of this type were indicated in § 2·11. The asymptotic solution for large b is given by $P(b) \sim C e^{-b\theta_0}$, where θ_0 $(= -\psi_0)$ is the positive root of the equation

$$\int_0^\infty \frac{\nu}{\mu} e^{\theta u} [1 - F(u)] \, du = 1,$$

or, on integration by parts, of

$$\left\{ \frac{\nu}{\mu\theta} e^{\theta u} [1 - F(u)] \right\}_0^\infty + \int_0^\infty \frac{\nu}{\mu\theta} e^{\theta u} \, dF(u) = 1,$$

i.e. of

$$\nu(M(\theta) - 1) - \mu\theta = 0,$$

which is equivalent to equation (2). To identify the two asymptotic solutions we have to show also that $C \sim 1$. Quoting again from the solution in § 2·11 we have under the conditions given there

$$C = \alpha(-\theta_0) \Big/ \frac{\partial \gamma(-\theta_0)}{\partial \theta_0},$$

where

$$\alpha(-\theta) = \int_0^\infty e^{\theta u} \left[\int_u^\infty \frac{\nu}{\mu} [1 - F(v)] \, dv \right] du,$$

which on successive integration by parts reduces to

$$\alpha(-\theta) = -\frac{\nu}{\mu\theta} \int_0^\infty v \, dF(v) + \frac{\gamma(-\theta)}{\theta},$$

$\gamma(-\theta)$ being given by $\nu[M(\theta) - 1]/(\mu\theta)$. Hence

$$\alpha(-\theta_0) = \frac{1}{\theta_0} - \frac{\nu\mu_1}{\mu\theta_0},$$

where μ_1 is the mean of the $F(u)$ distribution; and

$$\frac{\partial \gamma(-\theta_0)}{\partial \theta_0} = -\frac{\nu[M(\theta_0) - 1]}{\mu\theta_0^2} + \frac{\nu M'(\theta_0)}{\mu\theta_0}$$

$$= -\frac{1}{\theta_0} + \frac{\nu M'(\theta_0)}{\mu\theta_0}.$$

Thus

$$C = \frac{\mu - \nu\mu_1}{\nu M'(\theta_0) - \mu}. \tag{4}$$

To proceed further, we assume conditions under which as b increases the probability $P(b)$ is still of some interest as a non-vanishing quantity, so that θ_0 must become small. Hence we expand $M(\theta_0)$ in powers of θ_0 and equation (2) becomes, say,

$$\nu(\mu_1\theta_0 + \tfrac{1}{2}\mu_2\theta_0^2 + o(\theta_0^2)) - \mu\theta_0 = 0,$$

or
$$\theta_0 = \frac{\mu - \mu_1\nu + o(\theta_0)}{\tfrac{1}{2}\mu_2\nu}. \tag{5}$$

Hence
$$\nu M'(\theta_0) - \mu = \nu\mu_1 - \mu + \mu_2\nu\theta_0 + o(\theta_0)$$
$$= \mu - \nu\mu_1 + o(\theta_0).$$

In (5) the assumption that θ_0 is small is satisfied for finite $\mu_2\nu$ when the mean rate of increase in capital, $\mu - \mu_1\nu$, is small. We have then $\mu - \nu\mu_1 = O(\theta_0)$ and equation (4) reduces to $C = 1 + o(1)$.

It will be noticed that the first method gives the simpler solution (3), which as an *inequality* holds for all values of b; the second method may be compared with the integral equation methods of §§ 2·1 and 2·11, providing an exact solution if required.

The sampling theory of distribution functions. If a sample of n independent observations Z all following the same distribution function $F(z)$ is used to construct the empirical distribution function $F_n(z) = n_z/n$, where n_z is the number of observations whose values are less than or equal to z, it is known that $F_n(z) \to F(z)$, as $n \to \infty$, in the sense that†

$$P\{\lim_{n\to\infty} \max [|\, F_n(z) - F(z)\,|] = 0\} = 1.$$

A more useful result from a practical standpoint is Kolmogorov's asymptotic expression for

$$P_n \equiv P\{\max [|\, F_n(z) - F(z)\,|] < \lambda/\sqrt{n}\}$$

as n becomes large, in the case of $F(z)$ continuous. It is readily seen that the value of P_n is invariant to the shape of the increasing continuous function $F(z)$, for we may transform by a $(1,1)$ correspondence to the new random variable $Y \equiv F(Z)$, which has a rectangular distribution $G(y)$ in the interval $(0,1)$; thus, since $G_n(y) \equiv F_n(z)$, P_n need only be evaluated for a rectangular distribution.

† From now on we use $|x|$ for 'absolute value of x', and not $\|x\|$ as in Chapters 2 and 3 (from § 2·2).

Now if n independent observations Y are distributed at random in the interval $(0, 1)$, we cannot immediately apply the theory of the random walk with independent increments to the random function N_y/n, since $N_1/n = n/n = 1$ and the increments are obviously not independent. Consider, however, the joint distribution of $n+1$ independent values W_i from the same distributional law $H(w) = 1 - e^{-\alpha w}$ $(w \geqslant 0)$. These values have frequency law

$$\alpha^{n+1} \exp \left\{ -\alpha \sum_{i=1}^{n+1} w_i \right\} \prod_{i=1}^{n+1} dw_i.$$

The distribution of $U \equiv \Sigma_i W_i$ is well known to have frequency law

$$\alpha^{n+1} u^n e^{-\alpha u} du/n!,$$

whence the distribution of $W_1, \ldots, W_n$, given $U = u_0$, is

$$n! \prod_{i=1}^{n} dw_i/u_0^n \quad \left(\sum_{i=1}^{j} w_i \leqslant u_0 \text{ for all } j \leqslant n \right),$$

or for $u_0 = 1$,

$$n! \prod_{i=1}^{n} dw_i \quad \left(\sum_{i=1}^{j} w_i \leqslant 1 \text{ for all } j \leqslant n \right).$$

Transforming to the cumulative variables

$$U_j = \sum_{i=1}^{j} W_i \quad (j = 1, \ldots, n),$$

we obtain the equivalent frequency law

$$n! \prod_{j=1}^{n} du_j \quad (u_j \leqslant u_{j+1} \leqslant 1 \text{ for all } j < n),$$

which is the distribution of an *ordered* set of n independent variables uniformly distributed in the interval $(0, 1)$. We may thus interpret the random quantity $X(j) = U_j - j/n$ as a random walk conditional on $X(n+1) = -1/n = o(1/\sqrt{n})$.

To apply the results of §3·1 (equations (13) and (14)) we arrange that $E\{X\} = 0$ by choosing the mean $1/\alpha$ of W to be $1/n$. The increase in variance per unit 'time' is then $n/\alpha^2 = 1/n$. The standard deviation of fluctuations in X being of order $1/\sqrt{n}$, we take b to be of order $1/\sqrt{n}$ (b is thus large compared with $1/n$, the order of magnitude of individual increments of X). Putting

$t = 1$ and $b = \lambda/\sqrt{n}$ in equations (13) and (14) of §3·1, we have the asymptotic probability of the deviation $U_j - j/n$ not exceeding $\lambda/\sqrt{n}$, either in both directions or in one.

As the theoretical function $G(y) = y$ is at 45° to the y-axis, the boundaries a horizontal distance $\pm \lambda/\sqrt{n}$ from this function are also the same vertical distance; moreover, the 'corners' of the function $G_n(y)$ are the points $(U_1, 0)$, $(U_1, 1/n)$, $(U_2, 1/n)$, $(U_2, 2/n)$, ..., which include the points defined by $X(j)$ and also corresponding points a vertical distance $1/n$ below them. As $1/n \ll \lambda/\sqrt{n}$, all these points are included within the $\pm \lambda/\sqrt{n}$ boundary with the same asymptotic probability, which is thus

$$\sum_{s=-\infty}^{\infty} (-1)^s e^{-2\lambda^2 s^2}. \tag{6}$$

If we require the probability of not exceeding the deviation $\lambda/\sqrt{n}$ on one side only, we have similarly the result

$$1 - e^{-2\lambda^2}. \tag{7}$$

Sequential analysis. It is not the intention to discuss at length here the sampling technique in statistics known as sequential analysis (see Wald, 1947), which is a method of statistical sampling inspection in cases where the units sampled may conveniently be examined serially. Its main principles, and close relation to random-walk theory in the case of independent sample observations, can, however, be indicated. We consider the probability (or probability density in the case of a continuous distribution)

$$p_0 \equiv P\{S \mid \alpha_0\} = \prod_{i=1}^{n} p(x_i \mid \alpha_0) \tag{8}$$

of obtaining the independent sample observations

$$S \equiv (x_1, x_2, ..., x_n)$$

on a hypothesis α_0, and similarly for an alternative hypothesis α_1, and set up a rule: when $p_0/p_1 \leqslant A$, reject α_0 in favour of α_1; when $p_0/p_1 \geqslant B$, reject α_1.

Now as n increases, $X(n) = \log p_0 - \log p_1 \equiv L_0 - L_1$, say, is from (8) for α_0 (or α_1) true a cumulative random-walk sequence, with boundaries $a = \log A$, $b = \log B$; hence $P\{N = \infty\}$, where N

is the value of n for which $X(n)$ reaches one or other boundary, is zero. Let us stipulate that the chance of error in rejecting α_0 when in fact α_0 is true is ϵ_0, and that the chance of error in rejecting α_1 when in fact α_1 is true is ϵ_1. As the inequality $p_0 \leqslant A p_1$ is always satisfied when α_1 is accepted, the over-all probability for all samples for which α_1 is accepted when α_0 is true is at most A times the probability when α_1 is true. But the over-all probability for such samples is to be ϵ_0 if α_0 is true, and $1 - \epsilon_1$ if α_1 is true. Hence

$$\epsilon_0 \leqslant A(1 - \epsilon_1).$$

Similarly $\qquad\qquad 1 - \epsilon_0 \geqslant B\epsilon_1,$

and we must have

$$A \geqslant \frac{\epsilon_0}{1 - \epsilon_1}, \quad B \leqslant \frac{1 - \epsilon_0}{\epsilon_1}. \tag{9}$$

In the asymptotic case when the excess over the boundaries may be neglected, these inequalities become equalities.

The distributional theory of size of sample required for this sampling technique follows immediately from our earlier random-walk theory. For example, the average size of sample required is from equation (24) of § 2·1

$$E\{N\} = (P_a a + P_b b)/m, \tag{10}$$

where

$$a = \log A \sim \log\{\epsilon_0/(1 - \epsilon_1)\}, \quad b = \log B \sim \log\{(1 - \epsilon_0)/\epsilon_1\}$$

and $\qquad\qquad m = E\{\log p(x \mid \alpha_0) - \log p(x \mid \alpha_1)\}.$

The asymptotic case when α_0 and α_1 are 'close' enough to require a large sample for discrimination is worth examination. For definiteness suppose α_0 and α_1 denote alternative values of an actual parameter α of the distribution of an individual observation, where $\alpha_0 - \alpha_1$ is small. Then

$$L_0 - L_1 = (\alpha_0 - \alpha_1) L_0' - \tfrac{1}{2}(\alpha_0 - \alpha_1)^2 L''(\beta_0), \tag{11}$$

where dashes denote partial differentiation with respect to α, and β_0 is some value in the interval (α_0, α_1). Similarly

$$L_0 - L_1 = (\alpha_0 - \alpha_1) L_1' + \tfrac{1}{2}(\alpha_0 - \alpha_1)^2 L''(\beta_1). \tag{12}$$

Equation (11) is relevant when $\alpha = \alpha_0$ and (12) when $\alpha = \alpha_1$. In either case $L_0 - L_1$ is a cumulative sum of independent variables for each of which

$$m \sim (\alpha_0 - \alpha_1) E\{L_0'(x_r)\} - \tfrac{1}{2}(\alpha_0 - \alpha_1)^2 E\{L_0''(x_r)\}$$

for $\alpha_0 - \alpha_1 \sim 0$ (for L'' uniformly continuous in α in the neighbourhood of α_0, α_1) or

$$m \sim \tfrac{1}{2}(\alpha_0 - \alpha_1)^2 I_0 \quad (\alpha = \alpha_0),$$

as $\quad E\{L_0'(x_r \mid \alpha_0)\} = 0, \quad E\{[L_0'(x_r \mid \alpha_0)]^2\} = -E\{L_0''(x_r \mid \alpha_0)\} = I_0,$

say, under not very restrictive conditions on L (these results are well known in the theory of statistical estimation, I_0 being the information function introduced by R. A. Fisher; see, for example, Chapter 8). Further,

$$\sigma^2 \sim (\alpha_0 - \alpha_1)^2 E\{[L_0'(x_r \mid \alpha_0)]^2\} = (\alpha_0 - \alpha_1)^2 I_0 \quad (\alpha = \alpha_0).$$

We thus have the normal asymptotic case of § 2·1 with

$$2m = \sigma^2 = (\alpha_0 - \alpha_1)^2 I_0 \quad (\alpha = \alpha_0). \tag{13}$$

Similarly from equation (12) we find

$$-2m = \sigma^2 = (\alpha_0 - \alpha_1)^2 I_1 \quad (\alpha = \alpha_1). \tag{14}$$

Equation (19) of § 2·1 thus becomes

$$P_a \sim \frac{1 - e^{-b}}{e^{-a} - e^{-b}} \quad (\alpha = \alpha_0); \qquad \frac{1 - e^b}{e^a - e^b} \quad (\alpha = \alpha_1). \tag{15}$$

If, therefore, we require $P_b = 1 - \epsilon_0$ $(\alpha = \alpha_0)$, ϵ_1 $(\alpha = \alpha_1)$, we may solve equations (15) with these values inserted for P_b and obtain

$$e^b = \frac{1 - \epsilon_0}{\epsilon_1}, \quad e^a = \frac{\epsilon_0}{1 - \epsilon_1}, \tag{16}$$

in agreement with the general formula (9). The earlier theory for the normal case also provides at once the distributional theory for sample size in this general asymptotic case.

4·2 Simple renewal as a Markov process

The continual renewal or replacement of a single article which wears out in time with a frequency distribution $f(x)\,dx$ was discussed in Chapter 2 as an example of a random walk with increments following such a distribution, but we consider it again here from the standpoint of the theory of Markov processes. It is evident that the behaviour of the process at time t depends on more than a knowledge of whether a renewal occurred at some previous *chosen* time t_0, for whether an article wears out depends on its age, i.e. on the time from the previous *random* renewal time T_0. A specification of states in terms of the *age* at any time t is thus a Markov process specification, in fact, the process is an example of the multiplicative chains considered in § 3·42 in the very particular case of no 'multiplication'.

In terms of the formalism of that section, we must allow for the ageing in time, so that in the interval dt we have the transition

$$z(x) \to z(x+dt)\{1 - \mu(x)\,dt\} + z(0)\,\mu(x)\,dt,$$

whence
$$g(z(u)|x) = \mu(x)\,[z(0) - z(x)] + \frac{\partial z(x)}{\partial x} \tag{1}$$

and
$$\frac{\partial \Pi_t(z(u)|x)}{\partial t} = \mu(x)\,[\Pi_t(z(u)\,|\,0) - \Pi_t(z(u)\,|\,x)] + \frac{\partial \Pi_t(z(u)\,|\,x)}{\partial x}. \tag{2}$$

The 'death-rate' $\mu(x)$ in (1) and (2) now depends on the age, and is assumed finite for all x. Its connection with the 'generation distribution' $f(x)\,dx$ is seen by writing (2) equivalently in its renewal or integral equation form (cf. § 3·5)

$$\Pi_t(x) = \int_0^t \Pi_{t-u}(0)\,\mu(u+x) \exp\left\{-\int_0^u \mu(v+x)\,dv\right\} du$$
$$+ z(x+t) \exp\left\{-\int_0^t \mu(u+x)\,du\right\}, \tag{3}$$

giving
$$f(x) = \mu(x) \exp\left\{-\int_0^x \mu(u)\,du\right\}, \tag{4}$$

where it is assumed that the integral in the curly bracket is divergent as $x \to \infty$, so that

$$\int_0^\infty f(x)\,dx = 1.$$

Equation (2), or alternatively (3), can be solved in terms of the Laplace transform of Π_t, and provides, for example, *both* the characteristic function of the age distribution (put $z(u) = e^{i\theta u}$)

and the expected renewal density (put $z(u) = 1$ except in the age interval $(0, dt)$; note that the initial article is also included in the renewals at $t = 0$ if $x = 0$). However, such a 'portmanteau' solution is hardly required in this example, for the age distribution can alternatively and more directly be obtained from the renewal density, an article of age u being a survivor from a renewal at a time u previously.

Thus let the age-distribution function be $H(u \mid x, t)$ for an initial article age x at $t = 0$. Then clearly

$$H(u \mid x, t) = \int_0^{u \leqslant t} \exp\left\{ -\int_0^w \mu(v)\, dv \right\} r(t - w, x)\, dw$$
$$+ \left[\exp\left\{ -\int_x^{x+t} \mu(v)\, dv \right\} \right]_{u \geqslant x+t}, \quad (5)$$

where $r(t, x)$ is the renewal density, the last term on the right is zero if $u < x + t$, and the further increment to the first term on the right is zero if $u > t$. As t increases, this gives a limiting age-distribution function independent of x

$$H(u) = r \int_0^u \exp\left\{ -\int_0^w \mu(v)\, dv \right\} dw, \quad (6)$$

where r is the limiting constant renewal density (assumed here to exist; for weak conditions ensuring this, see Smith (1954)). Noting from (4) that

$$1 - F(u) = 1 - \int_0^u f(v)\, dv = \exp\left\{ -\int_0^u \mu(v)\, dv \right\},$$

we have from (6)

$$H(\infty) = 1 = r \int_0^\infty [1 - F(u)]\, du = r \int_0^\infty u f(u)\, du$$

(in agreement with equation (2) of §3·31). For simplicity we have assumed above that the density function $f(u)$ exists, but the modifications necessary if this is not so are not difficult.

4·21 Queues. A renewal problem of a rather different kind arises in the servicing or 'renewal' of machines when the machines have to await the attention of an operator before being again ready for use. This problem then becomes one of a number classifiable under the generic title of 'queue' problems. Another familiar example is the one of customers waiting to be served, including such situations as aircraft waiting to land at a single air-station.

A number of useful results can be deduced by an elementary application of the renewal type of argument. Thus suppose we have an 'input' feeding a queue of customers waiting to be served at one 'counter', the queue size at time t being $N(t)$. This includes the customer, if any, being served. Consider an instant τ when a customer is just leaving, so that

$$N(\tau + dt) = N(\tau) - 1 = Q,$$

say. Let the service-time for the next person be V, and suppose R new customers arrive during this time V. Then the size of queue when the next person leaves at time τ' is given by

$$Q' = Q - 1 + R + \delta(Q), \tag{1}$$

where $\qquad \delta(Q) = 0 \quad (Q \neq 0), \qquad \delta(Q) = 1 \quad (Q = 0).$

If the average rate of input is too high, no stationary process can exist. We shall, however, assume stationarity with $E(Q)$, $E(Q^2)$ finite. Averaging equation (1), we obtain under these assumptions

$$E(\delta) = 1 - E(R) = 1 - \rho, \tag{2}$$

say, where ρ (< 1) is the mean number of arrivals during the mean service time $E(V)$, or the 'traffic-intensity'.[†] Also by squaring (1) before averaging and remembering that $\delta^2 = \delta$, $\delta Q = 0$, we find

$$2E\{Q(1 - R)\} = E\{(1 - R)^2\} + E\{\delta(2R - 1)\}.$$

We shall now assume further a constant (stochastic) input rate, such that R is a Poisson variable independent of Q with mean λV, say. The last equation then gives

$$E(Q) = \rho + \tfrac{1}{2}[E(R^2) - \rho]/(1 - \rho).$$

Now $\qquad\qquad E(R^2) = E_V(\lambda V + \lambda^2 V^2)$
$$= \rho + \rho^2 + \lambda^2 \sigma_V^2,$$

as $E(V) = \alpha$, say $= \rho/\lambda$. Hence

$$E(Q) = \rho + \frac{\rho^2 + \lambda^2 \sigma_V^2}{2(1 - \rho)}. \tag{3}$$

[†] For a proof that $\rho < 1$ is the stationarity condition, see Lindley (1952).

Consider next the waiting-time W for a customer. The customer leaves after time $W + V$, and Q customers will form behind him in this time at an average rate $\lambda(W + V)$. Hence

$$\lambda E(W + V) = \rho + \tfrac{1}{2}[\rho^2 + \lambda^2 \sigma_V^2]/(1 - \rho),$$

and
$$\frac{E(W)}{\alpha} = \frac{\rho}{2(1 - \rho)}\left\{1 + \frac{\sigma_V^2}{\alpha^2}\right\}. \tag{4}$$

This formula shows that for maximum efficiency the variance σ_V^2 should be zero, and that the waiting-time will be doubled on the average if the service-time has the exponential distribution characteristic of random intervals.

Similar arguments may be used to obtain the complete frequency distribution for the waiting-time W, which it will be noticed will have a non-zero probability of being zero. Let

$$\Pi(z) = E(z^Q) = E\{z^{Q-1+\delta+R}\},$$
from (1); or
$$\Pi(z) = (1 - \rho)E_{\delta=1}\{z^R\} + \rho E_{\delta=0}\{z^{Q-1+R}\}, \tag{5}$$

as from (2) $1 - \rho$ is the chance that $\delta = 1$ or $Q = 0$. Now

$$E\{z^R\} = E_V\{e^{\lambda V(z-1)}\} = H(z),$$

say, as R is a Poisson variable with mean λV for given V, and further $H(z) = \beta\{\lambda(1 - z)\}$, where

$$\beta(\psi) = E_V\{e^{-\psi V}\} = \int_0^\infty e^{-\psi v}\,dB(v)$$

is the Laplace transform of the service-time distribution $B(v)$. From (5),
$$\Pi(z) = (1 - \rho)H(z) + \rho H(z) E_{\delta=0}\{z^{Q-1}\}.$$

But also
$$\Pi(z) = (1 - \rho) + \rho E_{\delta=0}\{z^Q\}.$$

From these two equations

$$\Pi(z) = \frac{(1 - \rho)(1 - z)}{1 - z/H(z)} = \frac{(1 - \rho)(1 - z)}{1 - z/\beta\{\lambda(1 - z)\}}. \tag{6}$$

For the waiting-time distribution function $C(w)$ define

$$\gamma(\psi) = E\{e^{-\psi W}\} = \int_0^\infty e^{-\psi w}\,dC(w).$$

From the stochastic relation between Q and $W + V$, we have

$$E_Q\{z^Q\} = E_{W, V}\{e^{\lambda(W+V)(z-1)}\},$$

whence
$$\gamma(\psi)\,\beta(\psi) = \frac{(1-\rho)\,\psi/\lambda}{1 - (1 - \psi/\lambda)/\beta(\psi)},$$

or
$$\gamma(\psi) = (1-\rho)/\{1 - \lambda[1 - \beta(\psi)]/\psi\}, \qquad (7)$$

a result due to Pollaczek. As a particular case, if the service-time is a purely random interval, so that its distribution is exponential, the continuous component of the waiting-time distribution has density

$$\rho\nu e^{-\nu w} \quad (\nu = \lambda/\rho - \lambda); \qquad (8)$$

this case was considered by Erlang with particular reference to telephone 'traffic'. For this case we have also

$$H(z) = \frac{1}{1 + \rho(1-z)}, \quad \Pi(z) = \frac{1-\rho}{1-\rho z}, \qquad (9)$$

and the size of queue left by a departing customer follows the geometric law $(1-\rho)\rho^q$ $(q = 0, 1, 2, \ldots)$.

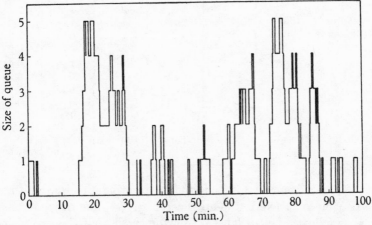

Fig. 5. Realization of a queue with a single server, starting with one customer at time zero (random intake and serving-time, with $\rho = \frac{1}{2}$).

In table 2 the results of an artificial queue realization for the case of random intake (mean interval 2 min.) and random serving time (mean interval 1 min., so that $\rho = \frac{1}{2}$) are shown. A portion of the realization is depicted in fig. 5. The agreement with theory appears fairly reasonable; it should be noticed, however, that a complete test of agreement would depend on the principles developed in Chapter 8 (see especially §8·21). The actual mean obtained for the arrival intervals was 1·96, but that for service-times came out by chance at the somewhat low figure of 0·82, which probably accounts for the rather low mean waiting-time of 0·78 for the customer.

Table 2. *Realization of a queue with random intake and serving time* ($\rho = \frac{1}{2}$)

Queue size (after service)	Frequencies		Waiting-times	Frequencies		
	Observed	Theoretical		Customer	Server	Theoretical
0	54	50·0	0	54	46	50
1	19	25·0	Non-zero:			
2	13	12·5	0– 1	20	23	19·7
3	9	6·3	1– 2	9	10	11·9
4	5	3·1	2– 3	8	7	7·2
5 or more	0	3·1	3– 4	6	7	4·4
			4– 5	2	4	2·7
Total	100	100	5– 6	0	1	1·6
Mean	0·92	1·00	6– 7	1	0	1·0
			7– 8	—	0	0·6
			8– 9	—	1	0·4
			9–10	—	0	0·2
			10–11	—	0	0·1
			11–12	—	0	0·1
			12 or more	—	1	0·1
			Total	100	100	100
			Mean	0·78	1·15	1·00

A more general method of investigating the waiting-time distribution (covering more general arrival-time distributions) is as follows. Let us label the customers so that the first waits time W_1, the second arrives at a time T later and waits W_2. Then if the serving-time for the first customer is V,

$$W_1 + V = W_2 + T$$

for $W_1 + V \geqslant T$, or if $V = U + T$,

$$W_2 = \begin{cases} W_1 + U & (W_1 + U \geqslant 0), \\ 0, & (W_1 + U < 0). \end{cases}$$

It is convenient to introduce a further variable Z which equals W for $W > 0$ and equals *minus* the time the server was previously unemployed if $W = 0$. Then

$$Z_2 = \begin{cases} Z_1 + U & (Z_1 > 0), \\ U & (Z_1 \leqslant 0). \end{cases}$$

Let $g(u)$ be the frequency function of U and $h(z)$ that of Z (this in general will exist, though that for W will not owing to the non-zero probability that $W = 0$). Then these functions must satisfy the integral equation

$$h_2(z_2) = \int_0^\infty h_1(z_1)\, g(z_2 - z_1)\, dz_1 + P_1 g(z_2),$$

where P_1 is the chance that $Z_1 \leqslant 0$. In the stationary case this is equivalent to the integral equation, of Wiener-Hopf type,

$$h(z) = \int_0^\infty h(y)\, g(z - y)\, dy + P g(z), \qquad (10)$$

where

$$P = \int_{-\infty}^0 h(y)\, dy.$$

This equation, due to Lindley, determines in principle $h(z)$ from the distribution of $U = V - T$, though it is rather awkward to handle for more general arrival-time distributions. Its solution has been discussed in this context by Smith (1953), who has shown in particular that if the service-time distribution is exponential so is that of the (non-zero) waiting-time, whatever the arrival-time distribution; this remarkable result was later shown by D. G. Kendall (1953) to apply even in the case of more than one server. Kendall has also pointed out that the preceding methods make use of the same general principle of extracting some convenient Markov sequence from the complete stochastic process, which is not Markovian in the general case unless extra variables are introduced (akin to the age in the simple renewal process of § 4·2).

Distribution of the busy period. Another feature of interest in the theory of queues is the size of the busy period, defined as the time T between instants when the queue was last zero and again becomes so. For independent arrival times it is a recurrence time in the sense of § 3·3. Let the probability density function for T be denoted by $g(t|s)$ when the service-time of the first customer in the busy period is s. The unconditional density function is thus

$$f(t) = \int g(t|s)\,dB(s). \tag{11}$$

The advantage of studying $g(t|s)$ is that the busy period 'generated' by $s_1 + s_2$ can be regarded as the sum of two independent periods generated by s_1 and s_2 separately. Thus if

$$\alpha(\psi|s) = \int e^{-\psi w} g(w|s)\,dw,$$

we have immediately

$$\alpha(\psi|s_1 + s_2) = \alpha(\psi|s_1)\,\alpha(\psi|s_2),$$

from which it follows that we can write

$$\alpha(\psi|s) = \exp\{-s\kappa(\psi)\}. \tag{12}$$

Consider now a renewal-type equation obtained by supposing, for a Poisson input of customers at a rate λ, that s for the initial customer is small. Then, with probability $1 - \lambda s$, $T = s$; and, with probability λs, T will take on the value appropriate to the distribution in (11), as the component from the first customer will be negligible. That is, as s becomes small enough

$$f(t|s) = (1 - \lambda s)\,\delta(t - s) + \lambda s f(t),$$

where $\delta(t - s)$ is the Dirac delta-function. Taking Laplace transforms, we thus find that for small s

$$e^{-s\kappa(\psi)} = (1 - \lambda s)\,e^{-\psi s} + \lambda s \alpha(\psi),$$

where, from (11), $\alpha(\psi) = \beta\{\kappa(\psi)\}$,

$\beta\{\psi\}$ being as before the Laplace transform of $B(s)$. Hence

$$\kappa(\psi) = \lambda + \psi - \lambda\beta\{\kappa(\psi)\},$$

or equivalently, $\alpha(\psi) = \beta\{\lambda + \psi - \lambda\alpha(\psi)\}. \tag{13}$

Equation (13) is valid whether or not the queueing process reaches equilibrium, though if it does not do so the distribution of the busy period may have a non-vanishing component at infinity. We will again examine the simplest case when the service-time distribution is exponential, so that

$$\beta\{\psi\} = 1/(1 + \rho\psi/\lambda).$$

Then (13) gives

$$\lambda\alpha^2(\psi) - (\psi + \lambda + \lambda/\rho)\,\alpha(\psi) + \lambda/\rho = 0.$$

Letting $\psi \to 0$ gives the equation for the probability P of the busy period terminating as

$$P^2 - (1 + 1/\rho)\,P + 1/\rho = 0, \tag{14}$$

which has a root $1/\rho$ for $\rho > 1$, but relevant value 1 for $\rho \leqslant 1$. The solution corresponding to $\alpha(0) = 1$ when $\rho \leqslant 1$ is

$$\alpha(\psi) = \frac{1}{2\rho}\left\{1 + \rho + \frac{\rho\psi}{\lambda} - \sqrt{\left(\left[1 + \rho + \frac{\rho\psi}{\lambda}\right]^2 - 4\rho\right)}\right\}, \tag{15}$$

corresponding to the probability density

$$f(t) = \frac{e^{-\lambda t(1+1/\rho)}}{t\sqrt{\rho}}\,I_1(2t\lambda/\sqrt{\rho}), \tag{16}$$

where $I_1(t)$ is the Bessel function of imaginary argument and first order.

The above method of deriving $f(t)$ is taken from Cox and Smith (1961), who show that the distribution of the total number N of customers served during the busy period may similarly be obtained; and also, in the case of more general service-time distributions, give asymptotic formulae for $f(t|s)$ and for the probability $P(t)$ that the busy period terminates after time t for large t. For an alternative method of approach see also Kingman (1962).

More than one server. For the case of more than one 'server', we shall make throughout the simplifying assumption of random input, e.g. telephone 'traffic' with a number s of lines. As $s \to \infty$, if we assume further a random time-interval for occupying a line, this problem is stochastically identical with that of equilibrium under a death-rate μ and an immigration rate λ. The equilibrium distribution for the number of occupied lines is thus

$$p_n = e^{-\lambda/\mu}(\lambda/\mu)^n/n!. \tag{17}$$

It should be noticed that this distribution can be obtained at once from the equation for $p_n(t)$

$$\partial p_n(t)/\partial t = -(\lambda + \mu n) p_n(t) + \lambda p_{n-1}(t) + (n+1) \mu p_{n+1}(t),$$

for in equilibrium

$$p_{n+1} = [(\lambda + \mu n) p_n - \lambda p_{n-1}]/[(n+1) \mu],$$

with $p_1 = \lambda p_0/\mu$.

However, if the number s of lines is finite, the number n of customers either making use of these lines or waiting to do so is governed by the modified equilibrium equation for $n \geqslant s$

$$0 = -(\lambda + \mu s) p_n + \lambda p_{n-1} + s\mu p_{n+1},$$

where
$$p_n = (\lambda/\mu)^n p_0/n! \quad (n \leqslant s), \tag{18}$$

but
$$p_n = (\lambda/\mu)^n p_0/[s! \, s^{n-s}] \quad (n > s). \tag{19}$$

The number of waiting customers is $n - s$ for $n \geqslant s$. The sum $\Sigma p_n/p_0$ converges if $\lambda/\mu < s$.

Finally, we shall consider a servicing problem where there are m operators available for repairing any of ms machines. We shall again consider only the simple case of random breakdowns (rate λ per machine) and random service-times (completed at rate μ). It is evident that a single group of m operators to ms machines is in principle more efficient than m separate groups of 1 operator to s machines, because the possibility of an operator being idle while a machine in another group requires attention is excluded if all operators are pooled. To investigate the gain in detail, we set up the equilibrium equations for the number of *machines* (not operators) idle:

$$0 = ms\lambda p_0 - \mu p_1,$$

$$[(ms-n)\lambda + n\mu] p_n = (ms-n+1)\lambda p_{n-1} + (n+1)\mu p_{n+1}$$
$$(1 \leqslant n < m),$$

$$[(ms-n)\lambda + m\mu] p_n = (ms-n+1)\lambda p_{n-1} + m\mu p_{n+1}$$
$$(m \leqslant n < ms),$$

$$\left. \right\} \tag{20}$$

whence for $0 \leqslant n < m$

$$(n+1)\mu p_{n+1} = (ms-n)\lambda p_n,$$

while for $n \geqslant m$,

$$m\mu p_{n+1} = (ms-n)\lambda p_n.$$

For $m = 1$, this gives

$$p_n = \frac{s!}{(s-n)!} \left(\frac{\lambda}{\mu}\right)^n p_0. \tag{21}$$

In the particular case when s becomes large, but the expected rate of breakdown remains finite, i.e. $s\lambda \to S$, say, the distribution (21) becomes

$$p_n = (S/\mu)^n p_0,$$

and the chance of no machine out of use is $1 - S/\mu$. This checks of course with the queue problem with the machines taking the role of customers. The expected number of machines in the waiting-line (excluding any machine being serviced) comes out to be $\rho^2/(1-\rho)$, where $\rho = S/\mu$.

In the case of general m, we shall consider the equations (20) in the same limiting situation of s large but $s\lambda \to S$. With $\rho = S/\mu$, the equations become

$$\left.\begin{aligned}
\rho m p_0 &= p_1, \\
(\rho m + n)\, p_n &= \rho m p_{n-1} + (n+1)\, p_{n+1} \quad (1 \leqslant n < m), \\
(\rho + 1)\, p_n &= \rho p_{n-1} + p_{n+1} \quad\quad\quad (n \geqslant m).
\end{aligned}\right\} \quad (22)$$

The last equation gives

$$p_n = \rho^{n-m} p_m \quad (n \geqslant m), \tag{23}$$

while for $1 \leqslant n \leqslant m$,

$$p_n = (\rho m/n)\, p_{n-1}. \tag{24}$$

These relations determine the form of the distribution for $n \geqslant m$ and $n < m$. For example, for $m = 2$,

$$p_1 = 2\rho p_0, \quad p_2 = 2\rho^2 p_0, \quad p_n = \rho^{n-2} p_2 = 2\rho^n p_0,$$

whence

$$p_0[1 + 2\rho/(1-\rho)] = 1,$$

or

$$p_0 = (1-\rho)/(1+\rho).$$

The expected number of machines in the waiting-line under these conditions is $2\rho^3/(1-\rho^2)$ and the ratio to twice the number for $m = 1$ is $\rho/(1+\rho)$, indicating the gain in efficiency already referred to. As m further increases, the ratio of the expected number of waiting machines to the total expected number for the $m = 1$ individual groups tends eventually to zero.

4·22 Theory of storage. Various stochastic processes are directly or indirectly related with queueing theory. The bunching of traffic along a road is often due to queues forming behind slow vehicles (cf. Table 7 in § 9·3). In storage and stock control, the *output* is to some extent analogous to the efflux of customers in a queue, and the *input* to their arrivals: though in the case of stock control the purchase of new stock is usually deliberate, in contrast with the random influx of water into a dam or reservoir. Moreover, the amount of stock, or water in the reservoir, is limited in amount, and would thus correspond to a queue of customers with a fixed upper size. The problem in insurance risk discussed in § 4·1 as an application of random walk theory is also obviously related, and we shall discuss one important aspect of the reservoir problem as an extension of this last problem.

Suppose (in place of the steady capital increase in the insurance problem) water is coming into a reservoir at a rate equivalent, at least asymptotically, to an additive process $I(t)$, and is leaving independently equivalent to the additive process $J(t)$. Suppose the capacity of the reservoir is C. Then the use of Wald's identity, if the initial volume of water is c, and we identify the boundaries of § 2·1 as
$$b = c, \quad a = c - C,$$
measuring output positively, gives
$$P_a e^{a\theta_0} + P_b e^{b\theta_0} \sim 1, \tag{1}$$
where θ_0 is the non-zero root of
$$K(\theta) \equiv K_J(\theta) + K_I(-\theta) = 0, \tag{2}$$
$K_I(\theta)$, $K_J(\theta)$ being the increments per unit time to the cumulant functions of $I(t)$, $J(t)$ respectively. In particular, if withdrawals are made at a rate ν and in amounts following a distribution with m.g.f. $M(\theta)$ (as in § 4·1) and contributions are received at a rate μ and m.g.f. $N(\theta)$, thus (2) becomes
$$\nu(M(\theta) - 1) + \mu(N(-\theta) - 1) = 0. \tag{3}$$
The probability of the reservoir first becoming empty, in contrast with first becoming full, is given by P_b; and the distribution of the time to first emptiness in the above sense may also be derived by the standard technique of § 2·1. But we are of course more interested in the distribution of the time to first emptiness, regardless of whether the reservoir is on occasions full. If the

capacity of the reservoir is effectively infinite, this further complication can be neglected. Otherwise, we must specify what happens when the reservoir is full, or nearly so; and we shall rather naturally assume that the volume is always limited by the capacity C.

Let the characteristic function of the time to emptiness be denoted by $\xi(\phi|c)$; and let the components of the characteristic function $\eta(\phi|c)$ to emptiness and fullness in the sense of §2·1 be given by the equation

$$\eta(\phi|c) = \eta_b(\phi|c) + \eta_a(\phi|c).$$

Then we have the relations

$$\left.\begin{array}{l} \xi(\phi|c) = \eta_b(\phi|c) + \eta_a(\phi|c)\,\xi(\phi|C), \\[2mm] \xi(\phi|C-\epsilon) = \eta_b(\phi|C-\epsilon) + \eta_a(\phi|C-\epsilon)\,\xi(\phi|C), \end{array}\right\} \tag{4}$$

whence, as $\epsilon \to 0$,

$$\left.\frac{\partial \xi(\phi|x)}{\partial x}\right|_C = \left.\frac{\partial \eta_b(\phi|x)}{\partial x}\right|_C + \xi(\phi|C)\left.\frac{\partial \eta_a(\phi|x)}{\partial x}\right|_C.$$

This last relation is most useful when Wald's identity is available in exact form even for short passages, i.e. when the passage to fullness at least is continuous. We shall moreover simplify the problem further by considering the normal diffusion case. Then diffusion from the boundary in a small time δt extends back to a distance of order $\sqrt{\delta t}$, and consistency is only possible if

$$\left.\frac{\partial \xi(\phi|x)}{\partial x}\right|_C = 0; \tag{5}$$

so that finally

$$\xi(\phi|c) = \eta_b(\phi|c) - \eta_a(\phi|c)\left[\frac{\partial \eta_b(\phi|x)}{\partial x}\bigg/\frac{\partial \eta_a(\phi|x)}{\partial x}\right]_C. \tag{6}$$

In this formula,

$$\eta_a(\phi|c) = \frac{e^{b\theta_1} - e^{b\theta_2}}{e^{b\theta_1+a\theta_2} - e^{b\theta_2+a\theta_1}}, \quad \eta_b(\phi|c) = \frac{e^{a\theta_2} - e^{a\theta_1}}{e^{b\theta_1+a\theta_2} - e^{b\theta_2+a\theta_1}},$$

where $\theta_1, \theta_2 = \{-m \pm \sqrt{(m^2 - 2i\sigma^2\phi)}\}/\sigma^2$

and $b = c$, $a = c - C$. Hence further

$$\left[\frac{\partial \eta_b(\phi|x)}{\partial x}\bigg/\frac{\partial \eta_a(\phi|x)}{\partial x}\right]_C = \frac{\theta_1 - \theta_2}{\theta_2 e^{C\theta_1} - \theta_1 e^{C\theta_2}}. \tag{7}$$

From the above formulae the mean time to emptiness is found to be

$$m(c) = \frac{c}{m} + \frac{\sigma^2}{2m^2} e^{\frac{-2mC}{\sigma^2}} \left(1 - e^{\frac{2mc}{\sigma^2}}\right). \tag{8}$$

As $m \to 0$, this gives in particular

$$m(c) = c(2C - c)/\sigma^2 \tag{9}$$

compared with the time $c(C-c)/\sigma^2$ to reach either boundary. The additional time Cc/σ^2 represents the mean time from fullness (C^2/σ^2) multiplied by the probability c/C of reaching this state. The reader may also find it instructive to derive the analogous results for the simple discrete step and time case, with probability p of one unit input (except that if the dam is full it remains in this state) and probability q of one unit output, and check his results with those obtained by Weesakul (1961) by a more direct method. (The equations involving $\xi(\phi | C - \epsilon)$ in this case must of course be replaced by appropriate difference relations involving $\xi(\phi | C - 1)$.) In particular, as $p \to q \to \frac{1}{2}$ (so that the equivalent $\sigma^2 \to 1$), the exact mean time to emptiness becomes $2Cc - c^2 + c$, with which result (9) agrees to a relative order of $1/c$. (An approach rather more related to the one here was used by Phatarfod (1963) when discussing inputs with negative exponential distribution.)

The above results for the asymptotic diffusion case can obviously be extended to the case of correlated increments by determining the effective diffusion per unit time in such a case (cf. § 2·23 on Wald's identity for Markov chains).

Another problem of interest is to find the equilibrium or stationary distribution of the reservoir content. This will in general, when input and/or output from the reservoir are allowed to contain Poisson components, consist of non-zero probabilities P_F and P_E of being full and empty respectively, as well as a density component for any intermediate volume c. However, the problem is much simplified if, as above, the purely diffusion case is considered, for then the stationary distribution consists merely of a density term (as is easily seen by consideration of the equilibrium situation at the boundaries). Moreover, as the distribution is stationary this density term must satisfy the equation

$$m \frac{\partial f}{\partial x} = \frac{1}{2}\sigma^2 \frac{\partial^2 f}{\partial x^2} \tag{10}$$

obtained from equation (9) of § 3·1 with $\frac{\partial f}{\partial t} = 0$. Hence

$$f = \frac{\dfrac{2m}{\sigma^2} e^{\frac{2mx}{\sigma^2}}}{e^{\frac{2mC}{\sigma^2}} - 1}, \tag{11}$$

where x denotes the reservoir content, m is the net mean gain per unit time and σ^2 the corresponding variance. Note that (11) is valid whether m is positive or negative, except that if $C \to \infty$ no limit exists unless m is negative.

Discussion of the more general case will be found in the literature; see for example, Prabhu's (1964) review. For other aspects of storage theory reference should also be made to Moran (1959); Moran points out that practical problems, where particular conditions may have to be introduced, will, like queueing problems, often have to be studied by means of Monte Carlo methods.

4·3 Population growth as a multiplicative process

In § 3·4 we considered a simple multiplicative birth-and-death process with constant birth- and death-rates λ and μ independent of time. This was extended to functions $\lambda(t)$ and $\mu(t)$ depending on the time, but for human and most animal populations it is essential to consider the variation of λ and μ with the *age* of the individual concerned, just as the energy of a particle of a cascade shower had to be introduced in § 3·42.

Before formulating such a process, we note briefly the usual treatment, which refers merely to *expected* numbers. To simplify the problem, we consider the standard case where the process is temporally homogeneous, and where the females only are enumerated. The expected female birth-rate $r(t)$ then satisfies an integral equation of renewal type, viz.

$$r(t) = r_0(t) + \int_0^t s(\tau)\, r(t-\tau)\, d\tau, \tag{1}$$

where $s(t)$ is the expected rate of female offspring at time t of a female born at time 0, and $r_0(t)$ depends on the value of $r(t)$ prior to $t = 0$. The function $s(t)$ is the product of the rate of offspring at time t to a female aged t and the chance of survival to time t, $1 - F(t)$. This equation may be solved either iteratively in terms of successive generations, or by an expansion of the type referred to in the theoretical solution quoted in § 2·11.

The linearity of equation (1) ensures its validity as an expectation equation under fairly general conditions, though it neglects the effect of a changing sex ratio, only females (or males) being considered, and is liable to give misleading results for comparatively new countries (like Australia) with a high immigration rate and anomalous sex ratio. In order to represent population growth as a *stochastic* process, it is necessary to give a more explicit and rather idealized specification of the statistical mechanism of growth. Before introducing the complication of the continuous age-structure, we may indicate the stochastic aspect by reference to the extinction question, treating this quite simply by keeping to 'generations' as a strictly discrete 'time' variable. The methods of § 2·3 may then be used to show that, on the assumption of independent female genealogical lines, a particular line will become extinct if R, the net reproduction rate defined as the mean of the female replacement distribution per female, is not greater than 1. Extinction similarly follows for any finite number of lines. This result implies the ultimate extinction of the population *in time* if we note that while generations overlap in time, the spread of generations (which for each line is a 'random walk' or additive process) is $O(\sqrt{t})$, and hence $o(t)$, as t increases.

When $R = 1$, the population mean size is ultimately stationary, but we see that this does not imply any more complete stochastic stationarity. Of course whether the divergence of the actual or expected size is practically important depends on the order of magnitude involved. The coefficient of variation, that is, the standard deviation divided by the mean, for a population with constant mean will be of order $\sqrt{(t/n_0)}$, where n_0 is the initial size (see § 2·3). For a human population of 50 million and t (measured in generations of about 25 years) equal to 2 units or 50 years, $\sqrt{(t/n_0)}$ is $\frac{1}{5000}$ and fluctuations are negligible; on the other hand, for an animal population of 100 with a 1-year life cycle, $\sqrt{(t/n_0)} \sim \frac{1}{3}$ in 10 years. When $R > 1$, the chance of extinction rapidly tends to zero with the increase in size (cf. the chance of extinction of favourable gene mutations referred to in § 2·3), but may still be appreciable for small numbers and in particular for $n_0 = 1$, i.e. the descendants of a single individual, as was illustrated in § 2·11.

Exact treatment of age-structure. We return now to the detailed specification problem, and make the following simplifying assumptions about the mechanism of the process:

(*a*) the sub-populations generated by any two co-existing individuals develop independently of each other;

(*b*) an individual of age x existing at time t has a chance $\lambda(x)\Delta t + o(\Delta t)$ of producing one new individual of age zero during the interval $(t, t+\Delta t)$, independently of previous events (multiple births, the planning of family size, etc., are ignored);

(*c*) an individual has a chance $\mu(x)\Delta t + o(\Delta t)$ of dying during the same interval.

The formalism of § 3·42 (cf. also § 4·2) gives for the transitions in the interval dt

$$z(x) \rightarrow z(x+dt)\{1 - [\lambda(x) + \mu(x)]\,dt\}$$
$$+ z(0)\,z(x+dt)\,\lambda(x)\,dt + \mu(x)\,dt,$$

whence the 'backward' equation for the characteristic functional of the number $N(u \mid x, t)$ of age less than or equal to u at time t, given one individual aged x at $t = 0$, expressed for convenience in terms of $z(u) \equiv \exp i\theta(u)$ as the probability-generating functional

$$\Pi_t(z(u) \mid x) \equiv E\left\{\exp\int i\theta(u)\,dN(u \mid x, t)\right\},$$

becomes

$$\frac{\partial \Pi_t(x)}{\partial t} = \lambda(x)\,[\Pi_t(0)\,\Pi_t(x) - \Pi_t(x)] + \mu(x)\,[1 - \Pi_t(x)] + \frac{\partial \Pi_t(x)}{\partial x}, \quad (2)$$

or in integral equation form

$$\Pi_t(x) = \int_0^t [\mu(x+u) + \lambda(x+u)\,\Pi_{t-u}(x+u)\,\Pi_{t-u}(0)]\,q(x+u \mid x)\,du$$
$$+ q(x+t \mid x)\,z(x+t), \quad (3)$$

where $q(x+t \mid x)$, representing the chance of no birth or death in the period t for an individual aged x at $t = 0$, is given by

$$q(x+t \mid x) = \exp\left\{-\int_x^{x+t} [\lambda(u) + \mu(u)]\,du\right\}. \quad (4)$$

The general solution of (2) or (3), at least in closed form, is unknown, though an iterative form of solution in terms of the contributions from successive generations is fairly readily obtained. In fact, if we write down in place of (3) the integral equation obtained by considering *deaths* alone as renewal or regenerative instants, we have

$$\Pi_t(x) = \int_0^t \mu(x+u)\, p(x+u\,|\,x)\, \Psi_t'(u;x)\, du$$
$$+ p(x+t\,|\,x)\, z(x+t)\, \Psi_t'(t;x), \quad (5)$$

where
$$p(x+t\,|\,x) = \exp\left\{ -\int_x^{x+t} \mu(u)\, du \right\} \quad (6)$$

is the chance of the initial individual not dying in the period t, and $\Psi_t'(u;x)$ denotes the functional at time t for the progeny born between 0 and u of an individual of initial age x *who does not die*. Now such an individual may be regarded as an 'immigration source' of progeny, by the principles of §3·41, so that

$$\Psi_t'(u;x) = \exp\left\{ \int_0^u \lambda(x+v)\, [\Pi_{t-v}(0) - 1]\, dv \right\}. \quad (7)$$

The new integral equation obtained by substituting (7) in (5) has the advantage that the functional in (7) is composed of contributions from new births. Thus if we write $\Pi^{(n)}$ as the contribution to Π of all generations up to the nth (the initial ancestor representing the zeroth generation), then on the right-hand side of (5) we must write $\Pi^{(n-1)}$, and the iterative form of solution of the first integral equation (3) is consequently

$$\Pi_t^{(n)}(x) = z(x+t)\, p(x+t\,|\,x) \exp\left\{ \int_0^t \lambda(x+u)\, [\Pi_{t-u}^{(n-1)}(0) - 1]\, du \right\}$$
$$+ \int_0^t \mu(x+u)\, p(x+u\,|\,x) \exp\left\{ \int_0^u \lambda(x+v)\, [\Pi_{t-v}^{(n-1)}(0) - 1]\, dv \right\} du. \quad (8)$$

In the particular case of $\lambda(x)$, $\mu(x)$ constant, the complete solution of (2) or (3) is known. We note from (3) that the general solution is linear in $z(x+t)$, and we may write

$$\Pi_t(x) = 1 + [z(x+t) - 1]\, U + V,$$

where U and V do not involve $z(x+t)$, while V vanishes and $U = 1$

when $t = 0$. We easily obtain, for $\lambda(x) \equiv \lambda$, $\mu(x) \equiv \mu$, the pair of equations for U, V

$$\frac{\partial U}{\partial t} = -\mu U + \lambda\{[z(t) - 1] U + V\} U,$$

$$\frac{\partial V}{\partial t} = -\mu V + \lambda\{[z(t) - 1] U + V\}(V + 1),$$

whence

$$\frac{1}{\lambda} \frac{\partial}{\partial t}\left(\frac{V e^{-\lambda t}}{U}\right) = [z(t) - 1] e^{-\lambda t},$$

$$\frac{1}{\lambda} \frac{\partial}{\partial t}\left(\frac{e^{-\mu t}}{U}\right) = -e^{-\mu t}\left\{[z(t) - 1] + \frac{V}{U}\right\}.$$

From the first of these equations,

$$\frac{V}{U} = e^{\lambda t} \int_0^t [z(u) - 1] \lambda e^{-\lambda u} du,$$

and substituting this result in the second equation, integrating by parts and using the initial value of U, we obtain

$$\frac{e^{-\mu t}}{U} = 1 - \int_0^t [z(u) - 1] \frac{\lambda}{\lambda - \mu} e^{-\mu u}[\lambda e^{(\lambda - \mu)(t-u)} - \mu] du.$$

Hence we find

$$\Pi_t(x) = 1 + \frac{(\lambda - \mu)\left\{[z(x+t) - 1] e^{-\mu t} + \lambda \int_0^t [z(u) - 1] e^{(\lambda - \mu)t} e^{-\lambda u} du\right\}}{(\lambda - \mu) - \lambda \int_0^t [z(u) - 1] e^{-\mu u}[\lambda e^{(\lambda - \mu)(t-u)} - \mu] du}.$$

$$(9)$$

Once the result (9) is obtained, it gives the simplest method of obtaining the moment formulae. Thus for ages less than t, the initial ancestor does not contribute (cf. the age distribution in the renewal problem of § 4·2). Putting $z(x+t) = 1$ in (9), otherwise $z(u) = e^{i\phi(u)}$, and taking logarithms, we find for the *cumulant functional*

$$K_t = \log\left(1 - \frac{\lambda}{\lambda - \mu} I_1\right) - \log\left(1 - \frac{\lambda}{\lambda - \mu} I_2\right), \tag{10}$$

where $\qquad I_1 = \int_0^t [e^{i\phi(u)} - 1] [\mu e^{(\lambda - \mu)t} e^{-\lambda u} - \mu e^{-\mu u}] \, du,$

$$I_2 = \int_0^t [e^{i\phi(u)} - 1] [\lambda e^{(\lambda - \mu)t} e^{-\lambda u} - \mu e^{-\mu u}] \, du.$$

Expanding the logarithms, and $e^{i\phi(u)}$, we obtain any required moment density function (§ 3·42). It is convenient to define in relation to the cumulant functional (10) the *cumulant densities*†

$$E\{dN(u,t)\} = \alpha(u,t) \, du = \mathrm{var}\,\{dN(u,t)\},$$
$$\mathrm{cov}\,\{dN(u,t), dN(v,t)\} = \beta(u,v,t) \, du \, dv \quad (u \neq v), \qquad (11)$$

(note that $\beta(u,v) + \alpha(u)\,\alpha(v)$ is the product-density of order two defined in § 3·42). Then, apart from excluding for convenience the anomalous direct contribution from the initial ancestor, we have in general

$$K_t = \int_0^t \alpha(u,t) \left[i\phi(u) - \tfrac{1}{2}\phi^2(u) \right] du$$
$$- \frac{1}{2} \int_0^t \int_0^t \beta(u,v,t) \, \phi(u) \, \phi(v) \, du \, dv + \dots. \qquad (12)$$

In the particular case (10), we find

$$\alpha(u,t) = \lambda e^{(\lambda - \mu)t - \lambda u} \quad (0 \leqslant u < t), \qquad (13)$$

$$\beta(u,v,t) = \frac{\lambda^2(\lambda + \mu)}{\lambda - \mu} e^{2(\lambda - \mu)t - \lambda(u+v)} - \frac{\lambda^2 \mu}{\lambda - \mu} e^{(\lambda - \mu)t} \left[e^{-\lambda u - \mu v} + e^{-\mu u - \lambda v} \right]$$
$$(u < v < t), \qquad (14)$$

the last formula holding also by symmetry for $v < u$.

These formulae can alternatively be derived directly by the methods developed in § 3·42 (cf. D. G. Kendall, 1949). The relevant moment equations will appear more familiar if derived from the 'forward' equation, which is (equation (11) of § 3·42)

$$d\Pi_t(z(u)) = \Pi_t(z^*(u)) - \Pi_t(z(u)),$$

---

† It is not perhaps immediately clear that the order of magnitude of the moments assumed in (11) is valid even when $u = x + t$ is excluded, especially in view of the special role of $u = 0$ in birth and renewal processes, but this can be checked if required, for example, from equation (5).

where

$$z^*(u) = z(u + dt) + \lambda(u)\,dt[z(0) - 1]\,z(u + dt) + \mu(u)\,dt[1 - z(u + dt)],$$

or, in terms of

$$K_t(i\phi(u)) \equiv \log E\left\{\exp\left[i\int_0^\infty \phi(u)\,dN(u, t)\right]\right\},$$

where

$$dK_t(i\phi(u)) = K_t(i\phi^*(u)) - K_t(i\phi(u)), \qquad (15)$$

$$i\phi^*(u) = i\phi(u + dt) + dt[\lambda(u)\,(e^{i\phi(0)} - 1) + \mu(u)\,(e^{-i\phi(u)} - 1)].$$

We shall not discuss these equations further here, except to check their consistency with the integral equation (1). We have from (15)

$$dK_t(i\phi(u)) = \int_0^\infty \alpha(u, t)\,[i\phi(u + dt) - i\phi(u) + dt(\lambda(u)\,i\phi(0)$$
$$- \mu(u)\,i\phi(u))]\,du + \dots,$$

whence

$$\left.\begin{array}{l} \dfrac{\partial \alpha}{\partial t} + \dfrac{\partial \alpha}{\partial u} + \mu\alpha = 0 \quad (u > 0), \\[2mm] \alpha(0, t) = \displaystyle\int_0^\infty \lambda(u)\,\alpha(u, t)\,du. \end{array}\right\} \qquad (16)$$

The first equation gives

$$\alpha(u, t) = p(u)\,\alpha(0, t - u) \quad (0 < u < t),$$

where (cf. equation (6)) $p(x) = \exp\left\{-\int_0^x \mu(v)\,dv\right\}$. The second equation in (16) then gives an integral equation identical with (1) if we write

$$r(t) \equiv \alpha(0, t), \quad s(t) = p(t)\,\lambda(t),$$

and $\quad r_0(t) \equiv \int_t^\infty \lambda(u)\,\alpha(u, t)\,du = \lambda(x + t)\exp\left\{-\int_x^{x+t} \mu(v)\,dv\right\}.$

The effect of immigration. The previous methods of allowing for immigration can be extended to cover the above specification. We shall consider only the age-independent case where $\lambda(u) \equiv \lambda$, $\mu(u) \equiv \mu$, and also we assume an immigration (constant) rate ν, the ages of immigrants being taken for simplicity all zero. As we shall be interested in the limiting population, we consider also an initial population of size zero. Then at time t the new moment-generating functional L_t will be given by

$$\log L_t = \nu \int_0^t [e^{K_t(u)} - 1] du,$$

or $L_t = \left[1 + \dfrac{\lambda}{\lambda - \mu} \int_0^t [e^{i\phi(u)} - 1] [\mu - \lambda e^{-(\mu - \lambda)(t-u)}] e^{-\mu u} du \right]^{-\nu/\lambda}.$ (17)

For $\mu > \lambda$, the limiting stationary distribution is given by

$$L_\infty = \left[1 - \frac{\lambda \mu}{\mu - \lambda} \int_0^\infty [e^{i\phi(u)} - 1] e^{-\mu u} du \right]^{-\nu/\lambda}. \tag{18}$$

Any required properties of the ultimate population can be obtained from (18). For example, if we write

$$i\phi(u) = \theta_1 + \theta_2 f(u),$$

we obtain the joint m.g.f. of the total number N of individuals and the total sum S of the function $f(U)$ over all the random ages U in the population. We find

$$M(\theta_1, \theta_2) = \left[\frac{\mu - \lambda}{\mu - \lambda e^{\theta_1} \psi(\theta_2)} \right]^{\nu/\lambda}, \tag{19}$$

where $\psi(\theta_2) = \displaystyle\int_0^\infty e^{\theta_2 f(u)} \mu e^{-\mu u} du.$ (20)

The marginal distribution of N is of course a negative binomial (as we have assumed the rates λ, μ, constant over the ages). Further, the conditional distribution of S/n for given $N = n$ is given by the m.g.f. $\psi^n(\theta/n)$ $(n > 0)$. This is exactly the same as if the n individuals were picked at random from an infinite population with the age distribution $\mu e^{-\mu x} dx$, a result which would be anticipated as all the individuals enter the population at age zero (whether by births or immigration) and die off at rate μ.

4·31   Growth and mutation in bacterial populations.

In the case of bacterial population growth, the standard model is rather different from that discussed in the last section, for the bacterium usually splits into two new offspring, so that, if the bacteria are in favourable conditions under which the death-rate can be neglected, the transition is represented by

$$z(x) \to z(x+dt)\{1 - \lambda(x)\,dt\} + z^2(0)\,\lambda(x)\,dt,$$

whence
$$\frac{\partial \Pi_t(x)}{\partial t} = \lambda(x)\left[\Pi_t^2(0) - \Pi_t(x)\right] + \frac{\partial \Pi_t(x)}{\partial x}, \tag{1}$$

or in integral equation form

$$\Pi_t(x) = \int_0^t \lambda(x+u)\,\Pi_{t-u}^2(0)\,P(x+u \mid x)\,du + z(x+t)\,P(x+t \mid x), \tag{2}$$

where
$$P(x+t \mid x) = \exp\left\{-\int_x^{x+t} \lambda(u)\,du\right\}.$$

These equations allow for any age-dependent probability of division, but assume that bacteria, once produced, are independent, so that their 'lifetimes' before division are not influenced either by other bacteria or by the previous values of lifetimes in their own genealogical history.

The solution of (1) or (2) is unknown in general, but if the lifetime U, with frequency distribution $\lambda(u)\,P(u \mid 0)\,du$, is approximately represented as the resultant of a number of independent stages or phases $U_1, U_2, \ldots, U_k$, each with exponential distribution of mean $1/\lambda_1, 1/\lambda_2, \ldots, 1/\lambda_k$ respectively, and we consider merely the total number $N(t)$ of individuals, beginning with one individual aged zero, the Laplace transform of (2) may be written

$$L(\psi) = M(\psi)\prod_{i=1}^{k}\left(1 + \frac{\psi}{\lambda_i}\right)^{-1} + z\left[\frac{1}{\psi} - \frac{1}{\psi}\prod_{i=1}^{k}\left(1 + \frac{\psi}{\lambda_i}\right)^{-1}\right],$$

where $M(\psi)$ is the transform of Π_t^2, or

$$\prod_{i=1}^{k}\left(1 + \frac{\psi}{\lambda_i}\right)L(\psi) = M(\psi) + \frac{z}{\psi}\left[\prod_{i=1}^{k}\left(1 + \frac{\psi}{\lambda_i}\right) - 1\right].$$

Transforming back and noting that $1 + \psi/\lambda_i \equiv 1 + 1/\lambda_i\,\partial/\partial t$, we obtain an equivalent differential equation

$$\prod_{i=1}^{k}\left(1 + \frac{1}{\lambda_i}\frac{\partial}{\partial t}\right)\Pi_t(z) = \Pi_t^2(z), \tag{3}$$

with initial conditions

$$\Pi_0(z) = z, \quad (\partial/\partial t)^i \Pi_0(z) = 0 \quad (1 \leqslant i < k).$$

From (3) soluble equations can be obtained at least for the moments. For example, in the case where all $\lambda_i = k\lambda$, we obtain for the first two factorial moments of N, m_1 and m_2, say,

$$\left.\begin{aligned} [(1 + D/(k\lambda))^k - 2] \, m_1 &= 0, \\ [(1 + D/(k\lambda))^k - 2] \, m_2 &= 2m_1^2, \end{aligned}\right\} \tag{4}$$

where $D \equiv \partial/\partial t$. In the particular case $k = 1$ we are back to the simple birth process

$$\left(1 + \frac{1}{\lambda}\frac{\partial}{\partial t}\right) \Pi_t = \Pi_t^2, \tag{5}$$

with complete solution

$$\Pi_t = ze^{-\lambda t}/[1 - z(1 - e^{-\lambda t})]. \tag{6}$$

From the more general equations (4) D. G. Kendall (1952) has found that, as t becomes large, $m_1 \sim 2^{1/k} \exp(\alpha_k \lambda t)/(2\alpha_k)$, where $\alpha_k = k(2^{1/k} - 1) \sim \log_e 2$ for large k, and the coefficient of variation $\sigma_N/m_1 \sim A_k/\sqrt{k}$, where $A_k \sim \sqrt{2}\log_e 2$ for large k. From experimental data by Kelly and Rahn for *Bacterium aerogenes*, with a mean 'lifetime' of about half an hour and a coefficient of variation about this mean of 15–30 %, a value of k about 30 appears to provide an appropriate model in this case.

Mutation. As an example of the growth of a heterogeneous population, we will consider the development of mutant forms in bacterial populations. This problem is of particular interest when the mutants are resistant to an environment (e.g. a nutrient medium impregnated with bacteriophage) unfavourable to the original type. The study and comparison with observation of particular stochastic models is important for understanding the causal mechanism involved—for example, in contrasting theories of (i) genetic random mutation occurring independently of the environment, and (ii) active interaction of the environment with the organism to produce the new resistant types.

The model we shall examine will assume only two possible bacterial types, normal and mutant, with a normal bacterium dividing into two new normals, or into one normal and one mutant, and a mutant dividing always into two mutants (i.e. 'back' mutation in this model is assumed negligible). We shall consider the growth prior to plating-out on a phage-impregnated medium, and assume that both normal and mutant types grow at equal rates (in the absence of mutations). We further assume that mutation occurs at the time of division, the chance of one of the two new bacteria after division being a mutant being p. Then equation (1) is replaced by

$$\left.\begin{aligned}
\frac{\partial \Pi_t^{(1)}(x)}{\partial t} &= \lambda(x)\{q([\Pi_t^{(1)}(0)]^2 - \Pi_t^{(1)}(x)) \\
&\quad + p(\Pi_t^{(1)}(0)\,\Pi_t^{(2)}(0) - \Pi_t^{(1)}(x))\} + \frac{\partial \Pi_t^{(1)}(x)}{\partial x}, \\
\frac{\partial \Pi_t^{(2)}(x)}{\partial t} &= \lambda(x)\{[\Pi_t^{(2)}(0)]^2 - \Pi_t^{(2)}(x)\} + \frac{\partial \Pi_t^{(2)}(x)}{\partial x},
\end{aligned}\right\} \tag{7}$$

where $\Pi_t^{(1)}(z_1(u), z_2(u)\,|\,x)$, $\Pi_t^{(2)}(z_1(u), z_2(u)\,|\,x)$ are the probability-generating functionals for one initial bacterium of normal and mutant type respectively (and $q = 1 - p$). In the simple stochastic model where we put $\lambda(x) = \lambda$, if further we put all $z_1(u) = z_1$, and all $z_2(u) = z_2$, these reduce to

$$\left.\begin{aligned}
\frac{\partial \Pi_t^{(1)}}{\partial t} &= \lambda q\{[\Pi_t^{(1)}]^2 - \Pi_t^{(1)}\} + \lambda p\{\Pi_t^{(1)}\Pi_t^{(2)} - \Pi_t^{(1)}\}, \\
\frac{\partial \Pi_t^{(2)}}{\partial t} &= \lambda\{[\Pi_t^{(2)}]^2 - \Pi_t^{(2)}\}.
\end{aligned}\right\} \tag{8}$$

As the second equation involves only $\Pi_t^{(2)}(z_2)$, we may solve it as for (5), and substitute in the first equation. Leaving $\Pi_t^{(2)}(z_2)$ for the moment arbitrary (for reference later), and putting for convenience

$$T = e^{-\lambda t}, \quad \Pi_t^{(2)} e^{\lambda t} = \Phi_T, \quad \Pi_t^{(1)} e^{\lambda t} = 1/\Psi_T,$$

we obtain
$$\frac{\partial \Psi_T}{\partial T} - p\Phi_T \Psi_T = q \tag{9}$$

with $\Psi_1(z_1, z_2) = 1/z_1$. From (9) we obtain

$$\Psi_T \Omega_T = \frac{1}{z_1} - q\int_T^1 \Omega_U\, dU, \tag{10}$$

where
$$\Omega_U = \exp \left\{ p \int_U^1 \Phi_V dV \right\}.$$

In particular, putting

$$\Phi_T = \Pi_t^{(2)}/T = z_2/[1 - z_2(1-T)],$$

we find

$$\Psi_T = [z_2 T + (1-z_2)]^p \left[\frac{1}{z_1} - \frac{1}{z_2} \right] + \frac{1}{z_2} [z_2 T + (1-z_2)]. \quad (11)$$

For the marginal distribution of mutants, we may put $z_1 \equiv 1$, $z_2 \equiv z$, and obtain

$$\Pi(z) = \frac{zT}{zT + (1-z) - (1-z)[zT + (1-z)]^p}, \quad (12)$$

starting from one normal bacterium, or

$$\Pi^n(z) = \left\{ \frac{zT}{zT + (1-z) - (1-z)[zT + (1-z)]^p} \right\}^n, \quad (13)$$

starting from n normal bacteria. In the usual case of p small, this is approximately

$$\Pi^n(z) \sim \left\{ 1 - \frac{m}{n} \frac{1-z}{z} \log [1 + z(T-1)] \right\}^{-n}, \quad (14)$$

where $m = np/T$, or when also T is small,

$$\Pi^n(z) \sim \left\{ 1 - \frac{m}{n} \frac{1-z}{z} \log (1-z) \right\}^{-n}. \quad (15)$$

It will be seen finally that for n fairly large, a condition under which stochastic fluctuations in the growth of the normal bacteria tend to be relatively small, this distribution is approximately

$$\Pi^n(z) \sim \exp \left\{ m \frac{1-z}{z} \log (1-z) \right\}. \quad (16)$$

This asymptotic distribution is rather remarkable in having no finite moments, though of course the more precise distribution (13) has. We should expect to arrive at the same asymptotic form under any reasonable hypothesis on the population growth of the normal type consistent with negligible fluctuation in its final size. For example, we may alternatively assume that the normal type grows 'deterministically' with exponential increase.

Confining our attention to the p.g.f. $\Pi(z)$ for the number of mutants, and using for convenience the 'forward' equation, we have (cf. §3·5)

$$\frac{\partial \Pi(z)}{\partial t} = n\lambda p e^{\lambda t}(z-1)\,\Pi(z) + \lambda(z^2-z)\frac{\partial \Pi(z)}{\partial z}, \qquad (17)$$

where it is assumed that $\lambda(x) = \lambda$ for the mutant type and the deterministic growth rate for the normal type is also now taken to be λ (on the previous assumptions it would be slightly less than λ, due to the occasional transfer to mutant type). Putting $m = npe^{\lambda t}$, we obtain from (17)

$$m\frac{\partial \Pi}{\partial m} = m(z-1)\,\Pi + (z^2-z)\frac{\partial \Pi}{\partial z}. \qquad (18)$$

Solving the partial differential equation (18) by standard methods, we have the auxiliary equations

$$\frac{dm}{m} = \frac{-dz}{z^2-z} = \frac{d\Pi}{m(z-1)\,\Pi},$$

with two independent solutions

(a) $m(1-z)/z = \text{constant}$,

(b) $\Pi(1-z)^{-m(1-z)/z} = \text{constant}$.

The general solution is thus

$$\Pi(1-z)^{-m(1-z)/z} = \psi\{m(1-z)/z\},$$

where, for $t = 0$ or $m_0 = np$, we have $\Pi = 1$, or

$$\psi\{np(1-z)/z\} = (1-z)^{-np(1-z)/z}.$$

This gives for $t \geqslant 0$,

$$\Pi(z) = (1-z+npz/m)^{m(1-z)/z}, \qquad (19)$$

which as $np/m \to 0$ also has the asymptotic form (16). In the form (19) the moments are still finite; we readily find the mean

$$n\lambda pte^{\lambda t} = \lambda mt = m\log(n_t/n), \qquad (20)$$

where n_t is the number $ne^{\lambda t}$ of normal bacteria. The variance is

$$2m(e^{\lambda t}-1) - \lambda mt \sim 2mn_t/n. \qquad (21)$$

For comparison the mean and variance from (13) are found to be

$$
\left.\begin{array}{l}
n(T^{-1} - T^{p-1}), \\
n(T^{p-1} - T^{-1} + 2pT^{p-2} - 2pT^{p-1} + T^{-2}[T^p - 1]^2),
\end{array}\right\} \tag{22}
$$

with the same asymptotic values as above when $p, T \to 0$ but $np/T \to m$.

The limiting distribution (16) has been tabulated by Lea and Coulson (1949) for varying values of m. They have also shown that for $m \geqslant 4$, say, a useful approximation is to take the quantity

$$
\frac{11 \cdot 6}{r/m - \log m + 4 \cdot 5} - 2 \cdot 02 \tag{23}
$$

as a normal variate with zero mean and unit standard deviation, where r is the observed number of mutants. Methods of statistical estimation (see Chapter 8) applied to such observed numbers from parallel cultures enable estimates of the mutation rate to be obtained. A simple though not fully efficient estimate is obtained from the median of the observed distribution. However, the use of the upper quartile, while nominally even less efficient, has the practical advantage of being more insensitive to the phenomenon of phenotypic delay.

This further complication is that the phenotypic appearance of the mutant type appears in some cases to be delayed a few generations, and it may be advisable to allow for this effect before the extent of agreement with observation is finally examined. The effect of smaller stochastic fluctuations in the reproduction of the mutant population should of course also be considered. This may be shown to halve the variance (21) in the limiting case of deterministic growth for both normal and mutant populations, but the asymptotic distribution (16), for which the variance has become infinite, is not affected.

With regard to the phenotypic delay, let us consider again the first of equations (8). This was solved first for arbitrary $\Pi_t^{(2)}$, and we are at liberty to interpret it in the general solution (10) as the p.g.f. of the number of *resistant* bacteria (i.e. bacteria in which the mutation is manifesting itself phenotypically) at time t, given one bacterium just mutated at time 0. If we are interested in the marginal distribution of resistant bacteria, we take

$\Pi_t^{(1)}$, $\Pi_t^{(2)}$ as functions merely of one further variable z, associated with the number of resistant bacteria, putting $z_1 = 1$ in (10). Suppose, for example, the multiplication of the mutated bacteria is, like the normal type, assumed to be of simple Markovian type with growth rate λ, but the chance that a mutated bacterium is immediately resistant is P and that if non-resistant, the chance of *one* of the two new mutant bacteria after its division being resistant is P, and so on. Then in place of $\Pi_t^{(2)}$ we put

$$P\Pi_t'(z) + (1-P)\,\Pi_t(z),$$

where $\Pi_t'(z)$ is $Tz/[1 - z(1-T)]$ and $\Pi_t(z)$ is given by (12) with P in place of p.

In the case of the general solution (10) (which assumes, however, a simple stochastic multiplicative process for the normal bacteria), if we put $z_1 = 1$, and regard $\Pi_t^{(1)}(z)$, $\Pi_t^{(2)}(z)$ as p.g.f.'s for the resistant bacteria only, D. G. Kendall (1952) has established that as $p, T \to 0$ but p/T remains constant,

$$\Pi_t^{(1)}(z) \to \left\{ 1 + pe^{\lambda t} \int_0^\infty [1 - \Pi_v^{(2)}(z)]\, e^{-\lambda v} \lambda\, dv \right\}^{-1}, \qquad (24)$$

and correspondingly, for n initial normal bacteria, the limit is this expression raised to the nth power. If, however, p and $1/n \to 0$ with np and t remaining constant, the resulting distribution starting from n initial bacteria tends to

$$\exp\left\{ -npe^{\lambda t} \int_0^t [1 - \Pi_v^{(2)}(z)]\, e^{-\lambda v} \lambda\, dv \right\}. \qquad (25)$$

When we substitute $\Pi_t^{(2)}(z) = Tz/[1 - z(1-T)]$, these equations become equivalent to earlier results. We may, moreover, as already noted for this special case, expect the second limit to be relevant for more general types of normal growth consistent with negligible fluctuations in the size of the normal bacterial population.

4·32 Population genetics. Several brief references to genetical problems have occurred in earlier sections, for example, to inbreeding in § 2·2, to natural selection in § 2·3, and, in relation to bacterial populations, in the last section; the theory of genetic recombination was also mentioned in § 2·11. The importance of the theory of stochastic processes to genetical and evolutionary theory should therefore have already become apparent, and the main purpose of the present section is to remind the reader of this by a further example in the domain of population genetics. Mathematical work on the genetical theory of evolution by natural selection is largely due to R. A. Fisher, J. B. S. Haldane and Sewall Wright, their work forming as technical a body of research as that in statistical mechanics, say, and requiring as detailed a study. More recently, the relation of their work to stochastic process theory has begun to be examined, in some cases putting it on a broader or more complete basis and in others bringing out more clearly the underlying assumptions.[†]

We shall consider again a mutation problem as in the last section, but in a more purely genetic context. We suppose that there is a population of size N_t, each individual of which carries a pair of genes AA, Aa or aa (cf. Example 2 in §2·2). The number of A genes in the total population will be denoted by A_t, and we write further $A_t/[2N_t] = P_t$. We shall assume that the bivariate process A_t, N_t is a Markov one in time t. This cannot be strictly true, for the exchange of genes will be related in individual genealogical lines to generation times, which we have seen must depend on age-dependent birth- and death-rates, but over the entire population it is a more realistic assumption, except perhaps for short-term effects if generation times are strictly periodic. Any effect of other genes has of course been ignored.

The detailed evolution of A_t and N_t will also depend on the phenotypic composition of the population at any instant, and its breeding characteristics. We can only hope to be able to neglect such factors in detail if we view the evolutionary changes in a rather broad manner. This means that only large changes compared with individual transitions can be considered, and we shall in effect be specifying these changes by an asymptotic diffusion process, though one of a Markov type (i.e. of the general type in equation (4), §3·5, or its extension to multivariate processes) rather than of the simple additive type.

We have seen that the behaviour of a finite population can be

† See especially Moran (1962).

fundamentally different according to whether its total size is expected to increase or decrease, and we may expect this to apply also to a population of mixed genetic constitution. In many situations it is perhaps reasonable to suppose that the total size N_t is fairly restricted by the environment, and keeps approximately constant, but, as Feller (1951) has stressed, this is a drastic assumption which severely limits the class of process studied. It would be useful, as in the processes of the last section, to study the joint distribution of A_t and N_t under wider conditions. Under general mutation and selection changes this is, however, not easy, and we consider here the solution under the special but usual assumption of N constant, following a treatment due to Malécot (1952).

As $N_t = N$ is constant, the Markov process is one in A_t only, or equivalently in $P_t = \frac{1}{2} A_t / N$. The equation for $M_t(\theta) \equiv E\{e^{\theta P_t}\}$ is then (equation (2), § 3·5, dropping t in Ψ for any homogeneous case)

$$\frac{\partial M_t(\theta)}{\partial t} = \Psi\left(\theta, \frac{\partial}{\partial \theta}\right) M_t(\theta), \tag{1}$$

where $$\Psi(\theta, P_t) = \lim_{\Delta t \to 0} \frac{\log E\{e^{\theta \Delta P_t} \mid P_t\}}{\Delta t}.$$

We now make the following simplifying assumptions:

(i) For small Δt, there is a chance $\kappa N P_t Q_t \Delta t$, where $Q_t = 1 - P_t$, of a single gene transition $A \to a$, and similarly of $a \to A$, due to the random shuffling of genes in the offspring of any mating. (Strictly speaking, these changes depend on the phenotypic male and female frequencies of the AA, Aa and aa gene pairs, even under random mating conditions, but it is in any case hardly possible to specify these changes more realistically under the artificial condition that N_t is constant.)

(ii) Selection is assumed to operate on the *ratio* P_t/Q_t of A to a genes, changing its value deterministically (again, for N_t constant, it seems difficult to formulate this more realistically) by an amount $\sigma \Delta t (P_t/Q_t)$ in Δt; this implies a change in P_t of $\sigma P_t Q_t \Delta t$.

(iii) There is a random mutation rate μ from A to a, i.e. the chance of such a mutation in Δt is $2 N P_t \mu \Delta t$; similarly, the rate from a to A is ν.

Assumptions (ii) and (iii) give a change in the mean of P_t of $\sigma P_t Q_t \Delta t + (\nu Q_t - \mu P_t) \Delta t$; assumption (i) gives no change in mean, but a variance $\frac{1}{2} \kappa P_t Q_t \Delta t / N$. The variance due to (iii) is

$$\frac{1}{2}(\nu Q_t + \mu P_t) \Delta t / N.$$

The third-cumulant contribution from (i) and (iii) is $O(1/N^2)$. Hence

$$\Psi'(\theta, P_t) = \theta[\sigma P_t Q_t + \nu Q_t - \mu P_t] + \frac{1}{2}\theta^2[\frac{1}{2}\kappa P_t Q_t + \frac{1}{2}(\nu Q_t + \mu P_t)]/N + O[\theta^3/N^2]. \quad (2)$$

The equation (1) thus becomes a diffusion equation if we assume we may neglect the terms of $O(1/N^2)$, so that θ only appears explicitly in Ψ' as a quadratic; equation (1) is, moreover, a partial differential equation of the second order in $\partial/\partial\theta$. If we assume that the solution of (1) has a limiting distribution, this must satisfy the equation

$$\Psi\left(\theta, \frac{\partial}{\partial\theta}\right) M(\theta) = 0. \quad (3)$$

For example, if the mutation rates ν and μ are zero, we have the equation

$$[\sigma\theta + \tfrac{1}{4}\kappa\theta^2/N]\left[\frac{\partial}{\partial\theta} - \frac{\partial^2}{\partial\theta^2}\right] M = 0. \quad (4)$$

This has the solution $Ae^\theta + B$. Moreover, from the full equation containing $\partial M_t/\partial t$, we see that $\partial M_t/\partial t = 0$ when $\theta = 0$ or $-4N\sigma/\kappa$. Hence for these values of θ, the limiting $M(\theta)$ is equal to its initial value $e^{r\theta}$, so that

$$A + B = 1, \quad Ae^{-4N\sigma/\kappa} + B = e^{-4N\sigma r/\kappa},$$

whence

$$M = \frac{1 - e^{-4N\sigma r/\kappa}}{1 - e^{-4N\sigma/\kappa}}(e^\theta - 1) + 1, \quad (5)$$

giving the complementary chances of extinction of the A or a genes in the absence of mutation. Malécot has shown further that this distribution is in fact the limiting solution of (1) and (2) for μ and ν zero.

As a second fairly tractable case, we suppose $\sigma = 0$ and μ, ν

are $O(1/N)$; then expanding M_i in powers of θ, we find the equation for $m_s = E\{P_i^s\}$

$$\frac{dm_s}{dt} = -s\nu(m_s - m_{s-1}) - \mu s m_s - \frac{\kappa s(s-1)}{4N}(m_s - m_{s-1}) + O\left(\frac{1}{N^2}\right), \quad (6)$$

and in particular, as $m_0 = 1$,

$$\frac{dm_1}{dt} = -\nu(m_1 - 1) - \mu m_1,$$

whence $\qquad m_1 = \dfrac{\nu[1 - e^{-(\mu+\nu)t}]}{\mu+\nu} + re^{-(\mu+\nu)t} \to \nu/(\mu+\nu).$

The equation (6) may be solved successively for $m_2, m_3, \dots$, and it is then evident that m_s has a limit as $t \to \infty$ for all s. This must therefore be given by the recurrence relation

$$m_s[s - 1 + 4N(\mu+\nu)/\kappa + O(1/N)]$$
$$= m_{s-1}[s - 1 + 4N\nu/\kappa + O(1/N)], \quad (7)$$

which defines the continuous distribution for P_∞ (independently of r)

$$f(p)\,dp \propto p^{4N\nu/\kappa - 1}(1 - p)^{4N\mu/\kappa - 1}\,dp. \quad (8)$$

This result may be obtained more simply from a general formula due to Wright

$$f(p) \propto \frac{1}{\sigma^2(p)}\exp\left\{2\int \frac{m(p)}{\sigma^2(p)}\,dp\right\}, \quad (9)$$

where $m(p)$ is the change in mean per unit time, and $\sigma^2(p)$ the change in variance. We put

$$m(p) = \nu(1-p) - \mu p \quad \text{and} \quad \sigma^2(p) = \tfrac{1}{2}\kappa p(1-p)/N,$$

and (8) follows. The formula (9) was, however, obtained merely by requiring the mean and variance of any stationary distribution with probability density to remain constant, and the solution (5) shows that it does not give the correct solution in all cases.

4·4 Epidemic models

A further important application of the techniques developed in Chapter 3 is to the mathematical theory of epidemics. We shall illustrate this by examining one or two simple but typical stochastic models associated with transmission of infection from person to person. As in the case of population growth a large part of the literature has been devoted to 'deterministic' forms of the equations (see, for example, Kermack and McKendrick, 1927–33; Soper, 1929).† Such discussions have been useful in often indicating sufficiently well the growth and spread of epidemics when numbers are large, but evidently may not be adequate when numbers are small and in particular in the early stages of an epidemic, so that it is essential to examine also the extent to which such studies are supported when more complete stochastic representations are employed.

Let us suppose that a population has at any time t a number $S(t)$ of individuals susceptible to a certain disease, and a number $I(t)$ of individuals actually infected. In problems for which recovered individuals are only temporarily immune from a further attack, we should need to specify such individuals also, but we suppose recovered individuals are permanently immune, and are similar to isolated (or dead) individuals in not giving rise to new infections. We shall for the moment assume that S (but not I) can be augmented by new susceptibles entering the population from outside; for example, in the case of measles, which is one of the most convenient epidemic diseases available for study, children are constantly growing up into the critical age period. To be precise, we assume the following idealized scheme of random transitions (cf. § 3·5), where the p.g.f. variable z corresponds to S and w to I:

Type of transition	Rate	Operator
$zw \to w^2$	λ	$\partial^2/\partial z\, \partial w$
$w \to 1$	μ	$\partial/\partial w$
$1 \to z$	ν	1

The equation for the p.g.f. $\Pi_t(z, w)$ is correspondingly

$$\frac{\partial \Pi}{\partial t} = \lambda(w^2 - zw)\frac{\partial^2 \Pi}{\partial z\, \partial w} + \mu(1 - w)\frac{\partial \Pi}{\partial w} + \nu(z - 1)\,\Pi. \qquad (1)$$

Differentiating this equation in turn with respect to w and z,

† See also, however, McKendrick (1926).

and then putting $z = w = 1$, we obtain the two equations

$$\left.\begin{aligned}
\frac{\partial x}{\partial t} &= \lambda E\{SI\} - \mu x, \\
\frac{\partial y}{\partial t} &= -\lambda E\{SI\} + \nu,
\end{aligned}\right\} \tag{2}$$

where $x = E\{I\}$, $y = E\{S\}$. In the deterministic model $E(SI)$ is replaced by xy, random variation in S and I being ignored.

The logistic process. A special well-known case is obtained by writing $\mu = \nu = 0$, $S = n - I$, the last condition corresponding to the assumption, that all individuals not infected in some finite constant population are susceptible. This gives the deterministic model

$$\frac{\partial x}{\partial t} = \lambda x(n - x), \tag{3}$$

with solution for the total number of infected up to time t,

$$x = n/[1 + e^{-\lambda n(t - t_0)}], \tag{4}$$

and rate of infection

$$\frac{\partial x}{\partial t} = \tfrac{1}{4}\lambda n^2 \operatorname{sech}^2[\tfrac{1}{2}\lambda n(t - t_0)], \tag{5}$$

where t_0 coincides with the time of maximum infection rate. The corresponding stochastic equation may be obtained (either by direct argument or from (1) by writing $\Pi(z, w) = z^n \Pi(z^{-1}w)$ and then putting $z = 1$) as

$$\frac{\partial \Pi}{\partial t} = \lambda w(w - 1)\left[(n - 1)\frac{\partial \Pi}{\partial w} - w\frac{\partial^2 \Pi}{\partial w^2}\right], \tag{6}$$

with $\Pi_0(w) = w^r$, say. Equivalently, in terms of the m.g.f. $M(\theta)$, equation (6) becomes

$$\frac{\partial M}{\partial t} = \lambda(e^\theta - 1)\left[n\frac{\partial M}{\partial \theta} - \frac{\partial^2 M}{\partial \theta^2}\right], \tag{7}$$

and from either equation, differentiating once with respect to w or θ and then putting $w = e^\theta = 1$, we easily obtain the equation

$$\dot{m}_1 = \lambda(nm_1 - m_2) \tag{8}$$

for the rate of change $\dot{m}_1$ of the mean $m_1 \equiv E\{I\}$, where m_2 is $E\{I^2\}$. It will be noticed that the non-linear nature of these models prevents a straightforward solution for the moments, though (8) allows m_2 to be calculated if m_1 is known. By writing (8) in the form

$$\dot{m}_1 = \lambda(nm_1 - m_1^2) - \lambda\sigma_I^2, \tag{9}$$

where σ_I^2 is the variance of I, and comparing it with the deterministic equation (3), we see that the effect of the variance σ_I^2 is to depress the initial rise of I on the average. The exact solution of (6) or (7) can in principle be obtained, being in fact a special case of the homogeneous birth process referred to in §3·2, but the detailed solution is rather complicated (see Bailey, 1950).

Deterministic approximations. Returning to the deterministic form of the equations (2), we may reduce these to equations for deviations about equilibrium, the latter corresponding to $y_0 = \mu/\lambda$, $x_0 = \nu/\mu$, by writing $x = x_0(1+u)$, $y = y_0(1+v)$, whence

$$\left.\begin{aligned} \tau\frac{\partial u}{\partial t} &= v(1+u), \\ \sigma\frac{\partial v}{\partial t} &= -(u+v+uv), \end{aligned}\right\} \tag{10}$$

where $\tau = 1/\mu$, $\sigma = \mu/(\nu\lambda)$. Some information on the nature of solutions of these equations is obtained by considering the case u and v small. Neglecting uv and eliminating v, we then obtain for u the equation

$$\frac{d^2u}{dt^2} + \frac{1}{\sigma}\frac{du}{dt} + \frac{1}{\sigma\tau}u = 0, \tag{11}$$

with solution

$$u = u_0 e^{-\frac{1}{2}t/\sigma}\cos\theta t \quad (\theta = \sqrt{[1/(\tau\sigma) - 1/(4\sigma^2)]}), \tag{12}$$

for suitable choice of time origin. The solution for v is then

$$v = u_0\sqrt{\beta}\,e^{-\frac{1}{2}t/\sigma}\cos(\theta t + \psi) \quad (0 \leqslant \psi \leqslant \pi), \tag{13}$$

where $\cos\psi = -\frac{1}{2}\sqrt{\beta}$, $\beta = \tau/\sigma$. These solutions represent damped harmonics with period $2\pi/\theta$. In the case of measles, Soper (1929) took τ as equivalent to an incubation period of a fortnight, and estimating σ for London as 68·2 weeks, calculated a period of 73·7 weeks from (12), and a damping factor from peak to peak of 0·58. These figures for small oscillations are comparatively insensitive to the oscillation amplitude, as may be ascertained either by direct arithmetic computation or by the development of the solution of the non-linear equations (7) as a power series in the solution of the linearized form. Up to the second-order terms we write

$$u = u_1 + a_{11}u_1^2 + a_{12}u_1 v_1 + a_{22}v_1^2, \left.\right\} \tag{14}$$
$$v = v_1 + b_{11}u_1^2 + b_{12}u_1 v_1 + b_{22}v_1^2, \left.\right\}$$

where u_1, v_1 are the linearized or first-order solutions. Then by straightforward investigation it is found that, if

$$a'_{ij} = (9 - 2\beta)\, a_{ij}, \quad b'_{ij} = (9 - 2\beta)\, b_{ij},$$

then

$$a'_{11} = 3 + 2\beta, \quad a'_{12} = 2\beta - 1, \qquad a'_{22} = 2 - 3/\beta, \left.\right\}$$
$$b'_{11} = \beta - 2\beta^2, \quad b'_{12} = 3 + 3\beta - 2\beta^2, \quad b'_{22} = 5 - 2\beta. \left.\right\} \tag{15}$$

In cases where β is small, so that $\psi \sim 90°$, this solution becomes approximately

$$u = u_1 + \tfrac{1}{3}u_0^2 e^{-t/\sigma} \cos 2\theta t, \left.\right\}$$
$$v = v_1(1 + \tfrac{1}{3}u_1), \left.\right\} \tag{16}$$

where u_1 is given by (12) and $v_1 = -u_0 \sqrt{\beta}\, e^{-\frac{1}{2}t/\sigma} \sin \theta t$.

These models may be extended to cover various modifications and additions to make them more realistic or more applicable to other situations—for example, (i) when immunity after infection and recovery is only partial or temporary, (ii) when a seasonal effect is present, or (iii) when the non-infectious and infectious parts of the incubation period are more precisely considered. The device used in § 4·31 of introducing a number of substates before the final transition to a new state takes place is convenient for modifying the lifetime distribution in any state in the stochastic model, though of course it increases the intractability of the equations. In the extreme case of a constant non-infectious period a immediately after infection, the deterministic form of (2) is readily seen to be

$$\frac{\partial x(t)}{\partial t} = \lambda x(t - a)\, y(t) - \mu x(t), \left.\right\}$$
$$\frac{\partial y(t)}{\partial t} = -\lambda x(t - a)\, y(t) + \nu. \left.\right\} \tag{17}$$

The equilibrium values of x and y are as before, and if $D \equiv \partial/\partial t$, we find

$$\left(D + \frac{1}{\sigma}\right)v + \frac{1}{\sigma}e^{-aD}u = 0,$$

$$-\frac{1}{\tau}v + \left(D + \frac{1}{\tau} - \frac{1}{\tau}e^{-aD}\right)u = 0.$$

If we suppose that the infectious period is short, we have μ and λ large, such that μ/λ remains as before. Thus $\tau \to 0$, and the equation for u then reduces to

$$\left[\left(D + \frac{1}{\sigma}\right)e^{aD} - D\right]u = 0. \tag{18}$$

The behaviour of u depends on the nature of the roots of

$$\left(D + \frac{1}{\sigma}\right)e^{aD} - D = 0.$$

For σ fairly large, we have

$$aD = -\log[1 + 1/(\sigma D)]$$

$$= -\frac{1}{\sigma D} + \frac{1}{2\sigma^2 D^2} \cdots,$$

with first approximation $D - i/\sqrt{(a\sigma)} = 0$, second approximation

$$aD = -\frac{1}{\sigma D} - \frac{a}{2\sigma},$$

which corresponds to the same type of solution as before with a in place of τ, and a damping term which has been halved (this may be confirmed more rigorously if required by taking the case of a finite number k of substates in the non-infectious part of the incubation period, and then letting k increase).

We have stressed that these deterministic formulations, while of some value in indicating the structural tendencies of the process in large populations, may be misleading if not supplemented by consideration of more complete stochastic models. This is so even for the simplest epidemic models such as the logistic process if it is remembered that owing to their non-linearity the equations are not independent of 'scale' even in their deterministic form. Thus the infectivity rate λ in (1), to be consistent with the numbers cited for London measles, has to be taken as 1/300,000. This is obviously too small except as an average of a highly variable chance of infection from children scattered over the whole of London. It is more relevant to consider a comparatively homogeneous group such as a school, for which stochastic fluctuations are relatively larger.

Contrasting stabilities in deterministic and stochastic models. However, a much more striking and important contrast between deterministic and stochastic epidemic models may be indicated. With the deterministic model associated with (1), the presence of the damping factor implies that an endemic stable equilibrium level should be reached; for infectious diseases such as measles this is not what is observed in practice. Modifications of the infectivity assumptions may reduce but do not eliminate this damping tendency, while seasonal changes in infectivity may be shown merely to induce regular annual oscillations about such an equilibrium level.

It will be recalled that *stochastically* this equilibrium level is precisely the one that is not permanent, as we saw in the theory of population growth; there the ultimate situation was either a very large population or extinction. This phenomenon appears to give a real meaning to the idea of a threshold level of susceptibles, below which a major epidemic will not occur. Consider, for example, an isolated susceptible population of size n, as in the logistic process, but where $\mu \neq 0$, so that the infected individuals are removed in due course. Equation (1) is then†

$$\frac{\partial \Pi}{\partial t} = \lambda(w^2 - zw)\frac{\partial^2 \Pi}{\partial z \, \partial w} + \mu(1-w)\frac{\partial \Pi}{\partial w}. \tag{19}$$

This situation can approximately represent (1) when ν is small, at the stage when one or two infected persons are introduced into a population previously free of infection. At the beginning of any epidemic, S will decrease from n as I increases, and the chance of a new infection will therefore be somewhat less than if we suppose S remains at the value n. But with this modification, (19) becomes simply the birth-and-death process

$$\frac{\partial \Pi}{\partial t} = \lambda n(w^2 - w)\frac{\partial \Pi}{\partial w} + \mu(1-w)\frac{\partial \Pi}{\partial w}. \tag{20}$$

Hence if $\lambda n \leqslant \mu$ the infection will fade out, possibly not until

† For exact solutions of equation (19) see Gani (1965), Siskind (1965).

after a 'minor outbreak' but without causing a 'major epidemic'. If $\lambda n > \mu$, there will be (see § 3·4) an approximate chance $[\mu/(\lambda n)]^j$ of the infection becoming extinct before it takes hold of the susceptible population, where j is the initial number of infected individuals. This chance tends to zero as n and/or j increase. (For a more exact investigation for finite n of the distribution of total size of an epidemic for the model (19), see Bailey (1953).)

Returning to the problem of the continuous epidemiological history of an isolated population, we see that in contrast to the deterministic endemic state, the infection will (under the above assumptions) disappear when the number of susceptibles is low enough, and can only enter the population again from outside. In a group which is not completely isolated (like a boarding-school) a continually recurring series of epidemics (with no damping) can, however, occur by a 'triggering-off' of each epidemic by fresh infection from outside whenever the number of susceptibles has reached a high enough level.

An approximate assessment of the distribution of the 're-newal' time of a major epidemic (under simplified assumptions) is as follows. Let the chance of one new infection from outside in t, $t + \delta t$ be $\epsilon \delta t + o(\delta t)$, and of none, $1 - \epsilon \delta t + o(\delta t)$. We suppose the susceptibles S are increasing at a much greater rate, and represent this increase by a *deterministic* rate ν. We suppose for definiteness that the initial number of susceptibles at $t = 0$ is negligible, so that the chance of 'extinction' of any new infection that enters is approximately $\mu/(\lambda \nu t)$ when $\lambda \nu t > \mu$. For ϵ small we neglect the delay in the growth of an epidemic (if any) from fresh infection, so that the chance of no epidemic up to time t is

$$\prod_{t > u > \mu/(\lambda\nu)} [1 - \epsilon \, du + \mu \epsilon \, du/(\lambda \nu u)] = \exp\left\{ -\epsilon \int_{\mu/(\lambda\nu)}^{t} [1 - \mu/(\lambda \nu u)] \, du \right\}$$

$$= \left(\frac{\lambda \nu t}{\mu} \right)^{\epsilon\mu/(\lambda\nu)} e^{-\epsilon[t - \mu/(\lambda\nu)]} \quad \left(t > \frac{\mu}{\lambda \nu} \right). \quad (21)$$

The equivalent frequency distribution, in terms of $T \equiv \lambda \nu t / \mu$, is

$$r(T - 1) \, T^{r-1} e^{-r(T-1)} \quad (r = \epsilon \mu/(\lambda \nu)), \quad (22)$$

with a mode at $T = 1 + 1/\sqrt{r}$. The above treatment is rather rough, but illustrates sufficiently well the stochastic mechanism; a more precise solution for the time of outbreak under the above assumptions may be obtained by means of the theory of the non-homogeneous birth-death-and-immigration process (§ 3·41) with birth-rate $\lambda \nu t$.

The distinction between 'minor outbreaks' (i.e. extinction after a small number of infection transmissions) and 'major epidemics' (i.e. spread of the infection through the susceptibles) will be less clear-cut for t in the neighbourhood of $\mu/(\lambda \nu)$, as minor outbreaks for $\lambda S/\mu$ approaching unity may be fairly prolonged, and major epidemics for $\lambda S/\mu$ not much larger than unity will be curtailed by the finite size of S; as t and S increase, however, the probable size of the epidemic will increase.

This type of stochastic mechanism has been demonstrated by means of artificial epidemic series constructed with the aid of random numbers. As infections multiply rapidly when an epidemic is under way and may be few and far between in quiescent periods between epidemics, it is usually convenient to employ the type of construction used for the simple birth-and-death process (fig. 4, Chapter 3), and determine the random times at which events take place rather than the values of the stochastic variables at fixed times. Thus for the last model (reverting for greater convenience to a *random* entry of susceptibles at average rate ν) the interval before a new occurrence has exponential distribution with mean $1/[\epsilon + \nu + \lambda IS + \mu I]$, and the relative chances of (i) $I \to I+1$, $S \to S$, (ii) $S \to S+1$, $I \to I$, (iii) $I \to I+1$, $S \to S-1$, (iv) $I \to I-1$, $S \to S$ are $\epsilon : \nu : \lambda IS : \mu I$. In fig. 6 (p. 133) part of an artificial series of this type is shown. In this particular series various modifications were introduced in order to make it more representative of real measles incidence in a boarding-school, with a 10 % seasonal variation of infectivity, and the entry of new susceptibles, and occasional infected children, restricted to the beginning of terms (see Bartlett, 1953); the essential features were, however, similar.

How far the observed quasi-periodic character of such epidemics as measles over a large continuous area such as a town tends to be maintained by a stochastic mechanism of the above kind requires further discussion. An examination of actual records confirms that infection may become temporarily absent from relatively large portions of an urban area; it can now, however, be transmitted from neighbouring districts, and a

complete stochastic model must include such possible transmissions. The extent of the effective 'isolation' of the various
local groups or districts will obviously be a relevant factor, for
with sufficient isolation each group would approximate to the
case treated above. We need first to set up a general mathematical
formalism for studying this rather complex problem. We
replace the p.g.f. Π of equation (1) by the functional $\Pi_t(z(\mathbf{r}), w(\mathbf{r}))$,
where $\mathbf{r}$ denotes the spatial coordinates of any individual; and
we assume possible transitions: $w(\mathbf{r}) \to 1$ at rate μ, $1 \to z(\mathbf{r})$ at
rate $\nu d\mathbf{r}$, $w(\mathbf{s}) z(\mathbf{r}) \to w(\mathbf{s}) w(\mathbf{r})$ at rate λ, where λ is a function
$\lambda(\mathbf{r} - \mathbf{s})$ of the separation† $\mathbf{r} - \mathbf{s}$. Then the principles of § 3·42
lead immediately to the generalization of the 'forward'
equation (1)

$$\frac{\partial \Pi_t}{\partial t} = \int \mu(1 - w(\mathbf{r})) \left\{ \frac{\partial \Pi_t}{\partial w(\mathbf{r}) \, d\mathbf{r}} \right\} d\mathbf{r} + \int \nu(z(\mathbf{r}) - 1) \, \Pi_t d\mathbf{r}$$

$$+ \iint \lambda(\mathbf{r} - \mathbf{s}) \, w(\mathbf{s}) \, (w(\mathbf{r}) - z(\mathbf{r})) \left\{ \frac{\partial^2 \Pi_t}{\partial w(\mathbf{s}) \, \partial z(\mathbf{r}) \, d\mathbf{s} \, d\mathbf{r}} \right\} d\mathbf{s} \, d\mathbf{r}. \quad (23)$$

In the situation when infection is almost absent in an area and
the number of susceptibles may be treated temporarily as
uninfluenced by the number of infected individuals, the process
degenerates to a multiplicative process in the sense of § 3·4, and
the 'backward' equation appears a little simpler. Thus in place
of (20) we may write for $\Pi_t(\mathbf{r})$, for *one* initial infected individual
at $\mathbf{r}$,

$$\frac{\partial \Pi_t(\mathbf{r})}{\partial t} = \mu(1 - \Pi_t(\mathbf{r})) + \int n(\mathbf{s}) \, \lambda(\mathbf{r} - \mathbf{s}) \, \Pi_t(\mathbf{r}) \, (\Pi_t(\mathbf{s}) - 1) \, d\mathbf{s}, \quad (24)$$

where $n(\mathbf{s})$ denotes the initial density‡ of susceptibles. In
particular, equation (24) includes the equation for the mean
density of infected individuals,

$$\frac{\partial m(\mathbf{r})}{\partial t} = -\mu m(\mathbf{r}) + \int n(\mathbf{s}) \, \lambda(\mathbf{r} - \mathbf{s}) \, m(\mathbf{s}) \, d\mathbf{s}. \quad (25)$$

This equation is an equation for the mean density $m(\mathbf{r})$ at any

---

† Cf. the distance function introduced, in a somewhat similar context, by
Rapoport (1951).

‡ Not, of course, strictly a density function, as the initial number of
susceptibles is finite.

arbitrary fixed point, given one initial individual at **r**, but with
assumptions of homogeneity and isotropy for n and λ, it is also
the equation for the mean density at **r** for one initial individual
at any arbitrary point, say the origin (or by superposition for any

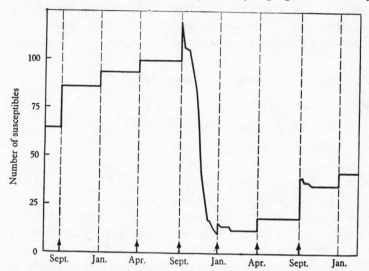

Fig. 6. Extract from mock epidemic series simulating successive measles
outbreaks in a boarding school. The graph shows a major epidemic beginning
in the autumn of the second year shown and ending in the New Year. A minor
outbreak also occurs the following autumn. The dotted lines indicate the
beginning of terms, when there are a number of new entrants to the 'school';
the arrows indicate dates when infection also entered. (Reprinted from
Applied Statistics, **2** (1953), 62.)

arbitrary initial distribution). To solve it in this case we may take
the bivariate Fourier transform $M(\theta_1, \theta_2)$ of $m(\mathbf{r})$ with respect to
r, and so obtain

$$M = e^{(n\Lambda - \mu)t}, \tag{26}$$

where Λ is the transform of $\lambda(\mathbf{r})$. While it has been previously
stressed that the mean density is not to be taken as necessarily
representative of the actual numbers, we may reasonably suppose
that for large n the chance of extinction is confined to an initial
period when the mean density is almost everywhere zero, and
under such conditions the solution (26) can be quite informative.
Asymptotically, in the particular case of purely local infection,
we may put $\Lambda t \sim \lambda t - \nu t(\theta_1^2 + \theta_2^2)$, and (26) represents the Gaussian
distribution solution of a diffusion equation, with multiplying

term $\alpha = n\lambda - \mu$. It is known (cf., for example, Fisher, 1937) that this solution implies a wave of propagation with velocity† $2\sqrt{(n\nu\alpha)}$. In the case when $n\lambda$ and μ are about equal, however, we can no longer expect the mean density solution (26) to be particularly useful, and a further study of equation (24) is desirable. Thus p_t, the chance of total extinction, may be obtained by putting $w(\mathbf{u}) = 0$ in (24). This gives, with the homogeneity condition $p_t(\mathbf{r}) = p_t$,

$$\dot{p}_t = \mu(1 - p_t) + n\lambda p_t(p_t - 1), \tag{27}$$

where $\lambda = \Lambda(\mathbf{0})$; whence

$$p_t = \mu[e^{(n\lambda - \mu)t} - 1]/[n\lambda e^{(n\lambda - \mu)t} - \mu]. \tag{28}$$

The solution for the second-order 'product-density' f_2, or other higher-order fluctuation formulae, may also be obtained, though more readily from the 'forward' equation equivalent to (24).

Further note on recurrent epidemics. Coming back to the theory of recurrent epidemics, the continuous history of the spread of infection over a large urban area is, as remarked earlier, a difficult problem; but it is now at least possible to comment on the overall picture. The stochastic mechanism illustrated in fig. 6 strictly applies to a comparatively small and sufficiently isolated community. For a larger community, new infection from outside the community may not be necessary for further epidemics; for the epidemic cycle may swing the increasing susceptible population back over its threshold before the tendency to extinction of infection below the threshold has had sufficient time to take complete effect. In the case of infection such as measles, this gives rise to a further concept, namely, a critical size of community above which the infection has a reasonable (say 50 %) chance of maintaining itself from one cycle to the next. With the model for measles, Monte Carlo studies have been carried out to supplement the rather crude theoretical results as yet available; these appear consistent with an *observed* critical size, either in England and Wales or in the United States during recent years, of the order of 250,000 total population (for further details, see Bartlett, 1960).

† It has, however, now been realized that this diffusion approximation is not adequate at the wavefront; see Daniels (1977), Mollison (1977) for further details.

Chapter 5

LIMITING STOCHASTIC OPERATIONS

5·1 Stochastic convergence

So far in this book we have hardly needed to consider direct limiting operations on the random variables and functions themselves, although of course we have encountered asymptotic and limiting properties for associated probabilities and probability distributions. Direct limiting operations on a stochastic process $X(t)$ include differentiation and integration, the possible equivalence of $X(t)$ to series expansions, and so on. A full discussion of these problems is included in Doob(1953)and our purpose in the present chapter is merely to indicate these methods and ideas to an extent sufficient for most applications.

In this first section we summarize the three main modes of convergence of a sequence of random variables $X_1, X_2, ..., X_n$ to a stochastic limit X, and define the particular mode of convergence mainly used in subsequent chapters (for a more detailed account, see Fréchet, 1937–8, Chapter 5).

Convergence in probability (*i.p.*). X_n converges to X in probability as $n \to \infty$ if, for any positive ϵ,

$$\lim_{n \to \infty} P\{|X_n - X| > \epsilon\} = 0. \tag{1}$$

A necessary and sufficient condition for such convergence is that for any positive ϵ and η there is an n_0 such that

$$P\{|X_n - X_m| > \epsilon\} < \eta \quad \text{for all } n, m \geqslant n_0. \tag{2}$$

An alternative necessary and sufficient condition is that $X_n \to X$ i.p. if the distribution function (d.f.) $G_n(y)$ of $Y_n = X_n - X$ tends to that of $Y \equiv 0$. This limiting d.f. of Y_n is sometimes denoted by $\epsilon(y)$, and is the proper function which is the integral of the improper Dirac δ-function $\delta(y)$. Note that it is necessary but not sufficient for $X_n \to X$ i.p. that the d.f. $F_n(x)$ of X_n tends to $F(x)$ uniformly at all points of continuity of $F(x)$.

The limit X is unique in the sense that if $X_n \to X$ i.p. and $X_n \to Z$ i.p. then $X \equiv Z$ almost certainly (i.e. $P\{X - Z = 0\} = 1$).

Almost certain (a.c.) or strong convergence. A stronger criterion of convergence (implying convergence i.p. but not conversely) is of particular importance in the probability limit theorems known as the 'laws of large numbers'. This is defined in terms of the entire sequence of random variables $X_1, X_2, ..., X_n, ...$ Regarding such a sequence as a new random variable with realized value $x_1, x_2, ..., x_n, ...$, we may say that this realized sequence either does or does not converge in the ordinary sense to a limit x. If the probability that it does so is unity, then we say that $X_n \to X$ almost certainly. This criterion is equivalent to what is called strong convergence in the theory of probability, defined by the condition that for any positive ϵ

$$\lim_{n \to \infty} P\{| X_m - X | > \epsilon \text{ for at least one } m \geqslant n\} = 0. \tag{3}$$

The equivalent condition not depending explicitly on X is that for any positive ϵ and η, there is an n_0 such that

$$P\{| X_n - X_m | > \epsilon \text{ for at least one } m \geqslant n\} < \eta \tag{4}$$

for all $n \geqslant n_0$. As a.c. convergence depends on the *simultaneous* behaviour of X_n for all $n \geqslant n_0$, it is obviously more difficult to handle, but the following sufficient criterion is useful. If

$$\Sigma_n E\{| X_n - X |^p\} < \infty \tag{5}$$

for some $p > 0$, then $X_n \to X$ a.c. An alternative condition is

$$\Sigma_n E\{[| X_{n+1} - X_n |/\epsilon_n]^p\} < \infty \quad \text{(where } \Sigma_n \epsilon_n < \infty). \tag{6}$$

Convergence 'in the mean' or 'in mean square' (m.s.). A third criterion we shall find particularly useful in stochastic process theory is analogous to the idea of 'convergence in the mean' in analysis, and is defined in general by the condition

$$\lim_{n \to \infty} E\{| X_n - X |^p\} = 0 \quad (p \geqslant 1), \tag{7}$$

though we shall only (unless otherwise stated) consider the case $p = 2$. We shall then sometimes write (7) in the notation employed in analysis:

$$\underset{n \to \infty}{\text{l.i.m.}} X_n = X. \tag{8}$$

However, owing to the danger of confusion in the use of the phrase 'in the mean' in statistics, we shall more often refer to (8) as convergence in mean square (or m.s. convergence).

The necessary and sufficient condition for m.s. convergence corresponding to the condition (2) for convergence i.p., or (4) for a.c. convergence, is

$$E\{|X_n - X_m|^2\} < \epsilon \text{ for all } n, m \geqslant n_0. \tag{9}$$

It may have been noticed by the reader that we have not here attempted to justify the existence of the limiting random variable X, though the equivalent conditions for convergence stated entirely in terms of the sequence variables X_n will suggest that this is possible. Moreover, it is easily shown that if $X_n \to X$, *either* a.c. *or* m.s., then $X_n \to X$ i.p. (but not conversely); as the i.p. limit X is unique, so is the a.c. or m.s. limit.

There is, however, no implication of a.c. convergence in m.s. convergence, nor conversely. This may be illustrated by an example. Let $X_n = n$ with probability $1/n^2$, and 0 with probability $1 - 1/n^2$. Then obviously

$$\lim_{n \to \infty} E\{|X_n - 0|^2\} = 1,$$

so that X_n does not converge in m.s. to 0 (there is clearly no other possible limit independent of n). But

$$P\{X_n > \epsilon \text{ for at least one } n \geqslant n_0\} \leqslant \sum_{n=n_0}^{\infty} P\{X_n > \epsilon\} = \sum_{n=n_0}^{\infty} \frac{1}{n^2}, \tag{10}$$

which converges, so that

$$\lim_{n_0 \to \infty} P\{X_n > \epsilon \text{ for at least one } n \geqslant n_0\} = 0,$$

and $X_n \to 0$ a.c. (The inequality in (10) follows from the addition law of probability for events $\mathscr{E}_1$ and $\mathscr{E}_2$, say, not mutually exclusive, namely,

$$P\{\mathscr{E}_1 \text{ or } \mathscr{E}_2\} = P\{\mathscr{E}_1\} + P\{\mathscr{E}_2\} - P\{\mathscr{E}_1 \text{ and } \mathscr{E}_2\} \leqslant P\{\mathscr{E}_1\} + P\{\mathscr{E}_2\}.)$$

Now suppose alternatively that $X_1, X_2, \ldots, X_n$ is a sequence of *independent* random variables such that $X_n = 1$ with probability $1/n$, and 0 with probability $1 - 1/n$. Then

$$\lim_{n \to \infty} E\{|X_n - 0|^2\} = 0,$$

and l.i.m. $X_n = 0$. But

$$P\{X_n = 0 \text{ for every } n \geqslant n_0\} = \left(\frac{n_0 - 1}{n_0}\right)\left(\frac{n_0}{n_0 + 1}\right)\left(\frac{n_0 + 1}{n_0 + 2}\right) \dots = 0,$$

whatever n_0, so that X_n does not converge a.c. to zero.

As a.c. convergence and m.s. convergence imply convergence i.p. they imply automatically the uniform convergence of the d.f. $F_n(x)$ to a limit $F(x)$ at all points of continuity of $F(x)$.

Application to stochastic sums. If we encounter a random variable defined as

$$X = \sum_{u=0}^{\infty} g_u Y_u, \tag{11}$$

the question of its convergence naturally occurs. The simplicity of m.s. convergence arises from its equivalence to ordinary limit properties for the second moment (always assumed finite when we make use of m.s. convergence). Thus condition (9) is then (for X_n real) equivalent to the condition

$$\lim_{n, m \to \infty} E\{X_n X_m\} \equiv \lim_{n, m \to \infty} \mu_{nm} = \mu \geqslant 0. \tag{12}$$

Applying this condition to (11), where we assume at present g_u, Y_u real, we have

$$\lim_{n, m \to \infty} \sum_{u=0}^{n} \sum_{v=0}^{m} g_u g_v E\{Y_u Y_v\} = \mu. \tag{13}$$

A sufficient condition for (13) to be true is that $\Sigma_u |g_u| \sqrt{(E\{Y_u^2\})}$ should converge. In the particular case when the Y_u are uncorrelated with zero mean and the same variance σ^2, (13) obviously becomes

$$\lim_{n \to \infty} \sum_{u=0}^{n} g_u^2 = \frac{\mu}{\sigma^2}, \tag{14}$$

so that the convergence of the sum in (14) is all that is needed.

Extension to random functions. We may extend these convergence definitions, as in analysis, to refer to functions, and consider the stochastic convergence of a random function $X(t)$, where t has a continuous range, to $X(\tau)$ as $t \to \tau$. For example, we say $X(t)$ is m.s. continuous at $t = \tau$ if

$$\lim_{t \to \tau} X(t) = X(\tau),$$

i.e. if

$$\lim_{t \to \tau} E\{|X(t) - X(\tau)|^2\} = 0. \tag{15}$$

This condition may be written (for real $X(t)$)

$$\lim_{t,s\to\tau} E\{X(t)\,X(s)\} \equiv \lim_{t,s\to\tau} \mu(t,s) = \mu(\tau,\tau) \equiv E\{X^2(\tau)\}. \qquad (16)$$

Notice that if (15) is true, we must necessarily have (as $|E\{Z\}|^2 \leqslant E\{Z^2\}$ for any random variable Z)

$$E\{\text{l.i.m.}_{t\to\tau} X(t)\} = E\{X(\tau)\} = \lim_{t\to\tau} E\{X(t)\}, \qquad (17)$$

so that the m.s. limiting operation and the expectation sign commute.

In the case of a.c. convergence, the extension implies a statement about 'almost all' realized functions $x(t)$ as $t\to\tau$, and the kind of difficulties referred to in § 1·3 will arise. In particular, the stochastic convergence of $X(t_n)$ to $X(\tau)$ for any arbitrary sequence $t_n\to\tau$, while the relevant property for investigating convergence i.p. or m.s. of $X(t)$ to $X(\tau)$, is no longer sufficient for a.c. convergence. Moreover, continuity i.p. (or m.s.) at every point t in an entire interval $(0,T)$ defines continuity i.p. (or m.s.) over the interval, but this is not true for a.c. continuity. An example is the additive process with unit jumps—this is a.c. continuous at every point, but the chance of at least one jump in any non-zero interval is greater than zero; it is a.c. continuity over an entire interval which is the more useful concept. A sufficient condition for this is

$$E\{|X(t)-X(t+h)|^2\} \leqslant C\,|h|^{\alpha+1} \quad (C,\alpha>0), \qquad (18)$$

for all t, $t+h$ in $(0,T)$.

5·11 Stochastic differentiation and integration. We next extend these ideas to differentiation and integration. Thus $\dot{X}(t)$ is said to be the m.s. differential coefficient of $X(t)$ if

$$\lim_{h\to 0} E\left\{\left|\frac{X(t+h)-X(t)}{h} - \dot{X}(t)\right|^2\right\} = 0, \qquad (1)$$

and it is easily shown from the principles indicated in § 5·1 that the equivalent condition for this is that $\mu(t,s)$ has partial derivatives $\partial\mu/\partial t$, $\partial\mu/\partial s$, $\partial^2\mu/\partial t\partial s$ (the latter in the symmetric or complete sense) at $t=s$. If $\dot{X}(t)$ exists at all points t, then these partial derivatives exist on the whole line $t=s$, and it may be shown further that they exist over the whole t, s plane. We then also have

$$E\{\dot{X}(t)\} = \frac{\partial}{\partial t} E\{X(t)\}, \quad E\{\dot{X}(t)\,X(s)\} = \frac{\partial \mu(t, s)}{\partial t},$$

$$E\{\dot{X}(t)\,\dot{X}(s)\} = \frac{\partial^2 \mu(t, s)}{\partial t\,\partial s}. \quad (2)$$

By repeating the argument for $\dot{X}(t)$, we obtain the further conditions for $\ddot{X}(t)$ to exist, and so on. If also $X(t)$ is stationary, so that $\mu(t, s)$ is a function only of $t - s$, it is necessarily differentiable for all t if differentiable at one point. For stationary stochastic processes with

$$\mu(t - s) = \cos |t - s|, \quad \exp\{-|t - s|\}, \quad \exp\{-(t - s)^2\}$$

(that such processes exist is shown in the next chapter), we see that the first and last are differentiable m.s. indefinitely, whereas the second is not even differentiable once. It is, however, m.s. continuous, corresponding to the ordinary continuity at $\tau = 0$ of $\exp\{-|\tau|\}$.

It is possible from (18) of § 5·1 to obtain sufficient conditions for a.c. differentiability; it is found that $X(t)$ is a.c. continuous if m.s. differentiable, a.c. differentiable once if m.s. differentiable twice, and so on. In particular, if $X(t)$ is m.s. differentiable indefinitely, it is a.c. differentiable indefinitely. We may go further and define *analytic* random functions $X(t)$ by taking t and hence $X(t)$ as complex. We then define

$$\mu(t, s) = E\{X^*(t)\,X(s)\}, \quad (3)$$

where X^* is the complex conjugate of X. $X(t)$ is said to be m.s. analytic in some region R of the t complex plane if it can be expanded in the m.s. convergent Taylor series

$$X(t) = \sum_{n=0}^{\infty} (t - t_0)^n \overset{(n)}{X}(t_0)/n! \quad (4)$$

for t, t_0 in R, where $\overset{(n)}{X}(t_0)$ is the nth m.s. derivative of $X(t)$ at t_0. The necessary and sufficient condition for (4) is that $\mu(t, s)$ is analytic in the corresponding region of both t and s. Moreover, $X(t)$ is now also a.c. differentiable to any order; it may be shown that the series for $X(t)$ in (4) also converges a.c. to $X(t)$, so that $X(t)$ is also a.c. analytic.

It has been usual in the statistical theory of turbulence to assume differentiability to any order for the random velocity $U(x)$ at a point x of a fluid, but has not always been realized that this automatically implies corresponding restrictions on the covariance or correlation function $\mu(x, y) = E\{U(x)\,U(y)\}$; thus we have already seen that we could not have

$$\mu(x, y) = \sigma^2 \exp\{-|x - y|\}.$$

It is possible also to define the Riemann m.s. integral

$$U = \int_a^b \phi(u)\,X(u)\,du, \tag{5}$$

where it is assumed that $\phi(u)$ is a real bounded function, piecewise continuous in the ordinary sense. The existence of such an integral is established in the usual way as the limit of a sum U_n obtained from a finite number of points $u_1, \ldots, u_n$, and as U_n is a random variable it may have a stochastic limit as the differences $u_r - u_{r-1} \to 0$. The necessary and sufficient condition for $U_n \to U$ m.s. is that

$$E\{U^2\} = \int_a^b \int_a^b \phi(u)\,\phi(v)\,\mu(u, v)\,du\,dv \tag{6}$$

exists in the ordinary Riemann sense. Then also

$$E\{UV\} = \int_a^b \int_a^b \phi(u)\,\psi(v)\,E\{X(u)\,Y(v)\}\,du\,dv$$

if

$$V = \int_a^b \psi(v)\,Y(v)\,dv$$

is similarly defined and exists. A Riemann-Stieltjes integral

$$U = \int_a^b \phi(u)\,dY(u) \tag{7}$$

may also be defined in the m.s. sense if and only if

$$E\{U^2\} = \int_a^b \int_a^b \phi(u)\,\phi(v)\,d\nu(u, v) \tag{8}$$

exists as an ordinary Riemann-Stieltjes integral, where

$$\nu(u, v) = E\{Y(u)\,Y(v)\}.$$

Another extension of (5) is the integral

$$U = \int_a^b X(u) \, d\Phi(u),$$ (9)

if the formal integral for $E\{U^2\}$ exists.

These definitions may readily be extended to infinite ranges of integration, to complex-valued functions ($\mu(u, v)$ being defined as in (3)), and also to random functions in more than one real variable. The equivalence of the m.s. property for $X(u)$ and the corresponding ordinary property for $\mu(u, v)$ in all these cases is primarily due to the linear character of differentiation and summation. It is also possible to define expressions like (5) and (7) as Lebesgue and Lebesgue-Stieltjes m.s. integrals, but the conditions (6) and (8) respectively, if now interpreted as Lebesgue integrals, are no longer sufficient, for the question of the m.s. measurability of $X(u)$ in the sense of measure-theory also arises, and is not necessarily covered by the measurability of $\mu(u, v)$. However, in practice the distinction between the two types of integral will hardly ever arise; a sufficient integrability condition usually adequate to cover this point is that the integral in (7) exists and has the same value in either the Riemann or Lebesgue sense if $Y(u)$ is of m.s. bounded variation in (a, b) (a condition equivalent to $\nu(u, v)$ being of bounded variation over the region $a \leqslant u \leqslant b$, $a \leqslant v \leqslant b$).

While the subsequent use of stochastic integrals or other stochastic limiting operations will often remain formal, the above remarks may help to indicate that a complete mathematical basis is available. Thus the rigorous existence of integrals like (7) or (9) enables a more complete definition of the characteristic functional defined in § 1·31 to be made. Results which hold for non-limiting linear operations, or for non-stochastic limiting operations will usually continue to hold, though sometimes perhaps requiring some further regularity conditions. For example, it may be proved that if $X(t)$ is a normal process, limiting linear combinations of $X(t)$, like $\dot{X}(t)$ or $\int X(t) \, dt$, are normally distributed. A special case of the integral (7), important in the theory of stationary processes (Chapter 6), occurs when

$\Delta Y(u)$ is uncorrelated with $\Delta' Y(u)$ for two disjoint intervals Δu, $\Delta' u$, so that

$$v(u,v) = \begin{cases} \sigma^2(u) & (u \leqslant v), \\ \sigma^2(v) & (u \geqslant v). \end{cases} \tag{10}$$

$Y(t)$ is then said to be an *orthogonal* process. The integral (8) then reduces to the single integral

$$E\{U^2\} = \int_a^b \phi^2(u) \, d\sigma^2(u).$$

More generally, when $\phi(u)$ and $Y(u)$ are complex, equation (8) becomes

$$E\{UU^*\} = \int_a^b \int_a^b \phi^*(u) \, \phi(v) \, dE\{Y^*(u) \, Y(v)\}, \tag{11}$$

and in particular for $Y(u)$ orthogonal, i.e.

$$E\{\Delta Y(u) \, \Delta'[Y^*(u)]\} = 0$$

for Δu, $\Delta' u$ disjoint,

$$E\{UU^*\} = \int_a^b \phi^*(u) \, \phi(u) \, dE\{Y^*(u) \, Y(u)\}. \tag{12}$$

Expansion theorems. In the above integrals we have supposed U defined in terms of a given random function $X(u)$ or $Y(u)$; if we more generally write $\phi(u,t)$ in place of $\phi(u)$, we should have a new random function

$$U(t) = \int_R \phi(u,t) \, dY(u), \tag{13}$$

where R denotes the range of integration of u. An important general expansion theorem states that a converse result holds, namely, that if the product or bilinear moment $\mu(t,s)$ of $U(y)$ is expressible in the form

$$\mu(t,s) = \int_R \int_R \phi^*(u,t) \, \phi(v,s) \, dv(u,v), \tag{14}$$

where $v(u,v)$ is a valid product-moment function, then there exists an expansion of $U(t)$ in terms of a random function $Y(u)$ of m.s. bounded variation given by (13), such that

$$v(u,v) = E\{Y^*(u) \, Y(v)\}.$$

(The integrals (13) and (14) are in general Lebesgue-Stieltjes integrals, but it has already been noted that if $\phi(u, t)$ is piecewise continuous in u equivalent Riemann-Stieltjes integrals exist.) The expansion (13) is, moreover, unique if the set of functions $\phi(u, t)$ possesses appropriate completeness properties (this means that $\iint \psi(v)\,\phi^*(u, t)\,d\nu\,(u, v) \equiv 0$ for all t implies

$$\iint \psi(v)\,\psi^*(u)\,d\nu(u, v) = 0);$$

$Y(u)$ can then be uniquely obtained by an inversion of (13). The most important case giving a unique expansion is $\phi(u, t) = e^{iut}$. A special case of the general theorem occurs if $Y(u)$ is orthogonal, so that (14) reduces to

$$\mu(t, s) = \int_R \phi^*(u, t)\,\phi(u, s)\,d\sigma^2(u), \tag{15}$$

where $\sigma^2(u) \equiv E\{Y^*(u)\,Y(u)\}$ is a non-decreasing positive function of u.

5·2   Stochastic linear difference and differential equations

It is often convenient to study a stochastic process defined directly by a stochastic equation. As a simple example consider the random sequence defined by the equation

$$X_r = \rho X_{r-1} + Y_r, \tag{1}$$

where Y_r is a sequence of uncorrelated variables with zero means and common variance σ^2. Solving (1) by iteration, we obtain

$$X_r = \rho^{r-u} X_u + \sum_{s=0}^{r-u-1} \rho^s Y_{r-s} \tag{2}$$

if $X_r = X_u$ at initial time u. As $u \to -\infty$, this gives a stable solution independent of X_u if $|\rho| < 1$. This solution is

$$X_r = \sum_{s=0}^{\infty} \rho^s Y_{r-s}, \tag{3}$$

which defines a valid (m.s.) process X_r with finite variance, as

$$\sum_{s=0}^{\infty} \rho^{2s} = 1/(1-\rho^2) < \infty.$$

A slightly more complicated case is

$$X_r + aX_{r-1} + bX_{r-2} = Y_r, \tag{4}$$

which is conveniently solved by putting

$$
\begin{aligned}
X_r &= \frac{1}{(E_r^2 + aE_r + b)} Y_{r+2} \\
&= \frac{1}{\mu_1 - \mu_2} \left\{ \frac{1}{E_r - \mu_1} - \frac{1}{E_r - \mu_2} \right\} Y_{r+2} \\
&= \frac{1}{\mu_1 - \mu_2} \{ E_r^{-1}(1 - \mu_1 E_r^{-1})^{-1} - E_r^{-1}(1 - \mu_2 E_r^{-1})^{-1} \} Y_{r+2},
\end{aligned}
$$

where $E_r = 1 + \Delta$ is the usual displacement operator in the calculus of finite differences and μ_1 and μ_2 are the roots of the equation $E_r^2 + aE_r + b = 0$. We thus obtain the general solution

$$X_r = A\mu_1^{r-u} + B\mu_2^{r-u} + \sum_{s=0}^{r-u-2} \frac{\mu_1^{s+1} - \mu_2^{s+1}}{\mu_1 - \mu_2} Y_{r-s},$$

where A and B are obtained from the values X_u, X_{u+1} at u, $u+1$. Again, if $|\mu_1|$, $|\mu_2| < 1$, and we let $u \to -\infty$, we obtain

$$X_r = \sum_{s=0}^{\infty} \frac{\mu_1^{s+1} - \mu_2^{s+1}}{\mu_1 - \mu_2} Y_{r-s}, \tag{5}$$

a valid process with variance

$$\sigma^2 \sum_{s=0}^{\infty} \left(\frac{\mu_1^{s+1} - \mu_2^{s+1}}{\mu_1 - \mu_2} \right)^2 = \frac{(1+b)\sigma^2}{(1-b)\{(1+b)^2 - a^2\}}. \tag{6}$$

The further statistical properties of X_r may also be studied either from its solution or from its defining difference equation. For example, for the covariance of X_r and X_s we obtain from (1) by multiplying by X_s and averaging ($E\{X_r\} = 0$ for (3), and also for (2) if $E\{X_u\} = 0$)

$$w(r, s) = \rho w(r-1, s) \quad (s < r), \tag{7}$$

whence
$$w(r, s) = \rho^{r-s} E\{X_s^2\}.$$

In case (3), $E\{X_s^2\}$ does not depend on s, and by squaring both sides of (1) and averaging,

$$\sigma_X^2 \equiv E\{X_s^2\} = \rho^2 \sigma_X^2 + \sigma^2,$$

whence
$$w(r, s) = \rho^{r-s} \sigma^2 / (1 - \rho^2), \tag{8}$$

a result alternatively obtainable from (3). Similarly from (4)

$$w(r, s) + aw(r-1, s) + bw(r-2, s) = 0 \quad (s < r), \qquad (9)$$

whence $\qquad w(r, s) = A(s)\mu_1^{r-s} + B(s)\mu_2^{r-s}.$

By putting $r = s$, we must have, corresponding to the solution (5), σ_X^2 not depending on s, so that $A(s)$ and $B(s)$ do not depend on s. By squaring (4) and averaging, we obtain

$$\sigma_X^2 = A + B,$$

where σ_X^2 is given by (6) To obtain another equation for A and B we take $s = r-1$, giving

$$(1+b)\,w(r, r-1) + a\sigma_X^2 = 0$$

$$-\frac{a\sigma_X^2}{1+b} = A\mu_1 + B\mu_2.$$

Hence we find, putting $a = -(\mu_1 + \mu_2)$, $b = \mu_1\mu_2$,

$$A = \frac{\sigma_X^2\mu_1(1-\mu_2^2)}{(1+\mu_1\mu_2)(\mu_1-\mu_2)}, \quad B = \frac{\sigma_X^2\mu_2(1-\mu_1^2)}{(1+\mu_1\mu_2)(\mu_2-\mu_1)}. \qquad (10)$$

The extension of these methods to higher-order linear difference equations of the type (4) is evident. The solutions are all of the type

$$X_r = \sum_{s=0}^{\infty} g_s Y_{r-s}, \qquad (11)$$

if we assume the process was 'started up' a long time ago. From the principles of the last section, this represents a well-defined solution (for $E\{Y_r\} = 0$, $E\{Y_r Y_s\} = 0$ when $r \neq s$) if

$$\sum_{s=0}^{\infty} g_s^2 < \infty. \qquad (12)$$

Equations of the type (1) and (4) are called *autoregressive* equations, and it is often assumed further that the Y_r's are entirely independent with the same distribution. With this further assumption (11) will be called a *linear* process. In this case let the joint distribution of $X_r, X_{r+1}, X_{r+2}, \ldots$ have characteristic function with logarithm

$$K\{i\phi_1, i\phi_2, i\phi_3, \ldots\} = \log E\{\exp[i(\phi_1 X_r + \phi_2 X_{r+1} + \ldots)]\}.$$

Substituting from (11) and remembering that the Y_r's are independent, we find

$$K = \sum_{v=-\infty}^{\infty} K_Y\{i\phi_1 g_{r-v} + i\phi_2 g_{r+1-v} + \ldots\} \quad (g_s = 0 \text{ for } s < 0),$$

where $K_Y(i\phi)$ is the cumulant function for Y. This is equivalent to

$$K = \sum_{u=-\infty}^{\infty} K_Y\{i\phi_1 g_u + i\phi_2 g_{u+1} + \ldots\}, \tag{13}$$

and, as it does not depend on r, indicates that X_r is a stationary process. In particular, we have the results for the second, third and fourth cumulants (the latter two when they exist) connecting X_r, X_{r+s}, X_{r+t}, X_{r+v},

$$\kappa_{11} \equiv E\{X_r X_{r+s}\} = \sigma_Y^2 \sum_{u=-\infty}^{\infty} g_u g_{u+s}, \tag{14}$$

$$\kappa_{111} \equiv E\{X_r X_{r+s} X_{r+t}\} = \kappa_3(Y) \sum_{u=-\infty}^{\infty} g_u g_{u+s} g_{u+t}, \tag{15}$$

$$\kappa_{1111} \equiv E\{X_r X_{r+s} X_{r+t} X_{r+v}\} - E\{X_r X_{r+s}\} E\{X_{r+t} X_{r+v}\}$$
$$- E\{X_r X_{r+t}\} E\{X_{r+s} X_{r+v}\} - E\{X_r X_{r+v}\} E\{X_{r+s} X_{r+t}\}$$
$$= \kappa_4(Y) \sum_{u=-\infty}^{\infty} g_u g_{u+s} g_{u+t} g_{u+v}. \tag{16}$$

The methods appropriate for continuous time are quite analogous, but now give rise to stochastic differential equations and integrals in place of differences and sums. There is one further complication to consider. Suppose for illustration we consider the concrete problem of a sensitive torsional galvanometer with a damped oscillation of the type

$$\ddot{x}(t) + \alpha\dot{x}(t) + \beta x(t) = 0. \tag{17}$$

This becomes a stochastic equation if we consider a term $Z(t)$ on the right-hand side representing the effect of additional random forces, so that (17) then becomes

$$\ddot{X}(t) + \alpha\dot{X}(t) + \beta X(t) = Z(t) \tag{18}$$

However, if these random forces are due to the random bombardment of the moving part of the galvanometer by air molecules, it is a more convenient assumption to suppose that the forces are of the nature of impulses, changing the velocity $\dot{X}(t)$

discontinuously. Even if we interpret the differential coefficients in the mean-square sense, it is easy to see that (18) is no longer an admissible equation, and must be replaced by an equation representing the changes in velocity $\dot{X}(t)$. We use as before the notation $df(t)$ for the change in $f(t)$ in dt, and as $Z(t)$ is replaced by the (m.s.) differential of the additive process $Y(t)$ of the accumulated impulse effects, (18) becomes†

$$d\dot{X}(t) + \alpha \dot{X}(t)\,dt + \beta X(t)\,dt = dY(t). \tag{19}$$

If we find the solution of this equation formally, we have

$$X(t) = A(u)\,e^{\lambda_1(t-u)} + B(u)\,e^{\lambda_2(t-u)} + \int_u^t \frac{e^{\lambda_1(t-v)} - e^{\lambda_2(t-v)}}{\lambda_1 - \lambda_2}\,dY(v), \tag{20}$$

where λ_1 and λ_2 are the roots of $\lambda^2 + \alpha\lambda + \beta = 0$, and $A(u)$, $B(u)$ are determined from the values of $X(t)$, $\dot{X}(t)$ at $t = u$. We may verify from (20) that this solution has a valid m.s. differential coefficient $\dot{X}(t)$, such that (19) is satisfied, thus checking the consistency of (19) with this solution. Analogously to the difference equation solution, if $|e^{\lambda_1}|$, $|e^{\lambda_2}| < 1$, $X(t)$ becomes independent of $X(u)$, $\dot{X}(u)$ as $u \to -\infty$, and the solution becomes

$$X(t) = \int_{-\infty}^t \frac{e^{\lambda_1(t-u)} - e^{\lambda_2(t-u)}}{\lambda_1 - \lambda_2}\,dY(u). \tag{21}$$

Again, the extension to higher-order linear equations is evident, and the solution (21) is more generally of the form

$$X(t) = \int_{-\infty}^t g(t-u)\,dY(u), \tag{22}$$

validly defined (for $E\{Y(u)\} = 0$ and $E\{Y^2(u)\} = \sigma^2$) if the piecewise continuous function $g(u)$ is such that

$$\int_0^\infty g^2(u)\,du < \infty. \tag{23}$$

For $Y(u)$ a homogeneous additive process, the process (22) will be called a linear process (cf. equation (11)), and it is readily found (cf. equation (13)) that its characteristic functional

$$C\{\Phi\} = E\left\{\exp\left[i\int X(t)\,d\Phi(t)\right]\right\}$$

---

† It is often useful to think of df, even if it may include a non-zero jump, as defined by $\delta f = df + o(\delta t)$, but a strict interpretation of (19) involves integration over some non-zero time interval (see, for example, Edwards and Moyal, 1955).

is given by

$$\log C\{\Phi\} = \int_{-\infty}^{\infty} K_Y \left\{ i \int_{-\infty}^{\infty} g(v)\, d\Phi(u+v) \right\} du \quad (g(v) = 0 \text{ for } v < 0), \tag{24}$$

where
$$K_Y\{i\phi\} = \log E \left\{ \exp \left[i\phi \int_0^1 dY(u) \right] \right\}. \tag{25}$$

From (24) the completely stationary character of the linear process (22) is apparent, and formulae parallel to (14), (15) and (16) may be deduced. In particular,

$$\sigma_X^2 = \sigma_Y^2 \int_0^{\infty} g^2(u)\, du, \tag{26}$$

where σ_Y^2 is the variance from (25), a result known as Campbell's theorem, and first deduced in the context of the 'shot effect' in vacuum tubes, the independent potential impulses $dY(u)$ due to stray electrons being damped at a later time t according to the coefficient $g(t-u)$ and giving a net result at time t represented by (22).

It is also clear from the principles of m.s. convergence that the above equations and solutions will hold in the m.s. sense as far as the second-order moments if $Y(u)$ is an orthogonal process rather than a completely additive process. This shows that the covariance $\mu(t,s) = E\{X(t)X(s)\}$ is a function $w(\tau)$ of $\tau = t-s$, including in particular that for $X(t)$ in (21). As in this case $X(t)$ is m.s. differentiable, we have $\partial^2 w(t-s)/\partial t\, \partial s = -\partial^2 w(\tau)/\partial\tau^2$, which therefore exists. Multiplying equation (19) by $X(s)$ $(s < t)$, and averaging, we find that $w(\tau)$ satisfies the equation

$$w''(\tau) + \alpha w'(\tau) + \beta w(\tau) = 0 \quad (\tau > 0), \tag{27}$$

where $w'(\tau) = dw(\tau)/d\tau$, etc., whence

$$w(\tau) = A e^{\lambda_1 \tau} + B e^{\lambda_2 \tau} \quad (\tau > 0).$$

We require also $w(0) = \sigma_X^2$, $w'(0) = 0$, the second condition following from

$$\frac{d}{dt} E\{X^2(t)\} = 2E\{X(t)\dot{X}(t)\} = 0.$$

This leads finally to

$$w(\tau) = \sigma_X^2 \left[\frac{\lambda_1 e^{\lambda_2 \tau} - \lambda_2 e^{\lambda_1 \tau}}{\lambda_1 - \lambda_2} \right] \quad (\tau > 0). \tag{28}$$

This result may alternatively be deduced from the general formula

$$w(\tau) = \sigma_Y^2 \int_0^\infty g(u) \, g(u + \tau) \, du. \tag{29}$$

Also by squaring both sides of (21) and averaging (equivalent to using formula (26)), we obtain

$$\sigma_X^2 = \sigma_Y^2 / (2\alpha\beta). \tag{30}$$

Multivariate autoregressive and linear processes. For reference later we note briefly some of the relevant extensions to multivariate or vector processes $\mathbf{X}(t)$. A difference equation of pth order may be written

$$[\mathbf{I} + \mathbf{A}_1 E_r^{-1} + \ldots + \mathbf{A}_p E_r^{-p}] \mathbf{X}_r = \mathbf{Y}_r, \tag{31}$$

where $\mathbf{A}_i$ are matrices, $\mathbf{I}$ is the unit matrix and $\mathbf{X}_r$ and $\mathbf{Y}_r$ are column vectors such that (a dash denoting a matrix transpose)

$$E\{\mathbf{Y}_r \mathbf{Y}_s'\} = \mathbf{W}\delta_{rs}, \quad E\{\mathbf{Y}_r\} = 0, \quad \delta_{rs} = \begin{cases} 1 & (r=s), \\ 0 & (r \neq s). \end{cases}$$

In the particular case $p = 1$, (31) becomes

$$[\mathbf{I} + \mathbf{A} E_r^{-1}] \mathbf{X}_r = \mathbf{Y}_r, \tag{32}$$

as a generalization of (1). Equation (32) gives for suitable $\mathbf{A}$ a well-defined stationary solution given symbolically by

$$\mathbf{X}_r = [\mathbf{I} + \mathbf{A} E_r^{-1}]^{-1} \mathbf{Y}_r, \tag{33}$$

which is included in the more general expression

$$\mathbf{X}_r = \sum_{s=0}^{\infty} \mathbf{G}_s \mathbf{Y}_{r-s}, \tag{34}$$

defined as a linear process if the $\mathbf{Y}_r$ are independent for different r with common distribution. The cumulant function (13) now becomes for $\mathbf{X}_r, \mathbf{X}_{r+1}, \ldots$ with associated vector variables $i\boldsymbol{\phi}_1, i\boldsymbol{\phi}_2, \ldots$

$$K = \sum_{u=-\infty}^{\infty} K_{\mathbf{Y}}\{i\mathbf{G}_u' \boldsymbol{\phi}_1 + i\mathbf{G}_{u+1}' \boldsymbol{\phi}_2 + \ldots\} \quad (\mathbf{G}_u = 0 \text{ for } u < 0), \tag{35}$$

where the cumulant function of $\mathbf{Y}$ is $K_{\mathbf{Y}}(i\boldsymbol{\phi})$. In particular,

$$\mathbf{V}_s \equiv E\{\mathbf{X}_r \mathbf{X}_{r+s}'\} = \sum_{u=-\infty}^{\infty} \mathbf{G}_u \mathbf{W} \mathbf{G}_{u+s}'. \tag{36}$$

where we require $\mathbf{V}_0 = \sum\limits_{u=-\infty}^{\infty} \mathbf{G}_u \mathbf{W} \mathbf{G}'_u < \infty.$ (37)

From the equation (32) we obtain by multiplying to the right by $\mathbf{X}'_{r-s}$ and averaging (for the stationary case)

$$\mathbf{V}'_s + \mathbf{A} \mathbf{V}'_{s-1} = 0 \quad (s > 0),$$

whence $\mathbf{V}_{-s} = \mathbf{V}'_s = (-\mathbf{A})^s \mathbf{V}_0,$ (38)

a result consistent with (36), as for (32) $\mathbf{G}_s = (-\mathbf{A})^s$. It should be noticed that (32), while of the first order, includes equations of higher order by appropriate definition of the variables of the process. For example, the second-order difference equation (4) may be defined as a first-order difference equation in the two variables $X_r, Z_r \equiv X_{r-1}$, as (4) is equivalent to

$$\begin{cases} X_r + aX_{r-1} + bZ_{r-1} = Y_r, \\ Z_r - \ \ X_{r-1} \qquad \ \ = 0. \end{cases}$$

For independent Y_r, this is equivalent to saying that linear processes resulting from higher-order autoregressive equations are vector linear Markov processes.

For continuous time the analogue of (32) is

$$d\mathbf{X}(t) + \mathbf{R}\mathbf{X}(t)\,dt = d\mathbf{Y}(t),$$ (39)

with stationary solution (for suitable $\mathbf{R}$)

$$\mathbf{X}(t) = \int_{-\infty}^{t} e^{-\mathbf{R}(t-u)}\,d\mathbf{Y}(u),$$ (40)

a special case of the linear process (for $\mathbf{Y}(u)$ additive)

$$\mathbf{X}(t) = \int_0^{\infty} \mathbf{G}(u)\,d\mathbf{Y}(t-u).$$ (41)

The characteristic functional $E\left\{\exp\left[\,i\int \mathbf{X}'(t)\,d\mathbf{\Phi}(t)\right]\right\}$ of (41) has logarithm

$$\int_{-\infty}^{\infty} K_{\mathbf{Y}}\left\{i\int_{-\infty}^{\infty} \mathbf{G}'(v)\,d\mathbf{\Phi}(u+v)\right\}du \quad (\mathbf{G}(v) = 0 \text{ for } v < 0),$$ (42)

where $K_{\mathbf{Y}}\{i\boldsymbol{\phi}\} = \log E\left\{\exp\left[\,i\boldsymbol{\phi}'\int_0^1 d\mathbf{Y}(u)\right]\right\}.$

In particular,

$$\mathbf{V}(\tau) \equiv E\{\mathbf{X}(t)\,\mathbf{X}'(t+\tau)\} = \int_{-\infty}^{\infty} \mathbf{G}(v)\,\mathbf{W}\mathbf{G}'(v+\tau)\,dv, \qquad (43)$$

where

$$\mathbf{W} = E\{\Delta\mathbf{Y}(u)\,\Delta\mathbf{Y}'(u)\} \quad (\Delta u = 1),$$

and

$$\mathbf{V}(0) = \int_{-\infty}^{\infty} \mathbf{G}(v)\,\mathbf{W}\mathbf{G}'(v)\,dv < \infty. \qquad (44)$$

From (39) we have

$$\frac{d\mathbf{V}'(\tau)}{d\tau} + \mathbf{R}\mathbf{V}'(\tau) = 0 \quad (\tau > 0),$$

$$\mathbf{V}'(\tau) = e^{-\mathbf{R}\tau}\mathbf{V}(0), \qquad (45)$$

a particular case of (43) with $\mathbf{G}(u) = e^{-\mathbf{R}u}$. As for discrete time, equation (39) includes higher-order differential equations by appropriate definition of variables. Thus for (19) we write $Z(t) = \dot{X}(t)$, and for additive $Y(t)$ it becomes a vector linear Markov process in $X(t)$, $Z(t)$.

5·21 Relations between direct stochastic equations and distribution equations.

In §5·2 distributional properties were often obtained either directly from the stochastic equation or deduced from the corresponding distributional solution. Both methods can be useful; in physical problems especially, where the velocity $U(t) \equiv \dot{X}(t)$ as well as $X(t)$ exists (as a continuous quantity in the m.s. sense, if any impulses acting on $U(t)$ are additive with total variance proportional to t) the two approaches will often reinforce each other.

As a simple example consider the (one-dimensional) motion of an unrestricted particle whose velocity is subjected to numerous 'small' random impulses, so that the distribution of the velocity satisfies the simplest diffusion equation

$$\frac{\partial f(u)}{\partial t} = \tfrac{1}{2}\sigma^2 \frac{\partial^2 f(u)}{\partial u^2}, \qquad (1)$$

with solution

$$f(u) = \frac{1}{\sqrt{(2\pi\sigma^2 t)}} \exp\left\{-\frac{1}{2}\frac{(u-u_0)^2}{\sigma^2 t}\right\}. \qquad (2)$$

The variance of u from (2) is $\sigma^2 t$. From the stochastic equation

$$dU(t) = dZ(t),$$

where $Z(t)$ is a normal additive process (i.e. a pure diffusion process with no non-zero transitions), we have

$$U(t) = U(0) + \int_0^t dZ(u), \tag{3}$$

and for $U(0)$ given as u_0, the variance of $U(t)$ in (3) is of course $\sigma^2 t$, where σ^2 is the variance increment per unit time of $Z(t)$ (a result more generally true than (1) and (2), as $Z(t)$ can in (3) be any homogeneous additive process with zero mean and finite variance).

Consider next the joint distribution of $U(t)$ and

$$X(t) = X(0) + \int_0^t U(u)\, du. \tag{4}$$

For $X(0) = x_0$, the mean of $X(t)$ is $x_0 + u_0 t$ and its variance

$$E\left\{\left[\int_0^t \int_0^v dv\, dZ(u)\right]^2\right\} = E\left\{\left[\int_0^t (t-u)\, dZ(u)\right]^2\right\}$$
$$= \tfrac{1}{3} t^3 \sigma^2. \tag{5}$$

Similarly, the covariance of $X(t)$ and $U(t)$ will be found to be $\tfrac{1}{2} t^2 \sigma^2$. To deduce these results from the diffusion equation, we require to extend this to the joint distribution of $X(t)$ and $U(t)$. Now for the characteristic function of $X(t)$ and $U(t)$ we have

$$C(\phi, \psi) = E\{e^{i\phi X + i\psi U}\},$$
$$\Delta C = E\{(e^{i\phi \Delta X + i\psi \Delta U} - 1)\, e^{i\phi X + i\psi U}\},$$

and for $\Delta X / \Delta t \to U$ (and ΔU independent of X),

$$\frac{\partial C}{\partial t} = \left[\Psi\left(i\psi, \frac{\partial}{\partial i\psi}\right) + i\phi \frac{\partial}{\partial i\psi} \right] C(\phi, \psi), \tag{6}$$

where $$\Psi\{i\psi, U\} \equiv \lim_{\Delta t \to 0} E\left\{ \left[\frac{e^{i\psi \Delta U} - 1}{\Delta t} \right] \middle| U \right\},$$

(cf. equation (2), §3·5). For the normal additive diffusion equation considered above we have $\Psi = -\tfrac{1}{2}\sigma^2 \psi^2$, whence

$$\frac{\partial C}{\partial t} = -\tfrac{1}{2}\sigma^2 \psi^2 C + \phi \frac{\partial C}{\partial \psi}. \tag{7}$$

In terms of the density function $f(u, x)$ equation (7) inverts to

$$\frac{\partial f}{\partial t} + u \frac{\partial f}{\partial x} = \tfrac{1}{2}\sigma^2 \frac{\partial^2 f}{\partial u^2}. \tag{8}$$

Equation (7) may be solved by writing

$$t' = t, \quad \psi' = \psi + \phi t, \quad \phi' = \phi, \quad \frac{\partial}{\partial t} - \phi \frac{\partial}{\partial \psi} = \frac{\partial}{\partial t'},$$

so that

$$\frac{\partial \log C}{\partial t'} = -\tfrac{1}{2}\sigma^2 (\psi' - \phi' t')^2,$$

$$\log C = -\tfrac{1}{2}\sigma^2 [(\psi')^2 t' - \psi' \phi'(t')^2 + (\phi')^2 (t')^3/3] + x_0(i\phi') + u_0(i\psi'),$$
$$= x_0(i\phi) + u_0 i(\psi + \phi t) - \tfrac{1}{2}\sigma^2 [\psi^2 t + \psi \phi t^2 + \phi^2 t^3/3], \tag{9}$$

representing a joint normal distribution with the first- and second-order moments agreeing with those previously found.

It has been pointed out that these moment formulae are true more generally for any additive process $Z(t)$. To solve the above distributional problem in this more general case, consider the characteristic functional

$$C(\Theta) = E\left\{ \exp \int_0^t i U(u)\, d\Theta(u) \right\}.$$

From the principles of the previous section we easily find from (3), by substituting for $U(u)$ in terms of $Z(u)$ and reversing the order of integration,

$$\log C\{\Theta\} = i u_0 \Theta(t) + \int_0^t K_Z\{i[\Theta(t) - \Theta(u)]\}\, du \quad (\Theta(0) = 0), \tag{10}$$

where

$$K_Z\{i\psi\} = \log E\left\{ \exp\left[i\psi \int_0^1 dZ(u) \right] \right\}.$$

To obtain the joint characteristic function of $U(t)$ and $X(t) - x_0$ we put $\Theta(u) = \psi \epsilon(u - t) + u\phi$ (where $\epsilon(u) = 1$, $u \geqslant 0$; 0, $u < 0$), and hence

$$\log C(\phi, \psi) = i u_0 (\psi + t\phi) + \int_0^t K_Z\{i[\psi + (t - u)\phi]\}\, du.$$

For $K_Z\{i\psi\} = -\tfrac{1}{2}\sigma^2\psi^2$, this agrees of course with (9).

A second well-known physical example is the Markov relation for the velocity $U(t)$ of a particle with damping proportional to α

$$dU(t) + \alpha U(t)\, dt = dZ(t), \tag{11}$$

which may be handled by similar methods. In fact, if we put $V(t) = U(t)\, e^{\alpha t}$, we obtain

$$dV(t) = e^{\alpha t} dZ(t),$$

$$V(t) = u_0 + \int_0^t e^{\alpha v} dZ(v),$$

and obtain in place of (10)

$$\log C\{\Theta\} = iu_0 \int_0^t e^{-\alpha u} d\Theta(u) + \int_0^t K_Z\left\{i\int_u^t e^{-\alpha(v-u)} d\Theta(v)\right\} du. \quad (12)$$

Thus for $X(t) - x_0$ and $U(t)$ we now obtain

$$\log C(\phi, \psi) = iu_0[\psi e^{-\alpha t} + (1 - e^{-\alpha t})\phi/\alpha]$$

$$+ \int_0^t K_Z\{i[\psi e^{-\alpha(t-u)} + (1 - e^{-\alpha(t-u)})\phi/\alpha]\} du.$$

In particular, for normal diffusion

$$\log C(\phi, \psi) = iu_0[\psi e^{-\alpha t} + (1 - e^{-\alpha t})\phi/\alpha]$$

$$- \frac{\sigma^2}{4\alpha}\left(\psi - \frac{\phi}{\alpha}\right)^2 (1 - e^{-2\alpha t}) - \sigma^2\left(\psi - \frac{\phi}{\alpha}\right)\frac{\phi(1 - e^{-\alpha t})}{\alpha^2} - \frac{\phi^2\sigma^2 t}{2\alpha^2}, \quad (13)$$

from which the variances and covariances of $U(t)$ and $X(t)$ may be deduced. The result (13) may alternatively be deduced directly from the equation for $C(\phi, \psi)$, which is

$$\frac{\partial C}{\partial t} = \phi\frac{\partial C}{\partial \psi} - \alpha\psi\frac{\partial C}{\partial \psi} - \tfrac{1}{2}\sigma^2\psi^2 C, \quad (14)$$

with equivalent equation for $f(u, x)$

$$\frac{\partial f}{\partial t} + u\frac{\partial f}{\partial x} - \alpha\frac{\partial}{\partial u}(uf) = \tfrac{1}{2}\sigma^2\frac{\partial^2 f}{\partial u^2}, \quad (15)$$

sometimes referred to as the Fokker-Planck equation.

Suppose now the motion is no longer unrestricted, but because of a *reflecting* barrier at $x = b$ ($x_0 = 0, b > 0$), the motion of a particle which hits the barrier is reversed. This implies that, corresponding to any particle very near the barrier and approaching it with velocity $U(t)$, there is just as likely to be one just reflected from it and receding with velocity $U(t)$. This gives the condition on the density, $f_1(u, x)$ say,

$$f_1(u, b) = f_1(-u, b). \quad (16)$$

The method of images referred to in §3·1 may conveniently be used to obtain a solution satisfying (8) or (15), and (16). The reflection of the particle means that the unrestricted density $f(u, x)$ at (u, x) is reinforced by an additional density emanating from an image point at $x = 2b$, with u_0 reversed in sign (to allow for the reflection); the net density at (u, x) is thus

$$f_1(u, x \mid 0, u_0) = f(u, x \mid 0, u_0) + f(u, x \mid 2b, -u_0), \qquad (17)$$

where $f(u, x)$ is the unrestricted solution of (8) or (15). To verify that (17) satisfies (16) it is sufficient to note that for either (8) or (15) $f(u, x)$ has the form $g(u - u_0 \eta(t), x - x_0 - u_0 \xi(t))$, where the normal distribution $g(w, z)$ is a function of a quadratic expression in w and z. This gives from (17)

$$\begin{aligned} f_1(u, b \mid 0, u_0) &= g(u - u_0 \eta, b - u_0 \xi) + g(u + u_0 \eta, -b + u_0 \xi) \\ &= f_1(-u, b \mid 0, u_0), \end{aligned}$$

as reversing the sign of u merely interchanges the two terms in g in this equation. This method may be extended to give the solution for two reflecting barriers on either side of the origin, the number of reversals of sign of u_0 corresponding to the order in the infinite series of images on either side. (Notice, however, that result (17) with the second term subtracted does not in general give the solution for an *absorbing* barrier at b, for the appropriate condition $f_1(u, b) = 0$ $(u < 0)$ is not satisfied.)

Obvious extensions of all the equations in this section are to motion in two or three dimensions. In physical problems further non-random forces (e.g. gravitation) may also need to be introduced (see, for example, Chandrasekhar, 1943; Moyal, 1949).

Distribution of $\dot{X}(t)$. The preceding equations were primarily in the random quantity $U(t)$, and the second quantity $X(t)$ appeared as a stochastic integral of $U(t)$. It may happen that $X(t)$ is the primary random quantity, and if $\dot{X}(t)$ exists in some well-defined sense, we may require the distribution of $\dot{X}(t)$, or the joint distribution of $X(t)$ and $\dot{X}(t)$. If we evaluate the joint distribution of $X(t + h)$ and $X(t)$, and the corresponding characteristic function is

$$C(\theta, \phi) = E\{\exp [i\theta X(t) + i\phi X(t + h)]\},$$

then we must have for $X(t)$, $\dot{X}(t)$ the characteristic function

$$\lim_{h \to 0} \left\{ E \exp \left[i\theta X(t) + i\psi \left(\frac{X(t + h) - X(t)}{h} \right) \right] \right\} = \lim_{h \to 0} C(\theta - \psi/h, \psi/h).$$

$$(18)$$

For example, for a stationary normal process with

$$\log C(\theta, \phi) = -\tfrac{1}{2}\sigma^2[\theta^2 + \phi^2 + 2\theta\phi e^{-\tfrac{1}{2}h^2}],$$

formula (18) gives $\quad -\tfrac{1}{2}\sigma^2(\theta^2 + \psi^2),$

consistently with $\dot{X}$ normal,

$$E\{X\dot{X}\} = 0, \quad E\{\dot{X}^2\} = -\sigma^2[\partial^2(e^{-\tfrac{1}{2}h^2})/\partial h^2]_{h=0}.$$

Systems of individuals. As an application of some of the results of this section, consider a system of n individuals or particles moving independently in some area or volume according to a well-specified stochastic process. Such individuals may be prevented from ultimate dispersion by restricting, e.g. reflecting, boundaries, so that the system becomes stationary. It is possible to demonstrate this final stationarity for, for example, the type of motion represented by equation (15); we allow also the boundaries to recede to infinity and n correspondingly to increase (the detailed proof is omitted). The spatial distribution becomes uniform, and the velocity distribution that consistent with (15), or its extension to more than one dimension, as $t \to \infty$. The joint distribution of velocity and position, given an initial velocity and position, together with the stationary spatial and velocity distribution, permits the calculation of the probability P_{12} of a particle in a region 1 at time 0 being in a region 2 at time t. For the two regions identical, $P_{12} = Q$, where $P \equiv 1 - Q$ is sometimes called the *probability after-effect*. In this last case, the mean number mQ being common to both regions, the joint p.g.f. for the numbers N_0 and N_t of individuals in the region at 0 and t respectively may be seen to be

$$\Pi(z_1, z_2) = \exp\{m(z_1 - 1) + m(z_2 - 1) + mQ(z_1 - 1)(z_2 - 1)\}. \quad (19)$$

Comparison of this distribution with that holding for the stationary form of the immigration-emigration process shows that they are identical. The differences between such an exact specification as the above and the use of the immigration-emigration process (§ 3·41) as an approximation are

(i) the above model will not be Markovian in the number of individuals, N_t;

(ii) the mathematical form of Q will not be the exponential function $e^{-\mu t}$.

The extent to which these differences become observationally important can be investigated. Of course if the mean density of individuals becomes large, the approximating assumption of independence retained above also breaks down. This last complication arises, for example, in statistical models of dense gases and liquids, but we shall not consider here the difficult problems arising in such a context, except to note that the motion of the entire system is then strictly only determined from the simultaneous detailed positions (and velocities) of all the particles. Logically the system is still classifiable as a point process in the sense of § 3·42; in particular, the simultaneous probability of a particle in an element $d\mathbf{r}$ of position-velocity phase-space and another in $d\mathbf{s}$ will be of the form $f_2 d\mathbf{r} d\mathbf{s}$, where f_2 is a 'product-density' of the second order. (For a detailed technical discussion of the molecular theory of fluids, see Green (1952).)

5·22 Equations for stochastic path integrals.

It is sometimes useful to extend the 'forward' and 'backward' equations for a Markov process $X(t)$ in continuous time to the more complicated process $\{X(t), S(t)\}$, where $S(t)$ is of the form†

$$\int_0^t U(X_u, dX_u, du) = \lim_{\Delta u \to 0} \Sigma U(X_u, \Delta X_u, \Delta u),$$

the notation X_u being conveniently used here for the value of the process $X(t)$ at $t = u$. The extended process being still Markovian, the operator-type equation of § 3·5, for example, may be correspondingly extended. We suppose

$$E\{\exp(i\theta \Delta S_t + i\phi \Delta X_t) | X_t\} = \Psi_S(i\phi, i\theta, X_t)\, \Delta t + o(\Delta t),$$

leading to the equation for

$$C(\phi, \theta, t) \equiv E\{\exp(i\theta S_t + i\phi X_t) | X_0\},$$

$$\frac{\partial C}{\partial t} = E\{\Psi_S(i\phi, i\theta, X_t) \exp(i\theta S_t + i\phi X_t) | X_0\}$$

$$= \Psi_S\left(i\phi, i\theta, \frac{\partial}{\partial i\phi}\right) C \tag{1}$$

---

† Stochastic path integrals for diffusion processes have been discussed by Darling and Siegert, Kac, Wiener and others, but for the somewhat more general type of integral considered here see Bartlett (1961).

operationally. A special case of this equation has been used in §5·21 for the integral

$$S(t) = \int_0^t X_u \, du$$

(with the interpretation of X_u velocity and hence S_u position), for which

$$\Psi_S\left(i\phi, i\theta, \frac{\partial}{\partial i\phi}\right) \equiv \Psi\left(i\phi, \frac{\partial}{\partial i\phi}\right) + i\theta\frac{\partial}{\partial i\phi}. \tag{2}$$

As another example, consider the integral

$$S(t) = \int_0^t (dX_u)^2. \tag{3}$$

For Brownian motion we obtain

$$\Psi_S(i\phi, i\theta, X_t) = \sigma^2(i\theta - \tfrac{1}{2}\phi^2) \tag{4}$$

and

$$C = \exp\{\sigma^2 t(i\theta - \tfrac{1}{2}\phi^2)\}; \tag{5}$$

and for the normal linear Markov process defined by

$$dX_t + \mu X_i \, dt = dZ_t,$$

where $Z(t)$ denotes Brownian motion,

$$\Psi_S(i\phi, i\theta, Xt) = -\mu X_t(i\phi) + \sigma^2(i\theta - \tfrac{1}{2}\phi^2) \tag{6}$$

and (in the case $X(t)$ stationary)

$$C = \exp\{\sigma^2(i\theta t - \tfrac{1}{4}\phi^2/\mu)\}. \tag{7}$$

In each of these two cases S_t shows a *certain* regular increase σ^2 per unit time (cf. §8·3).

However, for the simple birth-and-death process dX_u is either -1, 0 or 1, and S_t now denotes the total number of transitions from 0 to t. We easily find

$$\frac{\partial C}{\partial t} = \{\lambda[e^{i(\theta+\phi)} - 1] + \mu[e^{i(\theta-\phi)} - 1]\}\frac{\partial C}{\partial\phi}, \tag{8}$$

which can be solved. But if the distribution of S_t is required by itself, it is rather more convenient to consider the analogous 'backward' equation for S_t, as the properties of the latter equation allow us to put $\phi = 0$ in C_t *before* solving the equation.

As in §3·5, we shall for simplicity consider purely discontinuous Markov processes, for which we suppose that the chance of a non-zero transition $dX_t \leqslant y$ in $(0, dt)$ is $g(y|X_0)\,dt$, and write also

$$h(X_0) = \int_{-\infty}^{\infty} dg(y|X_0).$$

The integral equation for $C_t(\phi, \theta|X_0)$, if U is a function of X_u and dX_u only, becomes

$$C_t(\phi, \theta|X_0) =$$

$$\int_0^t \int_{-\infty}^{\infty} C_u(\phi, \theta|X_0 + y) \exp\left[i\theta U(X_0, y) - (t - u)\,h(X_0)\right] du\, dg\,(y|X_0)$$
$$+ \exp\left[i\phi X_0 - th(X_0)\right]. \quad (9)$$

In particular, for the birth-and-death process with

$$dg(1|X_0) = \lambda X_0, \quad dg(-1|X_0) = \mu X_0,$$

$$\frac{\partial C_t(X_0)}{\partial t} = X_0\{[\lambda C_t(X_0 + 1) + \mu C_t(X_0 - 1)]\, e^{i\theta} - (\lambda + \mu)\, C_t(X_0)\}. \quad (10)$$

In this example, we may still make use of the multiplicative property for the particular extended process considered, so that $C_t(X_0)$ is of the form $C_t^{X_0}$, where

$$\frac{\partial C_t}{\partial t} = [\lambda C_t^2 + \mu]\, e^{i\theta} - (\lambda + \mu)\, C_t. \quad (11)$$

Solving C_t in (11) for $\phi = 0$, when $C_t \equiv A_t(\theta)$, say, we find

$$A_t(\theta) = \frac{r_2\, e^{\lambda rt} - r_1 - (e^{\lambda rt} - 1)\, \mu/\lambda}{r_2 - r_1\, e^{\lambda rt} + e^{\lambda rt} - 1}, \quad (12)$$

where r_1 and r_2 are the roots of

$$\lambda z^2 - (\lambda + \mu)\, z\, e^{-i\theta} + \mu = 0,$$

such that as $\theta \to 0$, $r_1 \to 1$ and $r_2 \to \mu/\lambda$, and where $r = (r_1 - r_2)\, e^{i\theta}$. If we let $t \to \infty$ in (12), we obtain $A_t(\theta) \to A(\theta)$ where

$$A(\theta) = \begin{cases} r_2, & (\lambda > \mu), \\ r_1, & (\lambda < \mu). \end{cases} \quad (13)$$

Note that when $\lambda > \mu$ this does not hold for $\theta = 0$; though $r_1 = 1$ for $\theta = 0$. However,

$$r_2 = \frac{\mu\, e^{-i\theta}}{2\lambda}\left\{1 + \frac{\lambda}{\mu} - \frac{\lambda}{\mu}\, r\right\},$$

which is μ/λ when $\theta = 0$, and is also when divided by μ/λ identical with r_1 with λ/μ replacing μ/λ. This is consistent with the known results that for $\lambda > \mu$ (i) a fraction of the probability is ultimately absorbed at zero and the remainder at infinity, (ii) the process conditional on absorption at zero is equivalent to a process with birth-rate μ and death-rate λ (which is certain to be absorbed at zero), a result due to Waugh (1958).

The result (13) in the case $\lambda < \mu$ is related to a problem discussed by le Couteur (1949), in which effectively 'dying' individuals were divided into those strictly dying, with coefficient $\mu - \gamma \geqslant 0$, and emigrating individuals with coefficient γ. If the emigrating individuals are counted separately and we put $\gamma = \mu$, we have the number of downward transitions $\frac{1}{2}(S_t - X_t + X_0)$, which with the values X_0, X_t gives us the total number of transitions S_t. Another equivalent 'accounting system' in the birth-and-death process was used by D. G. Kendall (1948), who considered the number of births, $\frac{1}{2}(S_t + X_t - X_0)$.

5·3 Approximating and limiting solutions dependent on a parameter

In a number of the applications discussed earlier in this book, the stochastic process has depended on a parameter which may approach a limiting value for which the solution of the stochastic process equations is more tractable. The most common example is the number n in a population, this number being in simpler cases fixed or in others itself a random variable N_t of the process. Approximating diffusion equations, e.g. in population genetics, often arise in this way.

To illustrate in a special case, let us write down the *stochastic* equation (1) below (cf. §§ 5·2 and 5·21) for a single population of one type of individual, specified by the number of individuals N_t and the chance of a birth or death in time t, $t + dt$ of the form $\lambda(n)\,dt$ and $\mu(n)\,dt$ respectively when $N_t = n$. The process is assumed Markovian (but not in general multiplicative, as the individuals only act independently if $\lambda(n) = n\lambda$, $\mu(n) = n\mu$, the simple birth-and-death process).

$$dN_t = [\lambda(N_t) - \mu(N_t)]\,dt + dZ_t. \tag{1}$$

In (1) we have extracted

$$E\{N_{t+dt} - N_t | N_t\} = [\lambda(N_t) - \mu(N_t)] \, dt,$$

so that dZ_t, *defined* by equation (1), has $E\{dZ_t | N_t\} = 0$. Note that equation (1) has its deterministic analogue when dZ_t is neglected, viz.

$$\frac{dn_t}{dt} = \beta(n_t), \tag{2}$$

where $\beta(n_t) = \lambda(n_t) - \mu(n_t)$. Suppose in particular that the equilibrium equation for (2) has effectively one solution m such that $\beta(m) = 0$. The equilibrium distribution P_n for (1) when it exists is easily shown, from the forward equation, to be given by

$$P_{n+1} = \lambda(n) \, P_n / \mu(n+1). \tag{3}$$

For $n \sim m$, *where m is large*, write $x = (n-m)/\sqrt{m}$. Then if $P_n \equiv F(x)$, $\lambda(n)/\mu(n+1) \sim \lambda(n)/\mu(n) \equiv G(x)$, (3) becomes

$$\frac{F(x+\epsilon) - F(x)}{\epsilon F(x)} \sim \frac{G(x) - 1}{\epsilon}, \quad \text{where} \quad \epsilon = 1/\sqrt{m},$$

or

$$\frac{\partial F(x)}{F(x) \, \partial x} \sim \frac{x}{\epsilon} \left[\frac{\partial G(x)}{\partial x} \right]_{x=0},$$

whence

$$F(x) \sim C \exp\left(-\tfrac{1}{2} x^2 / \sigma_x^2 \right), \tag{4}$$

where $\sigma_x^2 \sim -\epsilon / [\partial G(x)/\partial x]_{x=0}$. Thus to a first approximation under appropriate conditions P_n is normal. The equation (1) may also be written

$$X_{t+dt} = X_t + \beta(N_t) \, dt/\sqrt{m} + dZ_t/\sqrt{m},$$

where $\beta(N_t) \equiv H(X_t) \sim X_t [\partial H(x)/\partial x]_{x=0}$, whence by squaring and averaging we verify in the stationary case

$$2[\partial H/\partial x]_{x=0} \, \sigma_x^2 \sim -[\lambda(m) + \mu(m)]/\sqrt{m} = -2\lambda(m)/\sqrt{m},$$

from the result $E\{(dZ_t)^2 | N_t\} = [\lambda(N_t) + \mu N(t)] \, dt$, whence

$$\sigma_x^2 \sim -\lambda(m) \bigg/ \left\{ \sqrt{m} \left[\frac{\partial H(x)}{\partial x} \right]_{x=0} \right\} = -\epsilon \bigg/ \left[\frac{\partial G(x)}{\partial x} \right]_{x=0},$$

as above. The variance of N_t is then $m\sigma_x^2$. Higher-order approximations may also be investigated. (Cf. Bartlett, 1960; and, for related discussions, see further references for this section.)

If we consider in particular

$$\lambda(n) = a_1 n - b_1 n^2, \quad \mu(n) = a_2 n + b_2 n^2,$$

then $m = (a_1 - a_2)/(b_1 + b_2)$, $\sigma^2 \sim (a_1 - b_1 m)/(b_1 + b_2)$. Thus for $a_1 = 0\cdot8077$, $a_2 = 0\cdot1145$, $b_1 = 0\cdot006932$, $b_2 = 0$, we find $m = 100$, $100\sigma_x^2 \sim a_2/b_1 = 16.52$. The mean and variance of N_t calculated by means of equation (3) are $99\cdot83$ and $16\cdot71$. The second-order approximation to this mean M, obtained by averaging equation (1), gives in this case

$$(a_1 - a_2)\, M - (b_1 + b_2)\, (M^2 + \sigma^2) = 0,$$

whence $M \sim 99\cdot83$, agreeing with the true value to this order of accuracy. (Note in this example that there is an alternative equilibrium at $n = 0$, but this may be neglected unless $N_0 \sim 0$; this would, however, no longer necessarily be so if $\sigma_x \gg 1/\sqrt{m}$.)

Chapter 6

STATIONARY PROCESSES

6·1   Processes stationary to the second order

In Chapter 1 we defined a stationary stochastic process as one whose distribution function $F_n(\mathbf{r})$ for any set of times $t_1, ..., t_n$ was dependent only on the intervals $t_r - t_s$ and independent of the absolute values of the t_r; in other words, it is invariant under translations of the t axis. We have already encountered such processes in earlier chapters. Thus if a homogeneous Markov process tends to a limiting distribution independent of initial

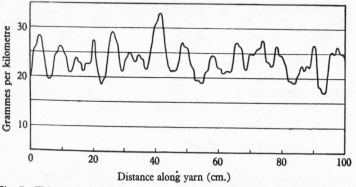

Fig. 7. This continuous time-series shows the variation in mass per unit length along a cotton yarn. (The graph was kindly supplied by Mr G. A. R. Foster and has been reprinted from *Applied Statistics*, **2** (1953), 52.)

conditions, it is evident that the process has become stationary in the above sense. Such an example was the 'emigration-and-immigration' process of § 3·41 (equation (5)), with its limiting marginal distribution a Poisson distribution. Another general class which can provide stationary processes was the linear process as defined in § 5·2.

In view of the importance of stationary processes in practice, representing a kind of stochastic equilibrium often typical of observed processes, e.g. in turbulence, communication systems, and time-series in general (a typical example is illustrated in fig. 7), their properties are further developed in this chapter.

We shall see that many of these properties will depend only on the structure of the process as specified by its first and second moments, always assumed finite (cf. §5·11), and hence will still hold even if the process is not completely stationary in the sense so far defined, but merely has first and second moments invariant under translation of the t axis. Processes stationary only to this extent will be called stationary to the second order. More generally, a process with finite moments will be called stationary to the rth order if all the moments of order r or less are invariant under translation (in §3·4 we had an example of a process stationary merely to the first order, with constant mean but increasing variance).

In the discussion of processes stationary at least to the second order, we shall usually find it convenient to take $E\{X(t)\} \equiv m = 0$, and define the covariance function in general for complex $X(t)$ as

$$w(s-t) \equiv E\{X^*(t)\,X(s)\} = \sigma^2\rho(s-t), \tag{1}$$

where $\sigma^2 \equiv E\{X^*(t)\,X(t)\}$ and $\rho(\tau)$ is called the autocorrelation function of $X(t)$, with $\rho(0) = 1$. For real $X(t)$, $\rho(\tau)$ is a real symmetric function of τ; for complex $X(t)$, $\rho^*(\tau) = \rho(-\tau)$. In the case of normal or Gaussian processes, i.e. processes whose distribution for any set $t_1, \dots, t_n$ is multivariate normal, the only parameters of the distribution are the first- and second-order moments; it at once follows that a normal process, stationary to the second order, is completely stationary. The normal process is often theoretically useful in providing an example of a process with given admissible first and second moments.

Consistency relations for product moments. Among the various consistency relations that must hold for the simultaneous distribution $F_n(\mathbf{r})$, the consistency relations for the correlations among $X_1, X_2, \dots, X_n$ follow from the non-negative character of $E\{ZZ^*\}$, where

$$Z = \lambda_1 X_1 + \lambda_2 X_2 + \dots + \lambda_n X_n.$$

This means that the Hermitian form

$$\Sigma_{u,v}\lambda_u^*\lambda_v\mu(t_u, t_v) \geqslant 0 \tag{2}$$

for any valid product-moment function $\mu(t, s)$, an important condition specifying the class of admissible product-moment functions. For processes stationary to the second order and with zero mean, (2) becomes

$$\Sigma_{u,v} \lambda_u^* \lambda_v \rho(t_v - t_u) \geqslant 0. \tag{3}$$

It follows that determinants formed from the coefficients of the form (3) are non-negative, e.g.

$$\begin{vmatrix} 1 & \rho(t_3 - t_2) & \rho(t_2 - t_1) \\ \rho(t_3 - t_2) & 1 & \rho(t_3 - t_1) \\ \rho(t_2 - t_1) & \rho(t_3 - t_1) & 1 \end{vmatrix} \geqslant 0,$$

or, if $t_3 - t_2 = t_2 - t_1 = h$, and X is real,

$$[1 - \rho(2h)][1 - 2\rho^2(h) + \rho(2h)] \geqslant 0,$$

or since $\rho(2h) \leqslant 1$, $\qquad 1 - 2\rho^2(h) + \rho(2h) \geqslant 0.$

Other particular relations could be deduced in this way, but we may obtain an equivalent comprehensive condition from the non-negative character of (3), since it at once follows from a theorem by Bochner (1936–7, Chapter 16) that if $\rho(\tau)$ is continuous at $\tau = 0$ (i.e. $X(t)$ is m.s. continuous and $\rho(\tau)$ is continuous for all τ) we may write $\rho(\tau)$ in the form

$$\rho(\tau) = \int_{-\infty}^{\infty} e^{i\tau\omega} dF(\omega), \tag{4}$$

where $F(\omega)$ is a never-decreasing function with $F(-\infty) = 0$, and as $\rho(0) = 1$, $F(+\infty) = 1$, and thus has the character of a distribution function. From (4) it follows that $\rho(\tau)$ for m.s. continuous stationary processes has the mathematical form of the characteristic function (Fourier transform) of a distribution function.

If $X(t)$ is real, we have $\rho(\tau)$ real and symmetric, and hence

$$\rho(\tau) = 2\int_0^{\infty} \cos\tau\omega \, dF(\omega). \tag{5}$$

The function $F_+(\omega)$, say, for real processes need only be defined from 0 to ∞, with the relation $\int_0^\omega 2dF(\omega) = \int_0^\omega dF_+(\omega).$

6·11 The spectral function. The function $F(\omega)$ is fundamental in the harmonic analysis of $X(t)$, as we shall see later. It will be called the spectral function for the stationary process

$X(t)$, and for real processes $F_+(\omega)$ will be called the integrated 'spectrum'.

In view of the importance of the result (4), a proof due to Loève (1945) not assuming Bochner's theorem (and in effect giving an alternative proof of Bochner's theorem) is added. We take in the non-negative form (3) λ_u proportional to $e^{-i\omega t u}$ and extend the sum to an integral (over a finite region for u and v), so that we have

$$\iint e^{-i\omega(s-t)} \rho(s-t)\, ds\, dt \geqslant 0. \tag{1}$$

We define (in the m.s. sense)

$$H(\omega) = \frac{1}{\sigma\sqrt{T}} \int_{-\frac{1}{2}T}^{\frac{1}{2}T} e^{-i\omega u} X(u)\, du,$$

so that

$$E\{H(\omega)\, H^*(\omega)\} = 2\pi f_T(\omega), \quad \text{say},$$

$$= \frac{1}{T} \int_{-\frac{1}{2}T}^{\frac{1}{2}T} \int_{-\frac{1}{2}T}^{\frac{1}{2}T} e^{-i\omega(u-v)} \rho(u-v)\, du\, dv$$

$$= \int_{-T}^{T} \left(1 - \frac{|\tau|}{T}\right) \rho(\tau)\, e^{-i\omega\tau}\, d\tau,$$

since $X(t)$ is stationary, this integral being, moreover, non-negative. Now define

$$\phi_T(\tau) = \begin{cases} \left(1 - \dfrac{|\tau|}{T}\right) \rho(\tau) & \text{for } |\tau| \leqslant T, \\[2mm] 0 & \text{for } |\tau| > T. \end{cases}$$

Then
$$2\pi f_T(\omega) = \int_{-\infty}^{\infty} \phi_T(\tau)\, e^{-i\omega\tau}\, d\tau \geqslant 0. \tag{2}$$

Multiply (2) by $(1 - |\omega|/\Omega)\, e^{it\omega}/(2\pi)$ and integrate with respect to ω from $-\Omega$ to Ω. We obtain

$$\int_{-\Omega}^{\Omega} \left(1 - \frac{|\omega|}{\Omega}\right) f_T(\omega)\, e^{it\omega}\, d\omega = \frac{1}{2\pi} \int_{-\infty}^{\infty} \phi_T(\tau) \left[\frac{\sin\{\frac{1}{2}\Omega(\tau-t)\}}{\frac{1}{2}\Omega(\tau-t)}\right]^2 \Omega\, d\tau.$$

The term multiplying $e^{it\omega}$ on the left-hand side is not negative, and hence the integral has the character of a characteristic function (apart perhaps from a multiplying constant). The right-hand side converges uniformly in every finite t-interval to $\phi_T(t)$ as $\Omega \to \infty$ (provided $\phi_T(t)$ is continuous, which follows from the *assumed* continuity of $\rho(t)$). Hence $\phi_T(t)$, for which

$$\phi_T(0) = \rho(0) = 1,$$

is mathematically a 'characteristic function' (using the limit property of characteristic function sequences, referred to in §1·31). From the same property it follows further that $\phi(t)$, the uniform limit of $\phi_T(t)$ in every finite t-interval as $T \to \infty$, is also a 'characteristic function'. As $\phi(t) \equiv \rho(t)$, the required result (4) of §6·1 follows.

This condition (4) (or (5) for real processes) is not only necessary but sufficient for the existence of a process with given auto-correlation function. For let

$$X(t) = e^{i(\Phi + \Omega t)},$$

where Ω has the distribution function $F(\omega)$, and Φ is independent of Ω and uniformly distributed over $(0, 2\pi)$. Then

$$E\{X(t)\} = 0, \quad E\{X(t)\,X^*(t)\} = 1,$$

$$E\{X^*(t)\,X(s)\} = E\{e^{i\Omega(s-t)}\}$$

$$= \int_{-\infty}^{\infty} e^{i\omega(s-t)}\,dF(\omega) = \rho(s-t).$$

Theorem (4) of §6·1 was first given by Khintchine (1934) for real processes, though its relation with the harmonic analysis of $X(t)$ was already implicit in earlier work by Wiener (1930). The assumption that $\rho(\tau)$ is continuous, equivalent to m.s. continuity of $X(t)$, will be made unless the contrary is stated.

Corresponding to the representation of any distribution function by two components, the 'discrete' and 'continuous' components (strictly speaking, the second is the absolutely continuous component, but we shall assume that there is no third singular component), we may write further

$$\rho(\tau) = |c_1|^2 \rho_1(\tau) + |c_2|^2 \rho_2(\tau), \quad (|c_1|^2 + |c_2|^2 = 1), \qquad (3)$$

where
$$\rho_1(\tau) = \Sigma_r e^{i\tau\omega_r} p_r, \qquad (4)$$

p_r being the discontinuities of $F(\omega)$, occurring at the points ω_r, and

$$\rho_2(\tau) = \int_{-\infty}^{\infty} e^{i\tau\omega} f(\omega)\,d\omega, \qquad (5)$$

where $f(\omega)$ is the spectral density. The inversion formulae for $F(\omega)$ and $f(\omega)$, while of course identical with those relating characteristic and distribution functions, are quoted for reference. For all points of continuity of $F(\omega)$, we have

$$F(\omega_2) - F(\omega_1) = \lim_{T \to \infty} \frac{1}{2\pi} \int_{-T}^{T} \frac{e^{-i\omega_2 \tau} - e^{-i\omega_1 \tau}}{-i\tau} \rho(\tau) \, d\tau, \qquad (6)$$

or, if $dF(\omega) = f(\omega) \, d\omega$,

$$f(\omega) = \frac{1}{2\pi} \int_{-\infty}^{\infty} e^{-i\omega\tau} \rho(\tau) \, d\tau. \qquad (7)$$

In the case $X(t)$ real the formula for $F_+(\omega)$ is

$$F_+(\omega) = \frac{2}{\pi} \int_0^{\infty} \rho(\tau) \frac{\sin \tau\omega}{\tau} \, d\tau. \qquad (8)$$

There are analogous theorems in the case of discrete time (with the equidistant intervals taken as unity). Thus Wold (1938) has shown that for real X_t, ρ_s is the autocorrelation coefficient between X_t and X_{t+s} only if it can be expressed in the form

$$\rho_s = \int_0^{\pi} \cos s\omega \, dF_+(\omega). \qquad (9)$$

More generally for complex $X(t)$, we have

$$\rho_s = \int_{-\pi}^{\pi} e^{is\omega} \, dF(\omega). \qquad (10)$$

No continuity assumption for X_t arises of course for stationary sequences. The inversion formulae in this case are

$$F(\omega) = \frac{\omega + \pi}{2\pi} + \frac{1}{2\pi} \lim_{n \to \infty} \left[\sum_{s=-n}^{-1} + \sum_{s=1}^{n} \right] \frac{e^{-is\omega}}{-is} \rho_s, \qquad (11)$$

$$F_+(\omega) = \frac{\omega}{\pi} + \frac{2}{\pi} \sum_{s=1}^{\infty} \frac{\sin s\omega}{s} \rho_s, \qquad (12)$$

and when $dF(\omega) = f(\omega) \, d\omega$,

$$f(\omega) = \frac{1}{2\pi} \sum_{s=-\infty}^{\infty} e^{-is\omega} \rho_s. \qquad (13)$$

It should perhaps be noticed that if a stationary sequence is extracted from a stationary process in continuous time by taking observations at intermittent intervals, the spectral function $dF(\omega)$ for the sequence does not coincide with that for the original process even over the range $-\pi$ to π, unless $dF(\omega)$ is zero for the original process outside this range; this is associated with the ambiguity of period of a harmonic series observed only at intervals, a higher order harmonic having the same observational characteristics.

Spectral moments. In the last chapter we saw that the m.s. differentiation properties of $X(t)$ were associated with the differentiation properties of $\rho(\tau)$. The connection between $\rho(\tau)$ and the spectral function $F(\omega)$ implies that there will also be associated properties of $F(\omega)$. Thus for real $X(t)$ we noted for stationary processes for which $\dot{X}(t)$ exists as a m.s. differential coefficient

$$E\{\dot{X}(t+\tau)\,X(t)\} = E\{\dot{X}(t)\,X(t-\tau)\} = \sigma^2\rho'(\tau), \qquad (14)$$

$$E\{\dot{X}(t)\,\dot{X}(t-\tau)\} = -\sigma^2\rho''(\tau), \qquad (15)$$

where dashes denote differentiation with respect to τ. Thus the autocorrelation function of $\dot{X}(t)$ is proportional to $-\rho''(\tau)$, which from (5) of § 6·1 has the form

$$-\int_0^\infty \omega^2 \cos\tau\omega\,dF_+(\omega). \qquad (16)$$

The relevant theorem, due to Slutsky, states that if the second moment of $F_+(\omega)$ is finite, then $X(t)$ is m.s. differentiable.

This result follows from the existence, when $\displaystyle\int_0^\infty \omega^2 dF_+(\omega)$ is finite, of the functions

$$\rho'(\tau) = -\int_0^\infty \omega\sin\tau\omega\,dF_+(\omega), \quad \rho''(\tau) = -\int_0^\infty \omega^2\cos\tau\omega\,dF_+(\omega),$$

whence $\rho'(\tau)$ is continuous with $\rho'(0) = 0$ and $\rho''(\tau)$ is continuous.

It may be shown further (cf. Doob, 1944, p. 236) that $X(t)$ is also a.c. differentiable.[†] For let $Y(t)$ be the stochastic process defined by the m.s. derivative $\dot{X}(t)$, and consider

$$Z \equiv X(t) - X(0) - \int_0^t Y(u)\,du.$$

It is readily shown that $E\{Z^2\} = 0$, whence $Z = 0$ a.c. Hence almost every realized function $x(t)$ is absolutely continuous with differential coefficient $y(t)$, the latter existing almost everywhere in $(0, t)$. Since $X(t)$ and hence $Y(t)$ are stationary, the qualification 'almost everywhere' may be omitted, and the result follows.

Among other relations between $\rho(\tau)$ and $F(\omega)$, we will note the obvious one from (7) that if $\rho(\tau)$ is absolutely integrable, $f(\omega)$ exists. As an example, consider the case $\rho(\tau) = \exp\{-\mu\,|\,\tau\,|\}$,

[†] There is a tacit 'regularity' condition on $X(t)$ assumed here (cf. also Doob, 1953, p. 536).

which we saw in §5·11 implied that $X(t)$ was m.s. continuous, but not m.s. differentiable. We have $\rho(\tau)$ absolutely integrable, and the corresponding spectral density is

$$f(\omega) = \frac{1}{2\pi} \int_{-\infty}^{\infty} e^{-i\tau\omega} e^{-\mu|\tau|} d\tau = \frac{1}{\pi} \frac{\mu}{\mu^2 + \omega^2}. \qquad (17)$$

However, the second moment of ω does not exist, corresponding to the non-existence of $\dot{X}(t)$ for this process.

An additive process $Z(t)$, or, what is equivalent for correlational properties, an orthogonal process, may have non-zero stationary increments $dZ(t)$ with autocorrelation function $\rho(0) = 1$, $\rho(\tau) = 0$ ($\tau \neq 0$). Such a process is not m.s. continuous, and has no proper spectrum, but its 'spectrum' may be thought of as a kind of limit with all 'frequencies' ω having equal weight, i.e. as a uniform distribution with a range tending to infinity. It is the analogue of a completely random stationary sequence of independent variables in discrete time, for which the spectrum is a uniform distribution from $-\pi$ to π (or 0 to π for $f_+(\omega)$).

6·12 Stationary point processes and covariance densities.

The stationary processes so far discussed have had some kind of continuity, apart from the increments of the orthogonal process just referred to. If, however, we have a point process $N(t)$ in time (cf. §3·42), such that $dN(t) = 1$ or 0, such processes may be stationary but not orthogonal. In fact, simple renewal processes (with a suitable random starting time for the renewals, or alternatively if started up a long time ago) will constitute examples, of which the simple Poisson additive process is only a very special case. For such processes we have seen that $E\{dN(t)\,dN(\tau)\}$ ($t \neq \tau$) is of a smaller order of magnitude than $E\{dN(t)\} = E\{[dN(t)]^2\}$, but, nevertheless, we may define a stationary covariance density (cf. equation (11), §4·3)

$$w(t - \tau)\,dt\,d\tau = \text{cov}\{dN(t), dN(\tau)\} = E\{dN(t)\,dN(\tau)\} - [E\{dN(t)\}]^2$$
$$(t \neq \tau),$$

such a density function $w(t - \tau)$ being only zero for orthogonal increments. That is, we assume

$$E\{dN(t)\,dN(\tau)\} = \begin{cases} \lambda\,dt & (t = \tau), \\ [w(t - \tau) + \lambda^2]\,dt\,d\tau & (t \neq \tau). \end{cases}$$

As an example, suppose the distribution of the time-interval between successive events in a simple renewal process is the distribution of example 2, § 2·11, namely (with $\mu = 2\lambda$),

$$f(t)\,dt = 4\lambda^2 t e^{-2\lambda t}dt \quad (0 \leqslant t < \infty).$$

This renewal distribution implies that the probability of an event in $t, t+dt$, given an event at $t=0$, but regardless of possible further events in the meantime, is (formula (6), § 3·3)

$$\lambda\,dt - \lambda e^{-4\lambda t}dt,$$

whence, letting $t \to \infty$, we see that the mean density $E\{dN(t)\}/dt$ is λ. We have further

$$E\{dN(t)\,dN(\tau)\} = \lambda\,d\tau[\lambda\,dt - \lambda e^{-4\lambda(t-\tau)}dt] \quad (t > \tau),$$

whence $$w(t-\tau) = -\lambda^2 e^{-4\lambda(t-\tau)} \quad (t > \tau). \tag{1}$$

This density is negative, implying an 'inhibition' by one event of an event too soon afterwards (this is why it was used as a possible model in the theory of genetic recombination). Consistently with this idea, we should expect the variance of the number of events occurring in any non-zero interval Δt to be less than the Poisson value $\lambda \Delta t$. We find in fact

$$E\left\{\left[\int_t^{t+\Delta t} dN(t)\right]^2\right\} - \lambda^2(\Delta t)^2$$
$$= \int_t^{t+\Delta t}\int_t^{t+\Delta t} w(u-v)\,du\,dv + \int_t^{t+\Delta t}\lambda\,du,$$

the second term in the last expression arising from the contribution for $u = v$ in the double integral. After substitution from (1) and some reduction we obtain

$$\tfrac{1}{2}\lambda\,\Delta t + \tfrac{1}{8}(1 - e^{-4\lambda\Delta t}), \tag{2}$$

which is less than $\lambda \Delta t$ for $\Delta t > 0$.

For stationary point processes, the covariance density $w(t-\tau)$ has no proper spectrum definable as for a non-zero covariance function in the sense of § 6·11, as its value for $t = \tau$ is of a different order of magnitude, but an extension of the theory of § 6·11 to cover it may be made (cf. also § 6·13). We may express the relation (4) of §6·1 more generally for the homogeneous integrated process

$$Y(t) = \int_0^t X(u)\, du \tag{3}$$

as $E\{Y(t)\,Y^*(\tau)\} = \lim_{\Omega \to \infty} \sigma^2 \int_{-\Omega}^{\Omega} \left(\frac{e^{it\omega}-1}{i\omega}\right)\left(\frac{e^{-i\tau\omega}-1}{-i\omega}\right) dF(\omega),$ (4)

and such a relation will exist also for a process such as

$$Y(t) = N(t) - \lambda t = \int_0^t dN(t) - \lambda t, \tag{5}$$

of the type considered above, even although $N(t)$ is not differentiable.

6·13 The spectra of stationary point processes.

The essential principles of the correlation and spectral theory of stationary point processes have been given in §6·12; but, in anticipation of the numerical application of this theory in §9·23, some further explicit results are developed below. While the spectral function $F(\omega)$ defined in equation (4) of §6·12 is not bounded as $\omega \to \infty$, it is nevertheless still a convenient specification of the covariance density $w(\tau)$. In fact, if we define the 'complete covariance density'
$$w_c(\tau) = \lambda\delta(\tau) + w(\tau), \tag{1}$$

where $\delta(\tau)$ is the Dirac delta-function, the spectral function $F(\omega)$ has the same Fourier relation with $w_c(\tau)$ as in the case of continuous time-series. For example, where $F(\omega)$ has merely a density component, we may define†

$$f(\omega) = \frac{1}{2\pi} \int_{-\infty}^{\infty} e^{-\omega i\tau} w_c(\tau)\, d\tau,$$

so that $g(w) \equiv 2\pi f(\omega)$

$$= \lambda + \int_{-\infty}^{\infty} e^{-i\omega\tau} w(\tau)\, d\tau. \tag{2}$$

For events occurring purely at random (the Poisson process), $w(\tau) = 0$ and $g(\omega)$ in (2) reduces to λ for all real ω. As with real continuous stationary processes, it is often convenient to define

† It is convenient to define $F(\omega)$ and $f(\omega)$, as we cannot scale $F(\infty)$ to 1, directly in relation to the generalized covariance.

a spectral density for non-negative ω by

$$g_+(\omega) = 2g(\omega).$$

Cases also arise where discrete spectral components exist, as shown in the following examples.

Example 1. Consider first a renewal process where the point process is defined by the times of renewal. Let the Laplace transform of the renewal distribution be $L(\psi)$. Then the renewal density $r(\tau)$ at τ given a renewal point at 0, has Laplace transform (cf. equation (6), §2·11)

$$R(\psi) = L(\psi)/[1 - L(\psi)]. \tag{3}$$

From the identifications $w(\tau) = \lambda r(\tau) - \lambda^2$, with Laplace transform $W(\psi) = \lambda R(\psi) - \lambda^2/\psi$, and $g(\omega) = \lambda + W(-i\omega) + W(i\omega)$, we obtain

$$g_+(\omega) = 2\lambda \left\{ 1 + \frac{L(i\omega)}{1 - L(i\omega)} + \frac{L(-i\omega)}{1 - L(-i\omega)} \right\}. \tag{4}$$

To particularize (4) further for an important class of renewal distributions, suppose

$$L(\psi) = [1 + \psi/(k\lambda)]^{-k},$$

corresponding to a χ^2-type distribution. Then (4) becomes

$$g_+(\omega) = 2\lambda \left\{ 1 + \frac{1}{[1 + i\omega/(k\lambda)]^k - 1} + \frac{1}{[1 - i\omega/(k\lambda)]^k - 1} \right\}. \tag{5}$$

For $k = 1$, this reduces to 2λ. For $k = 2$,

$$g_+(\omega) = 2\lambda - \frac{\lambda}{1 + \omega^2/(16\lambda^2)}, \tag{6}$$

which is of course consistent with equation (1) of §6·12.

For large k, (5) becomes approximately

$$g_+(\omega) \sim 2\lambda \left\{ 1 - \frac{1}{1 + \frac{1}{2}(e^{\omega^2/(k\lambda^2)} - 1)/(1 - \cos[\omega/\lambda])} \right\}, \tag{7}$$

which represents an oscillating curve starting near zero and rising to a maximum at 2λ, with the successive minima at heights gradually increasing to 2λ. It might be noted that as $k \to \infty$ the renewal process becomes one with renewals at more precise periodic intervals; but it is incorrect to put $k = \infty$, as there is an absence of long-term order in the renewal process for $k < \infty$ which implies a discontinuity in behaviour with the case $k = \infty$. Cases of strict periodicity are best treated separately.

Example 2. Consider a Poisson clustering process in one dimension where the 'parents' constitute a Poisson process and each have a number of 'offspring' distributed in some specified manner about the parent, and where all the 'family' contribute to the total process. For definiteness we suppose the offspring follow the parent individually in time, with an interval between the parent and the first offspring, and the subsequent intervals between adjacent offspring pairs from the same parent, all independently following the same distribution with characteristic function $C(\phi)$. We also suppose that the number of offspring to each parent follows the distribution p_r ($r = 0, 1, \ldots$), independently for each parent.

Contributions to the covariance density only arise within the same cluster or family. Let λ_0 be the density of parents. The first element $dN(t)$ in $dN(t)\,dN(t+\tau)$ may be any one of the family. Evaluation of all the possible terms gives, for $\tau > 0$,

$$w(\tau) = \lambda_0 \{ f_1(\tau)\,(p_1 + 2p_2 + 3p_3 + \ldots)$$
$$+ f_2(\tau)\,(p_2 + 2p_3 + \ldots)$$
$$+ f_3(\tau)\,(p_3 + 2p_4 \ldots) + \ldots \},$$

where $f_1(\tau)$ is the (formal) density corresponding to $C(\phi)$, $f_2(\tau)$ is the convolution for two consecutive intervals, corresponding to $C^2(\phi)$, etc. Thus the spectral density function associated with $w(\tau)$ is

$$\lambda_0 \{ [C(\omega) + C(-\omega)]\,E\{r\} + [C^2(\omega) + C^2(-\omega)]\,E'\{r-1\}$$
$$+ [C^3(\omega) + C^3(-\omega)]\,E'\{r-2\} + \ldots \},$$

where the prime denotes the contribution to the expectation for *positive* values of the variable. As also

$$\lambda = \lambda_0\,E\{r+1\},$$

we obtain

$$g_+(\omega) = 2\lambda \left\{ 1 + [C(\omega) + C(-\omega)]\frac{E\{r\}}{E\{r+1\}} + [C^2(\omega) + C^2(-\omega)]\frac{E'\{r-1\}}{E\{r+1\}} + \ldots \right\}. \tag{8}$$

If $f_1(\tau)$ is strictly a density function, $C(\omega)$ and $C(-\omega)$ tend to zero as $\omega \to \infty$, and $g_+(\omega) \to 2\lambda$. For small ω, $C(\omega)$ is continuous and tends to one, and as $\omega \to 0$

$$g_+(\omega) \to 2\lambda \{ 1 + E\{r(r+1)\}/E\{r+1\} \}. \tag{9}$$

This gives $g_+(0+)/g_+(\infty) = m + \sigma^2/m$ where m is $E\{r+1\}$, the mean cluster or family size and σ^2 is the variance of r. More generally, the ratio $g_+(0+)/g_+(\infty)$ may be shown to give the asymptotic ratio of the variance of $N(t)$ to its mean as t increases.

A degenerate case occurs if $f_1(\tau)$ tends to a Dirac delta-function centred at, say, $\tau = a$. Suppose for simplicity we have also $p_r = \delta_{r,1}$, i.e. 1 for $r = 1$, 0 otherwise. This corresponds to family pairs of fixed distance apart a. Equation (8) becomes

$$g_+(\omega) = 2\lambda \{ 1 + \cos a\omega \}, \tag{10}$$

a finite spectral density function still, although the covariance density function $w(\tau)$ is now $\frac{1}{2}\lambda\delta(\tau - a)$.

Example 3. If we define the quantity $J(\omega)$ calculated for a sample length T of a point process by the equation

$$J(\omega) = \sqrt{\left(\frac{2}{T}\right)} \int_0^T e^{it\omega}\, dN(t) \tag{11}$$

note that

$$E\{J(w)\, J^*(\omega)\} = 2\left\{\int_{-T}^{T} e^{i\tau\omega}\left(1 - \frac{|\tau|}{T}\right) w(\tau)\, d\tau + \lambda\right\}, \tag{12}$$

and as $T \to \infty$ tends to $g_+(\omega)$ (cf. equation (4), § 9·2).

Suppose in particular a point process consists of events occurring periodically at times $0, a, 2a, \ldots$ except for independent random shifts in each occurrence with distribution specified by the characteristic function $C(\phi)$. Then for a sample length T large enough we may (for appropriate $C(\phi)$) neglect any 'overlap' effect at the ends of the interval $(0, T)$ and write

$$J(\omega) \sim \sqrt{\left(\frac{2}{T}\right)}\{e^{i\omega\epsilon_1} + e^{i\omega(a+\epsilon_2)} + \ldots + e^{i\omega\{[n-1]\,a+\epsilon_n\}}\},$$

where $\epsilon_1, \epsilon_2, \ldots$ denote the random shifts. We have

$$E\{J(\omega)\, J^*(\omega)\} \sim \frac{2n}{T}\{1 - C(\omega)\, C^*(\omega)\} + \frac{2}{T} C(\omega)\, C^*(\omega)\, \frac{1 - \cos[na\omega]}{1 - \cos[a\omega]}. \tag{13}$$

The second term has large peaks at the frequencies $2\pi s/a$, of order $n^2 T = \lambda n$, and as T (and n) increase these peaks become discrete components in the spectrum with ordinates at $2\pi s/a$. In the case of random non-zero errors, the factor $C(\omega)\, C^*(\omega)$ provides a modulating factor on these ordinates, and the first term in (13) a non-zero spectral density component, so that the spectrum is a mixed one.

6·2   Generalized harmonic analysis

There is an important analysis of any stationary process $X(t)$ into orthogonal components, having an intimate relation with the properties of the correlation and covariance functions discussed in § 6·11. This follows from the general expansion theorem quoted at the end of § 5·11, this theorem indicating the existence of a harmonic analysis of $X(t)$ in the form

$$X(t) = \int_{-\infty}^{\infty} e^{it\omega}\, dZ(\omega), \tag{1}$$

where $Z(\omega)$ is an orthogonal process and the integral in (1) is interpreted in the m.s. sense. If we write $V(\omega) = E\{|Z(\omega)|^2\}$, then $V(\omega)$ is a never-decreasing function with

$$E\{|X(t)|^2\} = \sigma^2 = V(\infty) - V(-\infty),$$

so that for $V(-\infty)=0$, $V(\infty)=\sigma^2$. Then

$$E\{X^*(t)\,X(t+\tau)\}=E\left\{\int_{-\infty}^{\infty}\int_{-\infty}^{\infty}e^{i(t+\tau)u-itv}\,dZ(u)\,dZ^*(v)\right\}$$

$$=\int_{-\infty}^{\infty}e^{i\tau\omega}\,dV(\omega),\tag{2}$$

so that $V(\omega)=\sigma^2 F(\omega)$. Note that while $V(\omega)$ is bounded, $Z(\omega)$ need only be of m.s. bounded variation.

The expansion theorem quoted in §5·11 is most elegantly established by means of Hilbert space representations of stationary processes, but in view of the close relation of (1) to physical harmonic analysis and the theory of linear 'filters', it may be useful to sketch a more direct heuristic treatment of Blanc-Lapierre and Fortet (see Lévy, p. 125; cf. also the proof of Khintchine's theorem given in §6·11).

Consider the linear transformation or 'filter' operating on $X(t)$, giving

$$Y(t)=\int_{-\infty}^{\infty}X(t-u)\,g(u)\,du.\tag{3}$$

To see the effect of $g(u)$, suppose $X(u)$ can in fact be written in the form (1); then

$$Y(t)=\int_{-\infty}^{\infty}e^{it\omega}\,dZ(\omega)\int_{-\infty}^{\infty}g(v)\,e^{-i\omega v}\,dv$$

$$=\int_{-\infty}^{\infty}e^{it\omega}[h(\omega)\,dZ(\omega)],\tag{4}$$

where

$$h(\omega)=\int_{-\infty}^{\infty}g(v)\,e^{-i\omega v}\,dv=\psi(\omega)\,e^{i\phi(\omega)},\tag{5}$$

say, is called the 'gain' of the filter, the amplitude of each $dZ(\omega)$ being multiplied by ψ and the phase increased by ϕ. If we now take (formally) $h(\omega)=1$ in (ω_1,ω_2) and 0 elsewhere, then (4) would give

$$Y(t)=\int_{\omega_1}^{\omega_2}e^{it\omega}\,dZ(\omega);$$

the corresponding $g(v)$ from (5) is

$$g(v)=\frac{1}{2\pi}\left[\frac{e^{iv\omega_2}-e^{iv\omega_1}}{iv}\right],$$

whence (3) becomes

$$Y(t)=\frac{1}{2\pi}\int_{-\infty}^{\infty}\frac{e^{iv\omega_2}-e^{iv\omega_1}}{iv}\,X(t-v)\,dv.$$

As the function $h(\omega)$ just taken is discontinuous, it is strictly incompatible with integrability of $g(v)$, but this difficulty can be surmounted by a suitable passage to a limit, leading to the formula

$$\int_{\omega_1}^{\omega_2} e^{it\omega} dZ(\omega) = \frac{1}{2\pi} \underset{T \to \infty}{\text{l.i.m.}} \int_{-T}^{T} \frac{e^{iv\omega_2} - e^{iv\omega_1}}{iv} X(t-v) \, dv, \qquad (6)$$

or in particular when $t = 0$

$$\int_{\omega_1}^{\omega_2} dZ(\omega) = \frac{1}{2\pi} \underset{T \to \infty}{\text{l.i.m.}} \int_{-T}^{T} \frac{e^{-iv\omega_2} - e^{-iv\omega_1}}{-iv} X(v) \, dv. \qquad (7)$$

We have been led to the formula (6) starting from (1), but if we now *define* $Z(\omega)$ by (6), it is not difficult to show that

(i) $Y(t; \omega_1, \omega_2)$ and $Y(t; \omega_3, \omega_4)$ are uncorrelated for any non-overlapping intervals (ω_1, ω_2), (ω_3, ω_4), and hence that $e^{it\omega} dZ(\omega)$ and in particular $Z(\omega)$ represents an orthogonal process; and further,

(ii) $X(t)$ is identical with the sum of all such non-overlapping components $Y(t)$, whence (1) results. It may also be noted that if $X(t)$ is completely stationary, so also is each of its components of equation (6).

The integrated form of the inversion integral (7) should be compared with an integrated form of (1) which covers also the processes of § 6·12, viz.

$$Y(t) = \int_0^t X(u) \, du = \underset{\Omega \to \infty}{\text{l.i.m.}} \int_{-\Omega}^{\Omega} \left(\frac{e^{it\omega} - 1}{i\omega} \right) dZ(\omega). \qquad (8)$$

We note also the harmonic analysis for processes X_t in discrete time,

$$X_t = \int_{-\pi}^{\pi} e^{it\omega} dZ(\omega), \qquad (9)$$

corresponding to equation (10) of § 6·11.

A further theorem associated with the harmonic expansion (1) is that, if $X(t)$ is defined by an expansion of the form (1), $X(t)$ is stationary to the second order if and only if $Z(\omega)$ is orthogonal (it is assumed $E\{Z(\omega)\} = 0$). This follows at once from the covariance equation

$$E\{X^*(\tau) X(t)\} = \int_{-\infty}^{\infty} \int_{-\infty}^{\infty} e^{i(vt - u\tau)} E\{dZ^*(u) \, dZ(v)\}, \qquad (10)$$

for a necessary and sufficient condition for the covariance to be of the required form $w(t-\tau)$ is that each term given by the harmonic decomposition of the integral on the right-hand side has this property, whence $E\{\Delta Z^*(u)\,\Delta Z(v)\}=0$ for disjoint intervals Δu, Δv—the orthogonal property.

Discrete and continuous spectra. The discrete and (absolutely) continuous components of $F(\omega)$ (see equations (3)–(5), §6·11) imply a similar analysis of $X(t)$. For $V(\omega)$ the jump or *saltus* at ω is (see Cramér, 1937, p. 24)

$$\lim_{T\to\infty}\frac{1}{2T}\int_{-T}^{T}e^{-i\tau\omega}w(\tau)\,d\tau. \tag{11}$$

We should expect correspondingly, at each such point ω_n of discontinuity,

$$A_n=Z(\omega_n+0)-Z(\omega_n-0)=\underset{T\to\infty}{\text{l.i.m.}}\frac{1}{2T}\int_{-T}^{T}e^{-it\omega_n}X(t)\,dt, \tag{12}$$

an ergodic property which will follow from a result to be obtained in §6·21 (namely, that

$$A=\underset{T\to\infty}{\text{l.i.m.}}\frac{1}{2T}\int_{-T}^{T}X(t)\,dt \tag{13}$$

gives the jump in $Z(\omega)$ at $\omega=0$), by our regarding $e^{-it\omega_n}X(t)$ as a new stationary process $Y(t)$. Assuming this result for the moment, we may write

$$X(t)=\Sigma_n A_n e^{i\omega_n t}+\int_{-\infty}^{\infty}e^{i\omega t}dZ_2(\omega),$$

where $E\{A_n A_k^*\}=0$ $(n\ne k)$, and $dE\{Z_2 Z_2^*\}/d\omega$ is proportional to the spectral density component $f(\omega)$.

6·21 The ergodic property.

The value of the integral in (13) of §6·2 is a crucial one in the theory of stationary processes, for (with $E\{X(t)\}=0$) only if $A=0$ can the probability average $E\{X(t)\}$ be measured by means of the time average in this equation. The value of A readily follows from the Hilbert space theory already mentioned; here, however, we indicate Khintchine's original method of proof (1934).

Consider the expression for A before the limit is taken, U_T, say. Then A exists if

$$\xi\{T,V\}\equiv E\{|\,U_{T+V}-U_T\,|^2\}\to 0 \tag{1}$$

as $T \to \infty$, uniformly for all $V > 0$ (cf. equation (9), § 5·1). Now by evaluating $\xi\{T, V\}$ by expanding the term on the right we find

$$\xi\{T, V\} = \frac{\sigma^2}{4T^2(T+V)^2}$$

$$\times \left\{ T^2 \int_D \int_D \rho(u-v)\, du\, dv + V^2 \int_{-T}^T \int_{-T}^T \rho(u-v)\, du\, dv \right.$$

$$\left. - VT \int_D \int_{-T}^T [\rho(u-v) + \rho(v-u)]\, du\, dv \right\},$$

where D denotes a range of integration $-T-V$ to $-T$ and T to $T+V$. Making use of equation (4), § 6·1, we find further that

$$\xi\{T, V\} = \frac{V^2 \sigma^2}{(T+V)^2} \left\{ \int_{-\infty}^{\infty} \left[\left(\frac{\sin T\omega}{T\omega}\right)^2 + \left(\frac{\sin \frac{1}{2} V\omega}{\frac{1}{2} V\omega}\right)^2 \cos^2 (T\omega + \frac{1}{2} V\omega) \right. \right.$$

$$\left. \left. - 2 \left(\frac{\sin T\omega}{T\omega}\right) \left(\frac{\sin \frac{1}{2} V\omega}{\frac{1}{2} V\omega}\right) \cos (T\omega + \frac{1}{2} V\omega) \right] dF(\omega) \right\}, \quad (2)$$

an expression which tends to zero uniformly for all $V > 0$. This follows for bounded V from the initial factor; for arbitrarily large V it follows from an auxiliary mathematical lemma due to Khintchine, that *if in*

$$\eta = \int_{-\infty}^{\infty} \psi(\omega; T_1, T_2, \ldots)\, dF(\omega)$$

ψ *is a real continuous function of ω such that*

(i) $|\psi|$ *is bounded for all* $\omega, T_1, T_2, \ldots,$

(ii) $\psi \to C$ *as* $\omega \to 0$ *for any fixed* $T_1, T_2, \ldots,$

(iii) $\psi \to 0$ *as* $T_1, T_2, \ldots \to \infty,$ *uniformly in* $|\omega| \geqslant \delta > 0,$ *then*

$\eta \to C\{F(0+) - F(0-)\}$ as $T_1, T_2, \ldots \to \infty$. In the expression (2), $C = 0$, and hence $\xi(T, V) \to 0$.

It follows that the m.s. limit A exists. Only under certain conditions, however, is $A = 0 = E\{X(t)\}$. For

$$E\{|U_T|^2\} = \frac{\sigma^2}{4T^2} \int_{-T}^T \int_{-T}^T \rho(u-v)\, du\, dv$$

$$= \sigma^2 \int_{-\infty}^{\infty} \left(\frac{\sin T\omega}{T\omega}\right)^2 dF(\omega),$$

which from Khintchine's lemma tends to $\sigma^2\{F(0+)-F(0-)\}$, and is only zero provided $F(\omega)$ has no discontinuity at $\omega=0$. From (11) of the previous section, this condition is equivalent to

$$\lim_{T\to\infty} \frac{1}{2T} \int_{-T}^{T} w(\tau)\,d\tau = 0. \tag{3}$$

We have defined A above as the m.s. limit of the symmetrical time-average of $X(t)$, from $-T$ to T; but of course, as $X(t)$ is stationary, the time-average from 0 to T will also tend to the same limit.

6·3   Processes with continuous spectra

In equation (22), § 5·2, we defined the linear process

$$X(t)=\int_{-\infty}^{t} g(t-u)\,dY(u), \tag{1}$$

where $Y(u)$ was additive, and from its characteristic functional saw that $X(t)$ was completely stationary. (We also defined a linear process for discrete time with integration replaced by summation, but, as elsewhere in this chapter, the appropriate changes in the formulae for stationary sequences are usually evident.) If only second-order moment properties are being considered, it is convenient to allow $Y(u)$ and $g(u)$ in (1) to be possibly complex, and the condition that $X(t)$ is well-defined as a m.s. limit is then that its second moment

$$\sigma^2 = \sigma_Y^2 \int_0^{\infty} g(u)\,g^*(u)\,du < \infty. \tag{2}$$

The autocovariance function $w(\tau)$ is generalized to

$$w(\tau) = \sigma_Y^2 \int_{-\infty}^{\infty} g^*(u)\,g(u+\tau)\,du \quad (g(u)=0 \text{ for } u<0). \tag{3}$$

This last formula shows at once, if we write

$$h(\omega) = \int_{-\infty}^{\infty} e^{-iv\omega} g(v)\,dv, \tag{4}$$

that the spectral density exists and is given by

$$f(\omega) = \frac{\sigma_Y^2}{\sigma^2} \frac{1}{2\pi} \int_{-\infty}^{\infty} e^{-i\tau\omega} \left[\int_{-\infty}^{\infty} g^*(u) g(u+\tau) \, du \right] d\tau$$

$$= \frac{\sigma_Y^2}{2\pi\sigma^2} h(\omega) h^*(\omega). \tag{5}$$

This formula remains true if $Y(u)$ is orthogonal but not completely additive, and if $g(u)$ extends over $-\infty < u < \infty$ (provided of course the integral in (2) converges with $-\infty$ as its lower limit). The correlational and spectral properties of random functions of this kind are in fact included in those obtained by linear transformations of a stationary process (the 'filter' of § 6·2). In equation (3) of § 6·2 we had defined (with suitable change of notation for the random functions)

$$X(t) = L\{U(t)\} \equiv \int_{-\infty}^{\infty} g(t-u) \, U(u) \, du, \tag{6}$$

and found that the linear operator $L\{...\}$ defined by (6) was equivalent to an extra factor $h(\omega)$ in the harmonic analysis of $X(t)$ compared with $U(t)$, i.e.

$$U(t) = \int_{-\infty}^{\infty} e^{it\omega} dZ(\omega), \quad X(t) = \int_{-\infty}^{\infty} e^{it\omega} h(\omega) \, dZ(\omega).$$

$X(t)$ is necessarily stationary to the second order like $U(t)$, as $h(\omega) \, dZ(\omega)$ has similar orthogonal properties to $dZ(\omega)$. It is assumed that σ_X^2 is finite, i.e.

$$\int_{-\infty}^{\infty} h(\omega) h^*(\omega) \, dE\{Z(\omega) Z^*(\omega)\} \equiv \int_{-\infty}^{\infty} |h(\omega)|^2 dV(\omega) < \infty. \tag{7}$$

We have the further results

$$E\{U^*(t) X(t+\tau)\} = \int_{-\infty}^{\infty} e^{i\tau\omega} h(\omega) \, dV(\omega), \tag{8}$$

$$E\{X^*(t) X(t+\tau)\} = \int_{-\infty}^{\infty} e^{i\tau\omega} h(\omega) h^*(\omega) \, dV(\omega), \tag{9}$$

and if $\qquad dV(\omega) = \sigma_U^2 f_U(\omega) \, d\omega = v(\omega) \, d\omega,$

then from (9) $\qquad \sigma_X^2 f_X(\omega) = h(\omega) h^*(\omega) v(\omega). \tag{10}$

The result (5) above is the degenerate case of (10) as $v(\omega)$

approaches the uniform spectral 'distribution' of an orthogonal process. Particular cases of the linear operator $L\{...\}$ are a linear differential operator $\phi(d/dt) \equiv \phi(D)$, or, in cases of discrete time, a linear difference operator $\psi(E_t) \equiv \psi(e^D)$. The general formula (10) then gives formally, corresponding to the inverse linear operator L^{-1},

$$\sigma_X^2 f_X(\omega) = v(\omega)/|\phi(i\omega)|^2, \tag{11}$$

a result easily seen to be valid if $\phi(i\omega)$ does not vanish for any real ω. Examples referred to in §5·2 were:

(i) $e^D - \rho$, (ii) $e^{2D} + ae^D + b$, (iii) $D^2 + \alpha D + \beta$,

and (by implication) the Markov continuous time case (iv) $D + \mu$. For (iv), formula (11) gives

$$f_X(\omega) = \frac{1}{\sigma_X^2} \frac{v(\omega)}{\mu^2 + \omega^2} \to \frac{\sigma_Y^2}{2\pi\sigma_X^2} \frac{1}{\mu^2 + \omega^2} \tag{12}$$

in the orthogonal case. Similarly we find for (iii)

$$f_X(\omega) = \frac{\sigma_Y^2}{2\pi\sigma_X^2} \frac{1}{(\beta - \omega^2)^2 + \omega^2\alpha^2} \tag{13}$$

in the orthogonal case. The corresponding spectra for (i) and (ii) are referred to in Chapter 9.

A question which now arises is how general such a representation as (3) is for the covariance function. If the spectrum of a stationary process is assumed to be (absolutely) continuous, and we write

$$\alpha(\omega)\,\alpha^*(\omega) = \sigma_X^2 f(\omega), \tag{14}$$

then $$w(\tau) = \int_{-\infty}^{\infty} e^{i\omega\tau} \alpha(\omega)\,\alpha^*(\omega)\,d\omega,$$

which, from Parseval's theorem,

$$= \int_{-\infty}^{\infty} \beta(u + \tau)\,\beta^*(u)\,du, \tag{15}$$

where $$\beta(u) = \lim_{\Omega \to \infty} \frac{1}{\sqrt{(2\pi)}} \int_{-\Omega}^{\Omega} e^{i\omega u}\alpha(\omega)\,d\omega. \tag{16}$$

Equation (15) is identified with (3) by putting $g(u) = \beta(u)/\sigma_Y$, and implies (by appeal to the general expansion theorem quoted in § 5·11) the existence of an expansion

$$X(t) = \int_{-\infty}^{\infty} g(t-u) \, dY(u), \tag{17}$$

where $Y(u)$ is orthogonal. The representation (17) is thus general for any stationary process with spectral density $f(\omega)$, but it does not follow that an expansion of the form (1) is possible with $Y(u)$ orthogonal, for this requires further that $g(u)$ can be chosen so that $g(u) = 0$ for $u < 0$. This condition is given in a theorem by Paley and Wiener (1934, Theorem XII) that we must have

$$\int_{-\infty}^{\infty} \frac{|\log f(\omega)| \, d\omega}{1+\omega^2} < \infty. \tag{18}$$

To obtain $g(u)$ (or $\beta(u)$) from $\alpha(\omega)$, satisfying the condition $g(u) = 0$ for $u < 0$, we choose $\alpha(z)$ to have no singularities or zeros in the lower half-plane when regarded as a function of the complex variable z (the condition of no zeros avoids singularities in $1/\alpha(z)$, and is required for the use made of $\alpha(z)$ in Chapter 7).

For example, if

$$f(\omega) = \frac{\mu}{\pi} \frac{1}{\mu^2+\omega^2},$$

we take

$$\alpha(\omega) = \frac{\sigma_Y}{\sqrt{(2\pi)(\mu+i\omega)}},$$

giving

$$\beta(u) = \sigma_Y e^{-\mu u} \quad (u > 0).$$

More generally when $f(\omega)$ is of the form

$$\frac{C^2 \prod_{s=1}^{n} (\omega-\omega_s)(\omega-\omega_s^*)}{\prod_{s=n+1}^{m} (\omega-\omega_s)(\omega-\omega_s^*)} \quad (C \text{ real, } m > 2n, \, \mathscr{I}(\omega_s) > 0),$$

we may take

$$\alpha(\omega) = C \prod_{s=1}^{n} (\omega-\omega_s) \Big/ \prod_{s=n+1}^{m} (\omega-\omega_s). \tag{19}$$

As one example where (18) is not satisfied, we note the admissible process $X(t)$ for which

$$f(\omega) = \frac{1}{\sqrt{(2\pi)}} e^{-\frac{1}{2}\omega^2}, \quad \rho(\tau) = e^{-\frac{1}{2}\tau^2}. \tag{20}$$

This process is analytic (see § 5·11), and, as we shall see in the next chapter, it is thus not surprising that it cannot be represented by a formula like (1), which for $Y(t)$ additive is essentially a non-deterministic process, with new random increments $dY(t)$ arising as time proceeds. If $Y(t)$ is merely orthogonal, we can regard (1) as defining a process $X(t)$ at least non-deterministic 'to the second order', i.e. linear transformations of $X(u)$ for $u \leqslant t$ do not enable the increments $dY(u)$ for $u > t$ to be deduced.

6·31   Further examples of stationary processes.

Some further examples of stationary processes liable to arise in practice will be considered from the point of view of their correlational and spectral properties. In contrast with the class of processes of §6·3, a classical Fourier series has a discrete spectrum, and constitutes an example of a stationary process if the phases of the various harmonic components are independently and uniformly distributed over $(0, 2\pi)$. To note the simplest case,

$$X(t) = A \cos(\omega_1 t + \Phi) \qquad (1)$$

gives $\qquad E\{X^*(t) X(t + \tau)\} = \tfrac{1}{2}\sigma_A^2 \cos \omega_1 \tau,$

if Φ is uniformly distributed over $(0, 2\pi)$, and $E\{A\} = 0$.

For the Markov chain of equation (5), §3·41, recalled at the beginning of the present chapter, we had the conditional p.g.f., given $X(t) = n$ at $t = 0$,

$$\Pi_t(z) = [1 + T(z-1)]^n \exp\{\nu(z-1)(1-T)/\mu\}.$$

where $T = e^{-\mu t}$. Hence

$$\begin{aligned} E\{X(t) \mid n\} &= [\partial \Pi_t(z)/\partial z]_{z=1} \\ &= nT + \nu(1-T)/\mu, \end{aligned} \qquad (2)$$

showing that the regression of $X(t)$ on $X(0)$ is linear and that the autocorrelation function is $T = e^{-\mu t}$ $(t \geqslant 0)$. (Notice that this form of the autocorrelation function, contrary to an opinion sometimes held, does *not* necessarily hold for all stationary Markov processes—counter-examples are easily constructed.)

An example of some interest as providing a mixed spectrum is a renewal process involving two states, these being labelled for convenience as -1 and $+1$ and thus giving $E\{X(t)\} = 0$ if the probability of either state is $\tfrac{1}{2}$. The process is not Markovian in the simple sense as we suppose the chance of a transition from either state to the other in $t, t + \Delta t$ is $q(\tau) \Delta t + o(\Delta t)$, depending on its present lifetime τ in the state it is occupying. The chance $Q(u)$ of no transition in time u, measured from the last transition, is then

$$Q(u) = \exp \left\{ - \int_0^u q(t) \, dt \right\}. \tag{3}$$

The theory of the number of transitions occurring in any interval τ was developed in § 2·11. We have, if the probability of r transitions is $p_r(\tau)$ ($r = 0, 1, ..., not\ 1, 2, ...,$ as in § 2·11), and

$$P(\lambda, \tau) = \sum_{r=0}^{\infty} \lambda^r p_r(\tau),$$

then $\qquad P(\lambda, \tau) = 1 + (\lambda - 1) \int_0^\tau g(u, \lambda) \, du$

(equation (13), § 2·11), where

$$g(u, \lambda) = f(u) + \int_0^\tau \lambda f(u - y) \, g(y, \lambda) \, dy,$$

$f(u)$ here being $- \partial Q(u) / \partial u$. The correlation coefficient is easily seen to depend on the numbers of odd and even transitions; in fact, $\qquad \rho(\tau) = p_0(\tau) - p_1(\tau) + p_2(\tau) \ldots = P(-1, \tau). \tag{4}$

Two possibilities arise:

(a) the integral in (3) diverges, $Q(\infty) = 0$, and the lifetime distribution exists as a continuous distribution. Thus let

$$q(\tau) = \mu^2 \tau / (1 + \mu \tau), \tag{5}$$

which tends to μ for large τ (but to $\mu^2 \tau$ for small τ, so that a transition is inhibited by a previous one). Then from (3) we find

$$Q(u) = (1 + \mu u) \, e^{-\mu u}, \quad f(u) = \mu^2 u \, e^{-\mu u},$$

so that this case is the one treated at the end of § 2·11 and again in § 6·12. Comparing (4) with equation (19), § 2·11, we have

$$\rho(\tau) = e^{-\mu \tau} (\cos \mu \tau + \sin \mu \tau) \quad (\tau > 0). \tag{6}$$

However, this is the correlation when τ is measured from a transition point; in the *stationary* case we require τ to be measured from an arbitrary point, i.e. when the distribution of the *first* transition from this point is that given by equation (6) of § 4·2. The amendment of the formulae in § 2·11 to allow for this is not difficult (cf. Cox (1962), § 2·2), and leads to the modified result

$$\rho(\tau) = e^{-\mu\tau} \cos \mu\tau \quad (\tau > 0). \tag{7}$$

The spectral density for $\rho(\tau)$ in (7) is easily found by inversion.

(b) if the integral in (3) converges, there is a finite chance of no further transitions ever taking place. For example, let

$$q(\tau) = \mu/(1 + \mu\tau)^2, \tag{8}$$

so that
$$\int_0^u q(\tau)\, d\tau = \frac{\mu u}{1 + \mu u} \to 1 \quad \text{as} \quad u \to \infty.$$

Then $Q(\infty) = e^{-1}$. The chances of no transition, one transition, etc., as $\tau \to \infty$, are therefore e^{-1}, $(1 - e^{-1})e^{-1}, \ldots$.

Again, this refers to the situation when τ is measured from the time origin. Strictly speaking in this case, the ultimate stationary process is $X(t) = +1$ (or -1) with probability $\frac{1}{2}$, and corresponds to a concentration of the spectrum at zero frequency $\omega = 0$.

6·4 Complete stationarity

While the correlational and spectral properties of stationary processes depend merely on their second-order moments, we have had various examples of completely stationary processes, the stationary Markov chain, the normal stationary process and the stationary linear process. It may be asked whether there is any convenient alternative condition for complete stationarity, such as the Fourier relation required for the second-order moments in the case of second-order stationarity. This latter condition was most simply expressed in terms of the Fourier expansion of $X(t)$, for if

$$X(t) = \int_{-\infty}^{\infty} e^{it\omega}\, dZ(\omega), \tag{1}$$

then it was noted in § 6·2 that $Z(\omega)$ must be orthogonal, i.e.

$$E\{\Delta Z(\omega)\, \Delta Z^*(\omega')\} = 0 \quad (\Delta\omega,\, \Delta\omega' \text{ non-overlapping}). \tag{2}$$

From (2) it is seen that non-zero contributions to the total mean square $E\{|X(t)|^2\}$ only arise from $\omega = \omega'$. However, this is in terms of the double integral involving $dZ(\omega)$, $dZ^*(\omega')$ and for real processes, for which it may be seen that we require

$$-Z^*(\omega) = Z(-\omega),$$

the condition in terms of $Z(\omega)$ alone becomes $\omega + \omega' = 0$. This condition may be extended to the complete stationarity case. Suppose the characteristic functional of the real process $X(t)$ is

$$C\{\Theta\} \equiv E\left\{\exp\left[i\int_{-\infty}^{\infty} X(t)\,d\Theta(t)\right]\right\}. \tag{3}$$

Provided the 'scalar product' $\int X(t)\,d\Theta(t)$ of two functions $X(t)$, $d\Theta(t)$ is invariant, $C\{\Theta\}$ is invariant; the linear transformation represented by (1) is (for suitable regularity conditions on $\Theta(t)$) thus counteracted by the linear transformation

$$\phi(\omega) = \int_{-\infty}^{\infty} e^{it\omega}\,d\Theta(t), \tag{4}$$

whence the characteristic functional of $Z(\omega)$ is

$$C\{\Theta\} = c(\phi) \equiv E\left\{\exp\left[i\int_{-\infty}^{\infty} \phi(\omega)\,dZ(\omega)\right]\right\}. \tag{5}$$

If $X(t)$ is completely stationary, the properties of $X(t)$ are unaltered by writing $t + \tau$ for t. Making this change in (1), we see that $c(\phi)$ is unaltered if we replace $dZ(\omega)$ by $e^{i\tau\omega}dZ(\omega)$, or equivalently, if we replace $\phi(\omega)$ in (5) by $\phi(\omega)\,e^{i\tau\omega}$. In terms of the higher moments of $Z(\omega)$ of order r, if we assume that these are finite, it implies that non-zero contributions can only arise in the hyperplane

$$\omega + \omega' + \omega'' + \ldots + \omega^{(r-1)} = 0, \tag{6}$$

this being the appropriate extension of the condition $\omega + \omega' = 0$.

In terms of real functions, we may write the real process $X(t)$ as

$$X(t) = \int_{0}^{\infty} [\cos t\omega\,dU(\omega) + \sin t\omega\,dV(\omega)], \tag{7}$$

where $U(\omega) = Z(\omega) - Z(-\omega), \quad V(\omega) = i[Z(\omega) + Z(-\omega)]$,

so that degeneracy conditions on $U(\omega)$, $V(\omega)$ can be obtained from those on $Z(\omega)$.

A class of completely stationary processes. If $X(t)$ is a real process stationary to the second order with the representation (1), and $Z(\omega)$ is additive as well as orthogonal, is $X(t)$ completely stationary? By $Z(\omega)$ additive is meant that $U(\omega)$, $V(\omega)$ is a bivariate additive process with cumulant function

$$K(\alpha, \beta; \omega) = \int_{0}^{\omega} dK(\alpha, \beta; \omega), \tag{8}$$

say. With this additivity assumption, the cumulant (log characteristic) functional for $X(t)$ is

$$\log C\{\Theta\} = \int_0^\infty dK\left(i\int\cos u\omega\,d\Theta(u), i\int\sin v\omega\,d\Theta(v); \omega\right). \qquad (9)$$

Consider first the distribution of $X(t)$ at a single time-instant t, determined by its cumulant function

$$\int_0^\infty dK(i\theta\cos t\omega, i\theta\sin t\omega; \omega). \qquad (10)$$

This is independent of t (for all θ) if

$$K(\alpha, \beta) \equiv L(\alpha^2 + \beta^2) \equiv L([\alpha + i\beta][\alpha - i\beta]), \qquad (11)$$

and with this assumption (9) becomes

$$\int_0^\infty dL\left(-\int e^{iu\omega}\,d\Theta(u)\int e^{-iv\omega}\,d\Theta(v)\right), \qquad (12)$$

which is invariant to translation of the t-axis.

Notice that if it were assumed further that $U(\omega)$, $V(\omega)$ were independent of each other, then it would follow further that $Z(\omega)$ and hence $X(t)$ were normal processes, since the only solution of (11) satisfying also $K(\alpha, \beta) = K_1(\alpha) + K_2(\beta)$ is $\frac{1}{2}\sigma^2(\alpha^2 + \beta^2)$. However, the completely stationary process (12) need not be normal, arising in general from an additive process $U(\omega)$, $V(\omega)$ with a distribution function isotropic in the two variables (cf. §6·51).

If we assumed that $X(t)$ is normal, then it follows from the linear relation between $X(t)$ and $Z(\omega)$ that $Z(\omega)$ is normal (for stationarity to the second order it is easily shown from (7) that

$$E\{U^2(\omega)\} = E\{V^2(\omega)\} \quad \text{and} \quad E\{U(\omega)\,V(\omega)\} = 0,$$

and thus $U(\omega)$, $V(\omega)$ is a normal isotropic bivariate additive process). In turbulence theory it has been asked if a converse is true, that if $Z(\omega)$ is additive, is $X(t)$ and hence $Z(\omega)$ necessarily normal? The above more general class of completely stationary process shows that this does not follow from these conditions alone, even if it is assumed that $Z(\omega)$ has no fixed discontinuities, i.e. the spectrum of $X(t)$ is continuous. However, it is reasonable in this physical context to assume also that $X(t)$ is ergodic in regard to the measurement of its autocovariance. This would imply that the process

$$Y(t) = X(t)\,X(t+\tau) - w(\tau)$$

is ergodic, i.e. $$\operatorname*{l.i.m.}_{T\to\infty} \frac{1}{T}\int_0^T Y(t)\,dt = 0, \qquad (13)$$

an equivalent condition for which is seen to be

$$\lim_{T,\,U\to\infty} \frac{1}{TU}\int_0^T\int_0^U W(t,u)\,dt\,du = 0. \qquad (14)$$

Here $W(t,u)$ is the covariance function of $Y(t)$ and is a function of $t-u$ if $X(t)$ is completely stationary, but not necessarily if $X(t)$ is only given as stationary to the second order. It may be shown, even with this weaker condition, that (14) cannot be satisfied unless the additive process $Z(\omega)$ is normal. It then follows that $X(t)$ is normal and completely stationary. (A rigorous proof of this last result has been obtained by J. Eklind.)

6·41 Recurrence times for completely stationary processes.
In § 3·3 the distributional theory of recurrence and passage times was developed, but only in the case of 'renewal' processes, which included Markov processes, was a very complete theory practicable. Recurrence theory is especially of practical interest for completely stationary processes, even if they are not of the 'renewal' type, for they will in general exhibit continual recurrence, this phenomenon being linked with ergodicity properties. We shall consider stationary processes for which the ergodicity property holds in the most complete sense (not just for the time average of $X(t)$ or of $X(t)\,X(t+\tau)$), by which is meant that all probability distributions characterizing a process $X(t)$ are generated by any single realization $x(t)$. In terms of the characteristic functional (equation (3), § 6·4), defined, say, for $d\Theta(t) = 0$ outside an arbitrary range $(0,\tau)$, this will be obtained from the time-average

$$\lim_{T\to\infty} \frac{1}{T}\int_0^T \left\{ \exp\left[i\int_0^\tau x(u+t)\,d\Theta(u) \right] \right\} dt.$$

This is of course a strong ergodic property, but may reasonably be assumed for many physical applications—for example, in the theory of molecular density fluctuations. As first noted by

Smoluchowski, the extent to which the irreversibility represented by the Second Law of Thermodynamics continues to hold is dependent on the order of magnitude of the recurrence times and on the corresponding length of time-scale one is prepared to consider.

Even with these assumptions it only seems possible to give general formulae of a practicable kind for mean recurrence times. One powerful but simple method originally used by Smoluchowski (1915) is available when the recurring state S has a non-zero probability of occurrence at any instant.

It will be convenient to refer to the number of separate times a state is occupied in a 'long' interval of time T, on the understanding that all derived formulae will involve only ratios of such numbers, and hold in the limit as $T \to \infty$. We then denote, in the case of discrete time (or intermittent observation), the frequency of the specified state S being occupied continuously for an interval length k by N_k. Similarly, we denote by M_k the frequency for the composite state 'not S'. Then on our assumptions we necessarily have

$$P(S) = \frac{N_1 + 2N_2 + 3N_3 + \ldots}{N_1 + 2N_2 + 3N_3 + \ldots + M_1 + 2M_2 + 3M_3 + \ldots}, \quad (1)$$

$$P(S \mid S) = \frac{N_2 + 2N_3 + 3N_4 + \ldots}{N_1 + 2N_2 + 3N_3 + \ldots}, \quad (2)$$

$$1 = \frac{N_1 + N_2 + N_3 + \ldots}{M_1 + M_2 + M_3 + \ldots}, \quad (3)$$

where $P(S \mid S)$ denotes the conditional probability of S, given S at the preceding instant, and formula (3) follows from the equality in frequency of excursions from and to state S. If we now define the *lifetime* t_1 in state S as the total continuous interval in that state, we have by definition for the mean lifetime

$$T_1 \equiv E\{t_1\} = \tau \frac{N_1 + 2N_2 + \ldots}{N_1 + N_2 + \ldots} = \frac{\tau}{1 - P(S \mid S)}, \quad (4)$$

where τ is the interval between consecutive time-instants.

Similarly, if the *recurrence time* θ_1 is the total continuous interval out of state S, then

$$\Theta_1 \equiv E\{\theta_1\} = \tau \frac{M_1 + 2M_2 + \ldots}{M_1 + M_2 + \ldots} = T_1 \frac{1 - P(S)}{P(S)}. \quad (5)$$

To obtain the equivalent formulae as the time becomes strictly continuous (i.e. $\tau \to 0$), we assume that the limiting probability $P(\text{not } S \mid S)$ is of the form $\lambda \tau + o(\tau)$ (cf. the case of Markov chains § 3·3), so that formula (4) becomes in the limit

$$T_1 = 1/\lambda. \tag{6}$$

Formula (5) continues to hold; in the particular case when S becomes a rare state, so that $1 - P(S) \sim 1$, it becomes $\Theta_1 \sim T_1/P(S)$.

Application to the theory of density fluctuations. To illustrate this last formula we quote Smoluchowski's original results in the fluctuation theory of molecular concentrations. For a spherical volume of radius a, and independent motion of the molecules, the value λ may be calculated by standard arguments from the kinetic theory of gases, and is found to be

$$\lambda = \frac{3}{a}\left(\frac{kT}{2\pi m}\right)^{\frac{1}{2}} (n + \nu),$$

where the state S denotes n molecules in the volume, ν is the average number, and the expression in the square root has the usual thermodynamical interpretation. The distribution of n, in general Poissonian, is approximated for ν large by the normal law

$$P(n) \sim \frac{1}{\sqrt{(2\pi\nu)}} \exp\left\{-\frac{1}{2}\frac{(n-\nu)^2}{\nu}\right\}.$$

We thus finally obtain, for $n/\nu \sim 1$,

$$\Theta_1 \sim \frac{\pi a}{3}\left(\frac{m}{\nu kT}\right)^{\frac{1}{2}} \exp\left\{\frac{1}{2}\frac{(n-\nu)^2}{\nu}\right\}.$$

This formula is remarkable in its tremendous and rapid change for given n/ν as the size of volume changes. Thus for 1 % fluctuations in air molecules with density 3×10^{19}/c.c. and temperature $T = 300°$ K. (cf. Smoluchowski)

$$a = 1 \times 10^{-5} \text{ cm.}, \quad \Theta_1 \text{ of order } 10^{-11} \text{ sec.},$$

$$a = 3 \times 10^{-5} \text{ cm.}, \quad \Theta_1 \text{ of order } 10^{6} \text{ sec.},$$

$$a = 5 \times 10^{-5} \text{ cm.}, \quad \Theta_1 \text{ of order } 10^{68} \text{ sec.}$$

In spite of the essential reversibility of stationary processes, these results show how an abnormal deviation from the mean will be so long in recurring that a temporary impression of irreversibility is conveyed; a process may observationally appear reversible or irreversible according as the recurrence time of the initial state is short or long compared with the time for which the system is under observation.

Extension of Smoluchowski's formulae. We may derive a useful extension of Smoluchowski's mean-value formulae. We define

(i) the 'further lifetime' t_2 of state S, given S at an *arbitrary* instant;

(ii) the 'further recurrence time' θ_2 of state S, given 'not S' at an arbitrary instant.

Then by definition, for discrete time,

$$T_2 \equiv E\{t_2\} = \tau \frac{N_1 + (1+2)N_2 + (1+2+3)N_3 + \ldots}{N_1 + 2N_2 + 3N_3 + \ldots}, \tag{7}$$

$$\Theta_2 \equiv E(\theta_2) = \tau \frac{M_1 + (1+2)M_2 + (1+2+3)M_3 + \ldots}{M_1 + 2M_2 + 3M_3 + \ldots}. \tag{8}$$

However, we may define also the complete p.g.f.'s for t_1 and t_2,

$$\Pi_L' = \frac{zN_1 + z^2 N_2 + \ldots}{N_1 + N_2 + \ldots} \quad (\tau = 1);$$

$$\Pi_L'' = \frac{zN_1 + (z+z^2)N_2 + \ldots}{N_1 + 2N_2 + \ldots}$$

$$= \frac{z}{1-z}\left[\frac{N_1 + N_2 + \ldots}{N_1 + 2N_2 + \ldots}\right] - \frac{z}{1-z}\left[\frac{zN_1 + z^2 N_2 + \ldots}{N_1 + 2N_2 + \ldots}\right]$$

$$= \frac{z}{1-z}\left[\frac{1}{T_1} - \frac{\Pi_L'}{T_1}\right],$$

or $$\qquad\qquad T_1(1-z)\Pi_L'' = z(1-\Pi_L'). \tag{9}$$

A similar relation obviously holds between Π_R' and Π_R'' for the recurrence times θ_1 and θ_2, i.e.

$$\Theta_1(1-z)\Pi_R'' = z(1-\Pi_R'). \tag{10}$$

In the case of a Markov chain, the transitions of the process, given S at any instant, are independent of the state at previous instants, so that the variables t_1 and t_2 are stochastically equivalent. This immediately yields from (9), if we put $\Pi_L' = \Pi_L''$,

$$\Pi_L' = \frac{z}{z + (1-z)T_1}, \tag{11}$$

a result which could of course be obtained more directly. It should perhaps be stressed that the same result does *not* hold for θ_1, as θ_1 and θ_2 are not in general stochastically equivalent, even for a Markov process. In other words, a Markov process does not retain its Markovian character under grouping of its states (as we have noted earlier), and 'not S' cannot therefore be treated as a single state like $\dot{S}$, except in the trivial case of only two states in all.

When the interval τ (taken for convenience as unity in (9), (10) and (11)) tends to zero, (9) and (10) become

$$\psi T_1 M_L''(\psi) = 1 - M_L'(\psi), \tag{12}$$

$$\psi \Theta_1 M_R''(\psi) = 1 - M_R'(\psi), \tag{13}$$

and (11) becomes the exponential distribution. It is of interest to observe that the converse to (11) is also true and the relation (12) can lead to the exponential distribution for M_L' *only if* t_1 and t_2 are stochastically equivalent. For let $M_L' = 1/(1 + \psi T_1)$; then $M_L''(\psi)$ is also $1/(1 + \psi T_1)$.

If in (13) we do impose the condition that θ_1 and θ_2 are stochastically equivalent, we obtain the exponential distribution for $M_R'(\psi)$ and $M_R''(\psi)$. This may be used in the recurrence theory of finite Markov chains (§ 3·32) to obtain the condition that the recurrence distribution should reduce to this distribution. Denote the original transition-matrix by

$$\mathbf{R} = \left(\begin{array}{c|c} a & \mathbf{x}' \\ \hline \mathbf{b} & \mathbf{C} \end{array} \right),$$

where it should be noted that the vector $\mathbf{y}$, defined in § 3·32 as the initial probability vector for the remaining states, is proportional to $\mathbf{b}$ for θ_1, and for θ_2 to the asymptotic distribution of the individual 'not S' states given by the column latent vector of the zero latent root of $\mathbf{R}$ (in view of the ergodic assumption, there is only one zero root of $\mathbf{R}$). Hence the required condition is

$$\left(\begin{array}{c|c} a & \mathbf{x}' \\ \hline \mathbf{b} & \mathbf{C} \end{array} \right) \left(\begin{array}{c} \mu \\ \hline \mathbf{b} \end{array} \right) = \left(\begin{array}{c} a\mu + \mathbf{x}'\mathbf{b} \\ \hline \mu\mathbf{b} + \mathbf{C}\mathbf{b} \end{array} \right) = \left(\begin{array}{c} 0 \\ \hline \mathbf{0} \end{array} \right),$$

whence **b** must be a latent vector of **C**, with corresponding latent root $-\mu = \mathbf{x'b}/a$. A simple example is

$$\mathbf{R} = \begin{pmatrix} -2 & \vdots & 2 & 4 \\ \hdashline 1 & \vdots & -4 & 1 \\ 1 & \vdots & 2 & -5 \end{pmatrix}$$

for which $M'_R(\psi) = M''_R(\psi) = 1/(1 + \frac{1}{3}\psi)$.

Continuous infinity of states. When the probability of S becomes strictly zero, it is not very convenient to obtain the appropriate formulae for recurrence by direct passage from earlier formulae, as the concept of lifetime in a state breaks down; it may, in the case of a single variable X with continuous range, be replaced by the rate of change $\dot{X}$ of X, assuming this to exist in any well-defined sense, so that the simultaneous probability distribution $f(\xi, \eta) d\xi d\eta$ of $X, \dot{X}$ exists. As the process is stationary f will not depend on t. Thus for a given positive slope η the probability of $X = x$ in the time interval $t, t + \delta t$ as $\delta t \to 0$ is

$$\int_{x-\eta\delta t}^{x} f(\xi, \eta)\, d\xi \sim \eta f(x, \eta)\, \delta t,$$

and hence, for any slope, we have formally the density in time

$$\int_{-\infty}^{\infty} |\eta|\, f(x, \eta)\, d\eta, \tag{14}$$

and an expected recurrence time

$$\Theta_1 = 1 \bigg/ \int\!\!\int_{-\infty}^{\infty} |\eta|\, f(x, \eta)\, d\eta. \tag{15}$$

(The formula (14) may be rigorously justified if $f(\xi, \eta)$ is continuous in ξ, η and $\int f(x + \alpha\eta, \eta)\, d\eta$ converges uniformly with respect to α in some interval $-|\alpha_1| \leqslant \alpha \leqslant |\alpha_2|$ (see Rice, 1944–5).

As an example consider a normal process for which X and $\dot{X}$ are, being uncorrelated for a stationary process, independent, so that

$$f(\xi, \eta) = \frac{1}{2\pi\sigma_1\sigma_2} \exp\left\{ -\tfrac{1}{2}\left[\frac{\xi^2}{\sigma_1^2} + \frac{\eta^2}{\sigma_2^2} \right] \right\},$$

where we know from the general representation

$$X(t) = \int_{-\infty}^{\infty} e^{it\omega} dZ(\omega), \quad \dot{X}(t) = \int_{-\infty}^{\infty} i\omega e^{it\omega} dZ(\omega),$$

that
$$\sigma_2^2 = \sigma_1^2 \int_{-\infty}^{\infty} \omega^2 dF(\omega).$$

Now
$$\int_{-\infty}^{\infty} \frac{|\eta|}{2\pi\sigma_1\sigma_2} \exp\left\{-\frac{1}{2}\left[\frac{x^2}{\sigma_1^2} + \frac{\eta^2}{\sigma_2^2}\right]\right\} d\eta = \frac{\sigma_2}{\pi\sigma_1} \exp\left\{-\frac{1}{2}\frac{x^2}{\sigma_1^2}\right\},$$

so that
$$\Theta_1 = \frac{\pi}{\sigma_\omega} \exp\left\{\frac{1}{2}\frac{x^2}{\sigma_1^2}\right\}, \tag{16}$$

where σ_ω is the standard deviation of the spectral distribution of X. For a normal Markov process σ_ω is infinite and $\dot{X}$ does not exist, but the methods indicated at the end of § 3·3 may be used to confirm that in this case Θ_1 is strictly zero.

As an extension of equation (16), the formula for the mean recurrence time may be developed from the more general expansion for $f(x, \eta)$

$$\exp\left\{\sum_{r+s>2}\left(-\frac{\partial}{\partial x}\right)^r \left(-\frac{\partial}{\partial \eta}\right)^s \frac{\kappa_{rs}}{r!\,s!}\right\} \frac{1}{2\pi\sigma_1\sigma_2} \exp\left\{-\frac{1}{2}\left[\frac{x^2}{\sigma_1^2} + \frac{\eta^2}{\sigma_2^2}\right]\right\},$$

where for a completely stationary process $\kappa_{21}, = E\{X^2(t)\dot{X}(t)\}$ (for $E\{X(t)\} = 0$), $= 0$; $\kappa_{31} = E\{X^3(t)\dot{X}(t)\} = 0$, etc. We obtain by straightforward integration, retaining terms up to the fourth order, an average density of crossing the point $X(t) = x$ given by

$$\frac{1}{\pi}\frac{\sigma_2}{\sigma_1}\left\{1 + \frac{\gamma_{30}}{6}\left(\frac{x^3}{\sigma_1^3} - 3\frac{x}{\sigma_1}\right) + \frac{\gamma_{40}}{24}\left(\frac{x^4}{\sigma_1^4} - \frac{6x^2}{\sigma_1^2} + 3\right)\right.$$
$$\left. + \frac{\gamma_{12}}{2}\frac{x}{\sigma_1} + \frac{\gamma_{22}}{4}\left(\frac{x^2}{\sigma_1^2} - 1\right) - \frac{\gamma_{04}}{24}\right\}\exp\left\{-\frac{1}{2}\frac{x^2}{\sigma_1^2}\right\}, \quad (17)$$

where

$$\begin{cases} \gamma_{30} = E\{X^3\}/\sigma_1^3, & \gamma_{40} = E\{X^4\}/\sigma_1^4 - 3, \\ \gamma_{12} = E\{X\dot{X}^2\}/(\sigma_1\sigma_2^2), & \gamma_{22} = E\{X^2\dot{X}^2\}/(\sigma_1^2\sigma_2^2) - 1, \\ \gamma_{04} = E\{\dot{X}^4\}/\sigma_2^4 - 3. \end{cases}$$

In particular, if the marginal distribution of X is normal (as appears true if X denotes *velocity* in turbulent motion),

$$\gamma_{30} = \gamma_{40} = 0,$$

and (17) reduces to

$$\frac{1}{\pi}\frac{\sigma_2}{\sigma_1}\left\{1+\frac{\gamma_{12}}{2}\frac{x}{\sigma_1}+\frac{\gamma_{22}}{4}\left(\frac{x^2}{\sigma_1^2}-1\right)-\frac{\gamma_{04}}{24}\right\}\exp\left\{-\frac{1}{2}\frac{x^2}{\sigma_1^2}\right\}. \qquad (18)$$

For the average density of zeros, we put further $x = 0$.

6·5   Multivariate and multidimensional stationary processes

The extension of the correlation theory of stationary processes to multivariate processes does not present any particular new difficulties. The structure of the theory is evident if we relate the theory of the (column) vector process $\mathbf{X}(t) \equiv \{X_i(t)\}$ with that of the single process, considered for *arbitrary* λ_i,

$$X(t,\boldsymbol{\lambda}) \equiv \boldsymbol{\lambda}'\mathbf{X}(t) = \Sigma_i \lambda_i X_i(t).$$

This representation is adequate for second-order moment properties if $\boldsymbol{\lambda}$ is complex, allowing the possibility of separating the distinct terms $w_{ij}(\tau)$ and $w_{ji}(\tau)$ in the autocovariance of $X(t,\boldsymbol{\lambda})$. Such terms are not identical, as, even for real processes $X_i(t)$,

$$w_{ij}(\tau) \equiv E\{X_i(t)\,X_j(t+\tau)\}, \quad w_{ji}(\tau) \equiv E\{X_j(t)\,X_i(t+\tau)\},$$

when $E\{\mathbf{X}(t)\} = 0$. In the subsequent notation, it is convenient for complex vectors and matrices to let a transposed vector or matrix, still denoted by a dash, take on the conjugate values to the untransposed vector or matrix (e.g. we now write

$$\boldsymbol{\lambda}' \equiv (\lambda_1^*, \lambda_2^*, \dots)).$$

With this convention, we now define the covariance matrix

$$\mathbf{V}(\tau) \equiv E\{\mathbf{X}(t+\tau)\mathbf{X}'(t)\}, \qquad (1)$$

(it is convenient in the present section, involving complex quantities, to define $\mathbf{V}(\tau)$ equivalent to $\mathbf{V}(-\tau)$ of equation (43), §5·2).

From the general form of Khintchine's theorem for complex $X(t)$,

$$w(\tau,\boldsymbol{\lambda}) \equiv E\{X(t+\tau,\boldsymbol{\lambda})\,X^*(t,\boldsymbol{\lambda})\} = \int_{-\infty}^{\infty} e^{i\tau\omega}\,dW(\omega,\boldsymbol{\lambda}). \qquad (2)$$

But since also we may write, from equation (1) of § 6·2,

$$\mathbf{X}(t) = \int_{-\infty}^{\infty} e^{it\omega} d\mathbf{Z}(\omega),$$

$$X(t, \boldsymbol{\lambda}) = \boldsymbol{\lambda}'\mathbf{X}(t) = \int_{-\infty}^{\infty} e^{it\omega} \boldsymbol{\lambda}' d\mathbf{Z}(\omega), \tag{3}$$

where each item of $\mathbf{Z}(\omega)$, and also $\boldsymbol{\lambda}'\mathbf{Z}(\omega)$ for arbitrary $\boldsymbol{\lambda}$, are all orthogonal processes and hence $\mathbf{Z}(\omega)$ a vector orthogonal process, we obtain

$$\mathbf{V}(\tau) = \int_{-\infty}^{\infty} e^{i\tau\omega} d\mathbf{W}(\omega). \tag{4}$$

Hence $\qquad w(\tau, \boldsymbol{\lambda}) = \boldsymbol{\lambda}'\mathbf{V}(\tau)\boldsymbol{\lambda} = \int_{-\infty}^{\infty} e^{i\tau\omega} d[\boldsymbol{\lambda}'\mathbf{W}(\omega)\boldsymbol{\lambda}], \tag{5}$

and the Hermitian form $\boldsymbol{\lambda}'\mathbf{W}(\omega)\boldsymbol{\lambda}$ must consequently from (2) be the never-decreasing function $W(\omega, \boldsymbol{\lambda})$. This property of $\mathbf{W}(\omega)$ was first given by Cramér (1940).

The correlation properties of linear processes, which, as we saw in § 6·3, provide continuous spectra, were extended to vector processes at the end of § 5·2, so that we need merely note the associated spectral formulae. For the linear process

$$\mathbf{X}(t) = \int_{-\infty}^{\infty} \mathbf{G}(t-u) d\mathbf{Y}(u) \quad (\mathbf{G}(u) = 0 \text{ for } u < 0), \tag{6}$$

where $\mathbf{Y}(u)$ is additive, we obtain for the spectral density matrix $\mathbf{S}(\omega)$, say, the result

$$\mathbf{S}(\omega) = \frac{1}{2\pi} \mathbf{H}(\omega) \mathbf{V}\mathbf{H}'(\omega), \tag{7}$$

as the Fourier inverse of the formula

$$\mathbf{V}(\tau) = \int_{-\infty}^{\infty} \mathbf{G}(v) \mathbf{V}\mathbf{G}'(v-\tau) dv, \tag{8}$$

In (7),

$$\mathbf{H}(\omega) = \int_{-\infty}^{\infty} e^{-iv\omega} \mathbf{G}(v) dv.$$

The relations (7) and (8) also hold of course if $\mathbf{Y}(u)$ is orthogonal but not additive, and further if $\mathbf{G}(u) \neq 0$ for $u < 0$.

For the multivariate autoregressive equation of § 5·2,

$$d\mathbf{X}(t) + \mathbf{R}\mathbf{X}(t) dt = d\mathbf{Y}(t),$$

we obtain $\mathbf{H}(\omega) = (i\omega\mathbf{I} + \mathbf{R})^{-1}$, so that

$$S(\omega) = \frac{1}{2\pi}(\mathbf{R} + i\omega\mathbf{I})^{-1}\mathbf{V}(\mathbf{R}' - i\omega\mathbf{I})^{-1}. \tag{9}$$

To illustrate this rather condensed formula, consider the bivariate case ($\mathbf{R}$ real)

$$\mathbf{R} \equiv \begin{pmatrix} \alpha_{11} & \alpha_{12} \\ \alpha_{21} & \alpha_{22} \end{pmatrix}.$$

Then $(\mathbf{R} + i\omega\mathbf{I})^{-1}$

$$\equiv \begin{pmatrix} \alpha_{11} + i\omega & \alpha_{12} \\ \alpha_{21} & \alpha_{22} + i\omega \end{pmatrix}^{-1} = \frac{1}{\Delta} \begin{pmatrix} \alpha_{22} + i\omega & -\alpha_{12} \\ -\alpha_{21} & \alpha_{11} + i\omega \end{pmatrix},$$

where $\qquad \Delta = \alpha_{11}\alpha_{22} - \alpha_{12}\alpha_{21} + i\omega(\alpha_{22} + \alpha_{11}) - \omega^2,$

and

$$S(\omega) = \frac{\begin{pmatrix} \alpha_{22} + i\omega & -\alpha_{12} \\ -\alpha_{21} & \alpha_{11} + i\omega \end{pmatrix} \begin{pmatrix} \sigma_1^2 & \rho\sigma_1\sigma_2 \\ \rho\sigma_1\sigma_2 & \sigma_2^2 \end{pmatrix} \begin{pmatrix} \alpha_{22} - i\omega & -\alpha_{21} \\ -\alpha_{12} & \alpha_{11} - i\omega \end{pmatrix}}{2\pi\{(\alpha_{11}\alpha_{22} - \alpha_{12}\alpha_{21} - \omega^2)^2 + \omega^2(\alpha_{11} + \alpha_{22})^2\}}$$

$$= \frac{\begin{pmatrix} \beta_{11} & \beta_{12} \\ \beta_{21} & \beta_{22} \end{pmatrix}}{2\pi\{(\alpha_{11}\alpha_{22} - \alpha_{12}\alpha_{21} - \omega^2)^2 + \omega^2(\alpha_{11} + \alpha_{22})^2\}}, \tag{10}$$

where

$$\begin{cases} \beta_{11} = \sigma_1^2(\alpha_{22}^2 + \omega^2) - 2\rho\sigma_1\sigma_2\alpha_{12}\alpha_{22} + \sigma_2^2\alpha_{12}^2, \\ \beta_{12} = \beta_{21}^* = -\sigma_1^2\alpha_{21}\alpha_{22} + \rho\sigma_1\sigma_2(\alpha_{12}\alpha_{21} + \alpha_{11}\alpha_{22} + \omega^2) - \sigma_2^2\alpha_{11}\alpha_{12} \\ \qquad\qquad - i\omega[\sigma_1^2\alpha_{21} + \rho\sigma_1\sigma_2(\alpha_{22} - \alpha_{11}) - \sigma_2^2\alpha_{12}], \\ \beta_{22} = \sigma_1^2\alpha_{21}^2 - 2\rho\sigma_1\sigma_2\alpha_{11}\alpha_{21} + \sigma_2^2(\alpha_{11}^2 + \omega^2). \end{cases}$$

Multidimensional processes. Although a multidimensional process or 'field' $X(\mathbf{u})$, say, must not be confused with a vector process $\mathbf{X}(t)$, the method of studying the properties of a univariate process $X(t)$ by means of the properties of the vector random variable $\mathbf{X}(t_i) \equiv (X(t_1), X(t_2), ..., X(t_n))$, the set $t_1, ..., t_n$ being arbitrary, may evidently be extended to multidimensional processes, and implies a relation between the properties of, say, $X(u, v)$ and $X(u, v_j)$, the set v_j being arbitrary, as well as between $X(u, v)$ and the matrix or tensor variable $X(u_i, u_j)$.

The correlational properties of stationary processes developed in §§ 6·1 and 6·11 extend to multidimensional processes similarly to the extension of characteristic functions to more than one variable. For example, in the case of a two-dimensional stationary process (if the two dimensions are spatial, this is sometimes called a spatially homogeneous process), we write

$$\sigma^2 \rho(\tau, \nu) = E\{X^*(t, u)\, X(t + \tau, u + \nu)\},$$

where $\quad E\{X(t, u)\} = 0, \quad E\{X^*(t, u)\, X(t, u)\} = \sigma^2,$

and then have the relations

$$\rho(\tau, \nu) = \int e^{i(\tau\omega + \nu\mu)}\, dF(\omega, \mu), \tag{11}$$

$$X(t, u) = \int e^{i(t\omega + u\mu)}\, dZ(\omega, \mu), \tag{12}$$

where $\quad E\{Z(\omega, \mu)\, Z^*(\omega, \mu)\} = \sigma^2 F(\omega, \mu),$

and the integration is over the whole two-dimensional domain. These are the analogues of formulae (4) of § 6·1 and (1) of § 6·2, and $Z(\omega, \mu)$ is doubly orthogonal in ω and μ, i.e. there is no contribution to $E\left\{\int dZ(\omega, \mu) \int dZ^*(u, v)\right\}$ except along $\omega = u$, $\mu = v$.

In the statistical theory of turbulence the velocity at a fixed point $\mathbf{r} \equiv (x, y, z)$ is a random *vector* quantity $\mathbf{U} \equiv (U_1, U_2, U_3)$, depending on the space coordinates $\mathbf{r}$ and on t; it is thus both multivariate and multidimensional, and is an example of a random field vector changing with time. The further generalization of (11) and (12) to cover vector processes is evident.

6·51 Isotropy and other special conditions.

Any assumptions of symmetry in the case of field quantities impose special restrictions on the correlation and spectral functions. To take a simple case, a stationary process depending on two space coordinates x and y will, if circular symmetry is assumed, have a correlation coefficient depending only on r, the interval

$$\sqrt{\{(\Delta x)^2 + (\Delta y)^2\}}$$

between two points. Thus formula (11) of § 6·5 becomes

$$\rho(r) = \int e^{i(\omega \Delta x + \mu \Delta y)}\, dF(\omega, \mu). \tag{1}$$

When the derivative $f(\omega, \mu) = \partial^2 F/\partial \omega \, \partial \mu$ exists, the inverse formula becomes

$$f(\omega, \mu) = \frac{1}{(2\pi)^2} \int_{-\infty}^{\infty} \int_{-\infty}^{\infty} e^{-i\omega \Delta x - i\mu \Delta y} \rho(r) \, d\Delta x \, d\Delta y,$$

or, changing to polar coordinates, we have

$$f(\omega, \mu) = \frac{1}{(2\pi)^2} \int_{0}^{\infty} \int_{0}^{2\pi} e^{-ir(\omega \cos \theta + \mu \sin \theta)} \rho(r) \, r \, dr \, d\theta$$

$$= \frac{1}{2\pi} \int_{0}^{\infty} J_0(r\eta) \rho(r) \, r \, dr, \tag{2}$$

where $\eta = \sqrt{(\omega^2 + \mu^2)}$, whence it follows that $f(\omega, \mu) = g(\eta)$, a function of η only. Substituting in (1), we obtain the formula inverse to (2),

$$\rho(r) = 2\pi \int_{0}^{\infty} J_0(r\eta) g(\eta) \, \eta \, d\eta. \tag{3}$$

The function $\rho(r) = e^{-r}$ has been used as a possible correlation model for the varying fertility of an area of land from point to point. It might be noted that it does not appear possible to set up any isotropic analogue of the one-dimensional linear Markov process leading to the correlation function e^{-t} $(t > 0)$, in one time dimension, though two- and three-dimensional Markov processes with uniquely defined coordinate axes (corresponding, for example, to successive rows or layers of a crystal structure) do exist. This leads to some difficulty in constructing realizations of isotropic processes. In the case of isotropic point processes in two or three dimensions, arising, for example, in ecological models of plant communities or in models of stellar populations, one isotropic model which has been used is a purely random population of 'nuclei', surrounded by 'satellites' with an isotropic distribution of distances from their respective nuclei. Obviously, however, this is a model only appropriate in suitable contexts (cf. also §§ 6·52, 9·4 and 9·41).

Turning now to vector processes, we shall derive the form of the covariance matrix for a vector process in a space of three dimensions x_1, x_2, x_3. This has its most familiar application in isotropic turbulence models, where the vector variable refers to the velocity at a given point; we shall consequently denote the variable by $\mathbf{U}(\mathbf{x}; t) \equiv U_i(x_1, x_2, x_3; t)$ $(i = 1, 2, 3)$. The covariance matrix is assumed 'stationary' with respect to $\mathbf{x}$; its dependence on the times $t, t + \tau$ is left arbitrary, and need not be indicated. Spherical spatial symmetry then gives certain conditions on the covariance matrix, denoted by $V(\xi)$ for two points separated by the (column) vector $\xi \equiv (\xi_i)$. If $m \equiv (m_i)$, $n \equiv (n_i)$ are unit (column) vectors in arbitrary directions, the covariance between the velocities in these directions at the two points is $m'V(\xi)\, n$. Now the assumption of isotropy implies that this covariance depends only on the mutual configuration of ξ, m and n, and not on their orientations with respect to the Cartesian axes chosen. It follows (cf. Robertson, 1940) that the bilinear form $m'V(\xi)\, n$ must be of the general invariant form

$$m'V(\xi)\, n = A(\xi)\, m'\xi\xi'n + B(\xi)\, m'n + C(\xi)\, |\,\xi, m, n\,|, \qquad (4)$$

where $|\,\xi, m, n\,|$ denotes the determinant formed by the components of the three vectors, and $\xi = \sqrt{(\xi'\xi)}$. The assumption of isotropy also assumes, however, that reflection of the configuration does not affect the value of $m'V(\xi)\, n$ (i.e. the invariance of this scalar holds for the full rotation group, including improper as well as proper rotations). This excludes the last term in (4), as this changes sign on reflection, so that

$$m'V(\xi)\, n = A(\xi)\, m'\xi\xi'n + B(\xi)\, m'n, \qquad (5)$$

and

$$V(\xi) = A(\xi)\, \xi\xi' + B(\xi)\, I. \qquad (6)$$

Equation (6) gives the general form of the covariance matrix (or correlation matrix, as isotropy also implies that the variance of $m'U$ is independent of m; in fact from (6), $V(0) = B(0)\, I$).

A further relation is obtained if the velocities refer to an incompressible fluid, owing to the equation of continuity. The isotropic covariance matrix (6) (which is a second-order tensor) is then called solenoidal. With this further condition, which implies

$$\frac{\partial U_1(x)}{\partial x_1} + \frac{\partial U_2(x)}{\partial x_2} + \frac{\partial U_3(x)}{\partial x_3} = 0, \qquad (7)$$

we obtain

$$\left(\frac{\partial}{\partial x}\right)' V(\xi) = \left(\frac{\partial}{\partial \xi}\right)' V(\xi) = 0,$$

where $(\partial/\partial x)$ is the column vector $(\partial/\partial x_i)$, and $\xi = x - y$. Substituting for $V(\xi)$ from (6), and noting that

$$\frac{\partial A(\xi)}{\partial \xi_i} = \frac{\partial \xi}{\partial \xi_i}\, \frac{\partial A(\xi)}{\partial \xi} = \frac{\xi_i}{\xi}\, \frac{\partial A(\xi)}{\partial \xi},$$

we obtain
$$\left[4A(\xi) + \frac{\xi \partial A(\xi)}{\partial \xi} + \frac{1}{\xi} \frac{\partial B(\xi)}{\partial \xi} \right] \xi' = 0,$$

whence
$$4A(\xi) + \frac{\xi \partial A(\xi)}{\partial \xi} + \frac{1}{\xi} \frac{\partial B(\xi)}{\partial \xi} = 0. \tag{8}$$

For the isotropic forms of other moment tensors reference should be made to Robertson (1940).

In turbulence theory it is usual to define equivalent scalar functions $f(\xi)$, $g(\xi)$, in terms of which (6) is written

$$\mathbf{V}(\xi) = \frac{f - g}{\xi^2} \xi \xi' + g\mathbf{I}. \tag{9}$$

This implies $g \equiv B$, $A \equiv (f - g)/\xi^2$, so that (8) becomes after reduction
$$g = f + \frac{\xi}{2} \frac{\partial f}{\partial \xi}. \tag{10}$$

We shall also denote the trace $f + 2g$ of the matrix $\mathbf{V}$ in (9) by $R(\xi)$.

Consider now the corresponding forms for the spectral functions. We have in general

$$\mathbf{V}(\xi) = \int e^{i\xi'\omega} d\mathbf{F}(\omega); \tag{11}$$

assuming that the spectral function densities $\mathbf{f}(\omega) = d\mathbf{F}(\omega)/d\omega$ exist, we note further that the matrix or tensor spectral density

$$\mathbf{f}(\omega) = \frac{1}{(2\pi)^3} \int e^{-i\xi'\omega} \mathbf{V}(\xi)\, d\xi \quad (d\xi \equiv d\xi_1 d\xi_2 d\xi_3). \tag{12}$$

We then easily obtain, if we replace $\mathbf{V}(\xi)$ by the expression in (6),

$$\mathbf{f}(\omega) = \kappa(\omega)\, \omega\omega' + \eta(\omega)\, \mathbf{I} \quad (\omega = \sqrt{(\omega'\omega)}), \tag{13}$$

where, if $\alpha(\omega)$, $\beta(\omega)$ are the (three-dimensional) Fourier transforms of $A(\xi)$, $B(\xi)$ corresponding to equation (11),

$$\left. \begin{aligned} \kappa(\omega) &= \frac{1}{\omega^3} \frac{d\alpha}{d\omega} - \frac{1}{\omega^2} \frac{d^2\alpha}{d\omega^2}, \\ \eta(\omega) &= \beta - \frac{1}{\omega} \frac{d\alpha}{d\omega}. \end{aligned} \right\} \tag{14}$$

The further condition that $\mathbf{V}(\xi)$ is solenoidal as well as isotropic implies in the ω-space that $\mathbf{f}(\omega)$ in (13) also satisfies the relation

$\boldsymbol{\omega}'\mathbf{f}(\omega) = 0$, whence

$$\kappa(\omega)\,\omega^2 + \eta(\omega) = 0. \tag{15}$$

Relation (13) then becomes

$$\mathbf{f}(\omega) = \eta(\omega)\left(\mathbf{I} - \frac{\boldsymbol{\omega}\boldsymbol{\omega}'}{\omega^2}\right). \tag{16}$$

We define new functions in (13) analogously to (9) and write

$$\mathbf{f}(\omega) = \frac{a-b}{\omega^2}\boldsymbol{\omega}\boldsymbol{\omega}' + b\mathbf{I} \tag{17}$$

with $b \equiv \eta$ and $\omega^2\kappa = a - b$, whence (15) becomes

$$a(\omega) = 0. \tag{18}$$

Thus the function $a(\omega)$ vanishes for incompressible fluids.

Returning to the relations (14), and making use of the reduced formula for the three-dimensional transform of a scalar function like $A(\xi)$, viz.

$$\alpha(\omega) = \frac{4\pi}{(2\pi)^3}\int_0^\infty \frac{\sin\xi\omega}{\xi\omega}\,\xi^2 A(\xi)\,d\xi, \tag{19}$$

we obtain the formulae

$$\left.\begin{aligned} a(\omega) &= \frac{1}{2\pi^2\omega^2}\int_0^\infty \left[f\xi\omega\sin\xi\omega - 2(f-g)\left(\frac{\sin\xi\omega}{\xi\omega} - \cos\xi\omega\right)\right]d\xi, \\ b(\omega) &= \frac{1}{2\pi^2\omega^2}\int_0^\infty \left[g\xi\omega\sin\xi\omega + (f-g)\left(\frac{\sin\xi\omega}{\xi\omega} - \cos\xi\omega\right)\right]d\xi. \end{aligned}\right\} \tag{20}$$

Notice that $F(\omega)$, the trace $a + 2b$ of $\mathbf{f}(\omega)$, is related to $R(\xi)$ by

$$F(\omega) = \frac{1}{2\pi^2}\int_0^\infty \frac{\sin\xi\omega}{\xi\omega}\,\xi^2 R(\xi)\,d\xi. \tag{21}$$

Longitudinal and transverse components. If we consider the Fourier relation

$$\mathbf{U}(\boldsymbol{\xi}) = \int e^{i\boldsymbol{\xi}'\boldsymbol{\omega}}\,d\mathbf{Z}(\boldsymbol{\omega}) \tag{22}$$

for the random vector quantity $\mathbf{U}(\boldsymbol{\xi}) \equiv U_i(\boldsymbol{\xi})$ $(i = 1, 2, 3)$ itself, we may resolve $d\mathbf{Z}(\boldsymbol{\omega})$ into components along and perpendicular to $\boldsymbol{\omega}$, viz.

$$d\mathbf{Z}(\boldsymbol{\omega}) \equiv (\boldsymbol{\omega}/\omega)\,dZ_\omega(\boldsymbol{\omega}) + d\mathbf{T}(\boldsymbol{\omega}), \tag{23}$$

where, if $\boldsymbol{\alpha}$ and $\boldsymbol{\beta}$ are orthogonal unit vectors with $\boldsymbol{\omega}/\omega$, such that the corresponding components $dZ_\alpha(\boldsymbol{\omega})$, $dZ_\beta(\boldsymbol{\omega})$ are uncorrelated,

$$d\mathbf{T}(\boldsymbol{\omega}) \equiv \boldsymbol{\alpha}\,dZ_\alpha(\boldsymbol{\omega}) + \boldsymbol{\beta}\,dZ_\beta(\boldsymbol{\omega}). \tag{24}$$

The isotropy condition gives

$$E\{d\mathbf{T}(\boldsymbol{\omega})\,d\mathbf{T}'(\boldsymbol{\omega})\} \equiv (\boldsymbol{\alpha}\boldsymbol{\alpha}' + \boldsymbol{\beta}\boldsymbol{\beta}')\,E\{dZ_\alpha(\boldsymbol{\omega})\,dZ_\alpha^*(\boldsymbol{\omega})\}, \qquad (25)$$

and making use of the matrix relation

$$\boldsymbol{\alpha}\boldsymbol{\alpha}' + \boldsymbol{\beta}\boldsymbol{\beta}' + \boldsymbol{\omega}\boldsymbol{\omega}'/\omega^2 = \mathbf{I},$$

we note that the isotropy condition (25) implies further that

$$E\{d\mathbf{Z}(\boldsymbol{\omega})\,d\mathbf{Z}'(\boldsymbol{\omega})\} \equiv \mathbf{f}(\boldsymbol{\omega})\,d\boldsymbol{\omega}$$
$$= (\boldsymbol{\omega}\boldsymbol{\omega}'/\omega^2)\,(E\{dZ_\omega(\boldsymbol{\omega})\,dZ_\omega^*(\boldsymbol{\omega})\} - E\{dZ_\alpha(\boldsymbol{\omega})\,dZ_\alpha^*(\boldsymbol{\omega})\})$$
$$+ \mathbf{I}E\{dZ_\alpha(\boldsymbol{\omega})\,dZ_\alpha^*(\boldsymbol{\omega})\},$$

with $dZ_\omega(\boldsymbol{\omega})$ and $d\mathbf{T}(\boldsymbol{\omega})$ necessarily uncorrelated. This provides the identification

$$\left.\begin{aligned}E\{dZ_\omega(\boldsymbol{\omega})\,dZ_\omega^*(\boldsymbol{\omega})\} &\equiv a(\omega)\,d\boldsymbol{\omega},\\ E\{dZ_\alpha(\boldsymbol{\omega})\,dZ_\alpha^*(\boldsymbol{\omega})\} &\equiv E\{dZ_\beta(\boldsymbol{\omega})\,dZ_\beta^*(\boldsymbol{\omega})\} \equiv b(\omega)\,d\boldsymbol{\omega}.\end{aligned}\right\} \qquad (26)$$

In turbulence theory where $\mathbf{U}(\boldsymbol{\xi})$ is the velocity vector, we sometimes require to consider derived quantities such as the *vorticity* vector $\mathbf{v}$, where

$$v_i \equiv \frac{\partial U_j}{\partial x_k} - \frac{\partial U_k}{\partial x_j}.$$

Correspondingly, if $d\mathbf{W}(\boldsymbol{\omega})$ is the corresponding random Fourier component, we have

$$d\mathbf{W}(\boldsymbol{\omega}) = i\boldsymbol{\omega} \times d\mathbf{Z}(\boldsymbol{\omega}),$$

where $\times$ denotes a vector product. By definition $d\mathbf{W}(\boldsymbol{\omega})$ has no longitudinal component whether or not the fluid is incompressible. If we have isotropy, we have also the simple relation

$$E\{d\mathbf{W}(\boldsymbol{\omega})\,d\mathbf{W}'(\boldsymbol{\omega})\} = \omega^2 b(\omega)\left(\mathbf{I} - \frac{\boldsymbol{\omega}\boldsymbol{\omega}'}{\omega^2}\right)d\boldsymbol{\omega}. \qquad (27)$$

6·52 Two-dimensional point and line processes.

For a two-dimensional stationary *point* process such that

$$E\{dN(x,y)\} = \lambda\,dA \quad \text{and} \quad E\{dN(x,y)\,dN(x',y')\}$$

has the form $\quad \{\lambda^2 + w(x-x',y-y')\}\,dA\,dA',$ where

$$dN(x,y) = N(x+dx,y+dy) - N(x,y), \quad dA = dx\,dy.$$

we define

$$F(\omega_x, \omega_y) = \frac{1}{(2\pi)^2} \int_{-\infty}^{\infty} \int_{-\infty}^{\infty} e^{-i(x\omega_x + y\omega_y)} w(x, y) \, dx \, dy$$

and

$$g(\omega_x, \omega_y) = 4\pi^2 f(\omega_x, \omega_y).$$

There are obvious analogues of formulae (2) and (3) of § 6·51 in the isotropic case.

Example. As an example we consider a clustering model somewhat analogously to Example 2 of the previous § 6·13, except that now offspring may be distributed spatially around a parent, and for simplicity no 'second or further generation' distributed about each offspring will be introduced. The distribution p_r will still represent the distribution of number of offspring per parent, but each offspring in a family is distributed independently about its parent.

The contributions to the covariance density are compiled as before, but there are now no terms beyond the first convolution of the spatial distribution function. If this function is denoted by the density $f(\mathbf{R})$, where $\mathbf{R} = (x, y)$, and f_2 is the density for a *difference* $\mathbf{R} - \mathbf{R}'$, then

$$w(x, y) = \lambda_0 \{ [f(\mathbf{R}) + f(-\mathbf{R})] E\{r\} + f_2(\mathbf{R}) E\{r(r-1)\} \},$$

as for r offspring there are $\frac{1}{2}r(r-1)$ distances between pairs. Hence

$$g(\omega_x, \omega_y) = \lambda \bigg\{ 1 + [C(\omega_x, \omega_y) + C(-\omega_x, -\omega_y)] \frac{E\{r\}}{E\{r+1\}}$$

$$+ [C(\omega_x, \omega_y) C(-\omega_x, -\omega_y)] \frac{E\{r(r-1)\}}{E\{r+1\}} \bigg\}, \quad (1)$$

where $C(\theta, \phi)$ is the characteristic function of $f(\mathbf{R})$. Suppose, for example, that

$$C(\theta, \phi) = e^{-\frac{1}{2}\sigma^2(\theta^2 + \phi^2)}. \quad (2)$$

Then

$$g(\omega_x, \omega_y) = \lambda \bigg\{ 1 + \frac{2m_0 e^{-\frac{1}{2}\sigma^2(\omega_x^2 + \omega_y^2)}}{m_0 + 1} + \frac{(m_0^2 + v_0 - m_0) e^{-\sigma^2(\omega_x^2 + \omega_y^2)}}{m_0 + 1} \bigg\}, \quad (3)$$

where $m_0 = E\{r\}, v = E\{r^2\} - m_0^2$. In particular, if p_r is Poisson, so that $v = m_0$,

$$g(\omega_x, \omega_y) = \lambda \bigg\{ m_0 + \frac{(1 + m_0 e^{-\frac{1}{2}\sigma^2(\omega_x^2 + \omega_y^2)})^2}{m_0 + 1} \bigg\}. \quad (4)$$

Note that the mean number m per family (including the parent) is $1 + m_0$.

Line processes. It should be noted that in more than one dimension, as with quantitative variable distributions, the number of possibilities for point processes increases. Consider, in place of the assumption that the probability of an individual in $dA = dx\, dy$ is $\lambda\, dA$, the situation where 'particles' are moving about but preserving some permanence in time. For simplicity we consider particles on a line (e.g. vehicles on a road). Then there may be a valid one-dimensional point process in the space coordinate x or in time t, but not a two-dimensional point process, as in the two-dimensional space the elements are lines.

For these *line processes* the following relations are assumed.
  (i) First-order densities:

   (a) Space $f_x(x, t)\, dx = E\{d_x N(x, t)\}$

   (b) Time $f_t(x, t)\, dt = E\{d_t N(x, t)\}$

 (ii) Second-order densities:

   (a) Space $f_{xx'}(x, t; x', t)\, dx\, dx' = E\{d_x N(x, t)\, d_{x'} N(x', t)\}$

   (b) Time $f_{tt'}(x, t; x, t')\, dt\, dt' = E\{d_t N(x, t)\, d_{t'} N(x, t')\}$

Theoretically more than one kind of relation between the space and time densities could be envisaged, but for definiteness and with a view to applications we suppose that any particle has a definite velocity U at any moment. Then we have a two-dimensional point process in the phase-space variables x and u proceeding in time t, or (perhaps less naturally) one in t and u extending over x, so that we have the further densities:
  (i) First-order

   (c) Space-velocity $f_{x,\, u}(x, u, t)\, dx\, du = E\{d_{x,\, u} N(x, u, t)\}$

   (d) Time-velocity $f_{t,\, u}(x, u, t)\, dt\, du = E\{d_{t,\, u} N(x, u, t)\}$

 (ii) Second-order

   (c) Space-velocity $f_{xx',\, uu'}(x, u, t; x', u', t)\, dx\, dx'\, du\, du'$
$$= E\{d_{x,\, u} N(x, u, t)\, d_{x',\, u'} N(x', u', t)\}$$

   (d) Time-velocity $f_{tt',\, uu'}(x, u, t; x, u', t')\, dt\, dt'\, du\, du'$
$$= E\{d_{t,} N_u(x, u, t)\, d_{t',\, u'} N(x, u', t')\}$$

It is of course possible to define $f_{xx',\, uu'}$ at a pair of time instants t, t' instead of the same time t, and similarly for $f_{tt',\, uu'}$ at x, x'; but the definitions as listed above will be used unless otherwise indicated.

Various relations obviously hold, e.g.

$$f_x = \int_{-\infty}^{\infty} f_{x,u}\, du,$$

$$f_t = \int_{-\infty}^{\infty} f_{t,u}\, du;$$

but note also

$$f_{t,u} = |u| f_{x,u},$$

$$f_t = \int_{-\infty}^{\infty} |u| f_{x,u}\, du = \overline{|u(x)|} f_x,$$

where $\overline{|u(x)|}$ is the space-conditional mean (absolute) velocity at time t. Moreover,

$$|u| f_{t,u}\, du = u^2 f_{x,u}\, du$$

so that the time-conditional mean (absolute) velocity at x is

$$\overline{|u(t)|} = \int_{-\infty}^{\infty} u^2 f_{x,u}\, du / [\overline{|u(x)|} f_x] \geqslant \overline{|u(x)|}.$$

For the second-order densities,

$$f_{u',uu'} = |uu'| f_{xx',uu'}.$$

With the further assumption that any particle has a definite acceleration A and conservation holds for the particles, various continuity equations hold. Let $h(a|x,u,t)$ be the space-velocity conditional density function for a, where any non-zero

$$d_{x,u} N(x,u,t+dt)$$

must have arisen from some non-zero

$$d_{x,u} N(x-u\,dt,\, u-a\,dt,\, t)$$

so that

$$d_{x,u} N(x,u,t+dt) = \int_{-\infty}^{\infty} h(a|x,u,t)\, dN_{x,u}(x-u\,dt,\, u-a\,dt,\, t).$$

Averaging over $N(x,u,t)$ provides the continuity equation in phase-space, viz.

$$\frac{\partial f_{x,u}}{\partial t} + u\frac{\partial f_{x,u}}{\partial x} + \frac{\partial}{\partial u}(\bar{a} f_{x,u}) = 0, \tag{5}$$

where a is the space-velocity conditional mean acceleration. On integration with respect to u this provides the more familiar continuity equation in ordinary space

$$\frac{\partial f_x}{\partial t} + \frac{\partial}{\partial x}\{\bar{u}(x)f_x\} = 0, \tag{6}$$

where $\bar{u}(x)$ is the space-conditional mean velocity.

In the theory of traffic flow Lighthill and Whitham (1955) have made use of equation (6) combined with an empirical relation between f_x and f_t, this relation being roughly parabolic in shape, owing to $\bar{u}(x)$ being found to be roughly linear in its dependence on f_x (constraints between vehicles depressing $\bar{u}(x)$ as f_x increases). In more recent developments the situation has been considered in finer detail as a stochastic process, analogously to investigations at the molecular level in hydrodynamics (see, for example, Newell (1955), Bartlett (1957), Miller (1962), Prigogine *et al.* (1963); cf. also § 9·23).

The spectrum of a random line process. In certain physical problems (for example, in the study of two-dimensional sheets of fibres or filaments) it may be relevant to consider the generalized spectra of line processes, as we have already done for point processes. Before, however, considering the limiting case of a line process, we note the general formula for the spectrum of the process (assumed 'stationary', i.e. homogeneous, in $\mathbf{r}$)

$$X(\mathbf{r}) = \int \xi(\mathbf{s} - \mathbf{r})\,dN(\mathbf{s})$$

generated by linear superposition from a plane point process $N(\mathbf{r})$ viz.

$$\sigma_x^2 f_x(\boldsymbol{\omega}) = h(\boldsymbol{\omega})\,h^*(\boldsymbol{\omega})\,g_N(\boldsymbol{\omega}), \tag{7}$$

where $$h(\boldsymbol{\omega}) = \int e^{-i(\omega_x x + \omega_y y)}\,\xi(\mathbf{r})\,d\mathbf{r}$$

and $g_N(\boldsymbol{\omega})$ is the complete (generalized) spectrum of $N(\mathbf{r})$. For $N(\mathbf{r})$ a Poisson process in the plane, $g_N(\boldsymbol{\omega})$ is constant over $\boldsymbol{\omega} \equiv (\omega_x, \omega_y)$.

If the $\xi(\mathbf{r})$ are also random independently of $N(\mathbf{r})$, with an independent sample function for each jump in $N(\mathbf{r})$, then

$$E\{X(\mathbf{r})\,X(\mathbf{r}')\} = \iint_{\mathbf{s}+\mathbf{s}'} E_\xi\{\xi(\mathbf{s}-\mathbf{r})\}\, E_\xi\{\xi(\mathbf{s}'-\mathbf{r}')\}\, E\{dN(\mathbf{s})\,dN(\mathbf{s}')\}$$

$$+ \int_{\mathbf{s}} E_\xi\{\xi(\mathbf{s}-\mathbf{r})\,\xi(\mathbf{s}-\mathbf{r}')\}\, E\{dN(\mathbf{s})\}$$

Taking Fourier transforms of both sides, and including for convenience in (7) the delta-function components at $\boldsymbol{\omega}=0$ due to the retention of the mean of $X(\mathbf{r})$, we obtain

$$\sigma_x^2 f_x(\boldsymbol{\omega}) = E_\xi\{h(\boldsymbol{\omega})\}\, E_\xi\{h^*(\boldsymbol{\omega})\}\,\{g_N(\boldsymbol{\omega})-\lambda\} + \lambda E_\xi\{h(\boldsymbol{\omega})\,h^*(\boldsymbol{\omega})\}, \qquad (8)$$

where $E\{dN(\mathbf{r})\}=\lambda\,d\mathbf{r}$.

Consider in particular a system of rectangles centred at the jump-points of $N(\mathbf{s})$, with length l and width d, and random orientation θ, say. Then for given θ

$$h(\boldsymbol{\omega}) = \int_{-\frac{1}{2}l}^{\frac{1}{2}l} \int_{-\frac{1}{2}d}^{\frac{1}{2}d} \exp\{-i\omega_x(u\cos\theta - v\sin\theta)$$
$$-i\omega_y(u\sin\theta + v\cos\theta)\}\,du\,dv$$

and
$$E_\theta\{h(\boldsymbol{\omega})\} = \int_{-\frac{1}{2}l}^{\frac{1}{2}l} \int_{-\frac{1}{2}d}^{\frac{1}{2}d} J_0(\omega t)\,du\,dv,$$

where $\omega^2 = \omega_x^2 + \omega_y^2$, $t^2 = u^2 + v^2$. In the case d very small, we may write

$$E_\theta\{h(\boldsymbol{\omega})\} \sim 2d \int_0^{\frac{1}{2}l} J_0(\omega u)\,du.$$

For the same case,

$$E_\theta\{h(\boldsymbol{\omega})\,h^*(\boldsymbol{\omega})\} \sim 2ld^2 \int_0^l (1 - z/l)\, J_0(\omega z)\,dz.$$

Hence
$$\sigma_x^2 f_x(\boldsymbol{\omega}) \sim 4d^2 \left(\int_0^{\frac{1}{2}l} J_0(\omega u)\,du \right)^2 \{g_N(\boldsymbol{\omega})-\lambda\}$$

$$+ 2\lambda l d^2 \int_0^l (1 - z/l)\, J_0(\omega z)\,dz. \qquad (9)$$

Notice that under the same assumption of d very small, it is immaterial whether overlap of the rectangles is allowed for or ignored.

In the case of a Poisson process $N(\mathbf{r})$, $g_N(\boldsymbol{\omega})=\lambda$ except for the discrete component at the origin due to the mean. Hence in this case, apart from the point $\omega = 0$.

$$\sigma_x^2 f_x(\omega) \sim 2\lambda l d^2 \int_0^l (1 - z/l) J_0(\omega z) \, dz. \tag{10}$$

For a true line process, for which $d \to 0$, this formula becomes exact, provided of course the measure is first re-scaled by dividing by d^2. In the further special case $l \to \infty$, we obtain

$$\sigma_x^2 f_x(\omega) \sim 2\lambda l d^2 / \omega. \tag{11}$$

For the case l small (or l finite and ω small)

$$\sigma_x^2 f_x(\omega) \sim \lambda l^2 d^2 \left(1 - \frac{\omega^2 l^2}{24} + \frac{\omega^4 l^4}{960} \cdots \right); \tag{12}$$

and is the average of this expression over l if l is random.

If, instead of $N(\mathbf{r})$ being purely random, a two-dimensional clustering model were more appropriate, the form (11) for large ω would remain valid, but the general expression (8) would include the first term in (9) if orientation were random (integration being to $\frac{1}{2}E(l)$ if l were random also). Further standardization of the spectrum might be an advantage in appropriate applications; for example (for constant l),

$$\frac{\sigma_x^2 f_x(\omega)}{4d^2 \left\{ \int_0^{\frac{1}{2}l} J_0(\omega z) \, dz \right\}^2} \sim \{g_N(\omega) - \lambda\} + \frac{\frac{1}{2}\lambda l \int_0^l (1 - z/l) J_0(\omega z) \, dz}{\left\{ \int_0^{\frac{1}{2}l} J_0(\omega z) \, dz \right\}^2}$$

$$= \{g_N(\omega) - \lambda\} + \lambda \left\{ 1 - \frac{\omega^4 l^4}{4608} + \cdots \right\}. \tag{13}$$

In the case of random l, however, the second part of (13) would no longer be independent of a term in ω^2.

Of course in some contexts the above spectrum in terms of Cartesian co-ordinates would not be very relevant. Thus for a random stream of vehicles with a distribution of velocities but free overtaking, an analysis of position (or time) as one co-ordinate, and of speed (i.e. the slope of the position-time graph for each vehicle) would be more convenient (see Bartlett, 1965).

6·53 Stationary processes on the circle and sphere. In the sampling theory of stationary time-series the artificial device

is sometimes used of imposing the constraint $x_{n+1} = x_1$. Again, Hannan (1960) discusses 'circular' processes where the discrete time sequence $1, 2, \ldots n$ may be considered to be situated on a circle. The theory of stationary processes in continuous space constrained to a circle or sphere is perhaps of particular relevance; for, although not so common in occurrence as the straightforward generalizations to two or three dimensions, there are obvious possible physical applications. We shall indicate the spectral theory for processes in a quantitative variable, though a similar theory would apply also to point processes.

For the circle at least the theory is rather simpler than the analogous infinite one-dimensional parameter, as the implied periodicity restricts expansions to Fourier series instead of integrals, and all spectra are discrete. Writing $E\{X(t)\} = 0$ and t varying from 0 to 2π for convenience, we may expand $X(t)$ in the form

$$X(t) = \sum_{s=-\infty}^{\infty} Z_s e^{ist}, \tag{1}$$

where

$$Z_s = \frac{1}{2\pi} \int_0^{2\pi} e^{-ist} X(t) \, dt, \tag{2}$$

and are uncorrelated when $X(t)$ is stationary.

The autocovariance function is

$$w(t' - t) = E\{X^*(t) X(t')\} = \sum_{s=-\infty}^{\infty} e^{is(t'-t)} E\{Z_s Z_s^*\}, \tag{3}$$

the corresponding (unstandardized) spectrum being

$$\sigma_s^2 = E\{Z_s Z_s^*\}. \tag{4}$$

For the sphere the theory is a little clumsier but equally straightforward, making use of the orthogonal expansions familiar in physics. The parameter is now the end-point of a vector $\mathbf{r}$ which rotates over a sphere of fixed radius, say unity. Then in place of (1) we may now write

$$X(\mathbf{r}) = \sum_{s=0}^{\infty} \sum_{n=-s}^{s} Z_{sn} Y_s^n(\mathbf{r}) \tag{5}$$

in terms of spherical harmonics $Y_s^n(\mathbf{r})$. Stationarity to the second order is here defined as the dependence of the covariance function merely on the angle θ_{12} between two vectors $\mathbf{r}_1$ and $\mathbf{r}_2$, and the constancy of the mean $E\{X(\mathbf{r})\}$, which is again assumed zero for convenience. (In this context isotropy of $X(\mathbf{r})$ is perhaps a more natural term than stationarity.)

The random coefficients Z_{sn} are doubly orthogonal for stationarity, i.e.
$$E\{Z_{sn} Z_{rm}^*\} = 0 \tag{6}$$

unless $s = r$, $n = m$. It may also be shown that
$$E\{Z_{sn} Z_{sn}^*\} = \sigma_s^2 \tag{7}$$

depending on the suffix s only, and
$$w(\theta_{12}) = E\{X^*(\mathbf{r}_1) X(\mathbf{r}_2)\} = \frac{1}{4\pi} \sum_{s=0}^{\infty} (2s+1) \sigma_s^2 P_s(\cos\theta_{12}), \tag{8}$$

where $P_s(\cos\theta_{12})$ is the Legendre polynomial of degree s. For further relevant formulae and remarks on their possible application in meteorology, see Jones (1963).

6·54 Lattice models and Markov fields.

Before we discuss the analogues of *discrete* time-series in two (or more) spatial dimensions, i.e. stochastic models defined on a rectangular grid or lattice of sites, it is advisable to clarify the one-dimensional case further, for we shall see that, in addition to the complication of more dimensions, there is the switch in emphasis from temporal processes developing in one direction and spatial processes, which may be laterally symmetric (cf. also §8·22). With the latter, the analogue of the simple linear Markov sequence (equation (1), §5·2) might be some symmetric relation
$$X_r = \beta(X_{r-1} + X_{r+1}) + Y_r. \tag{1}$$

Now if in (1) the Y_r are assumed *independent*, this is an example of Whittle's *simultaneous* models (Whittle, 1954). From the principles of obtaining spectral functions for linear models, it is readily seen that the spectral density for the model (1) will be proportional to
$$[1 - 2\beta\cos\omega]^{-2}. \tag{2}$$

However, it is recalled from §2·22 that the one-sided Markov sequence, which has spectrum proportional to
$$[(1 - \rho e^{i\omega})(1 - \rho e^{-i\omega})]^{-1} = [1 + \rho^2 - 2\rho\cos\omega]^{-1},$$

is equivalent to a nearest-neighbour model which, from the regression relation

$$E\{X_r | X_{r-1}, X_{r+1}\} = \frac{\rho}{1+\rho^2}(X_{r-1}+X_{r+1}), \tag{3}$$

is in danger of being confused with (1), although clearly its spectral function is, in contrast with (2), proportional to

$$[1-2\beta\cos\omega]^{-1}, \quad (\beta = \rho/(1+\rho^2)). \tag{4}$$

Indeed, equation (3) implies that

$$X_r = \beta(X_{r-1}+X_{r+1})+Z_r \tag{5}$$

where Z_r is uncorrelated with X_r (independent of X_r for normal sequences). Equation (5) is different in meaning from (1), and in particular it is easily shown that the Z_r sequence is not, like the Y_r in (1), an uncorrelated one.

In two dimensions, the analogous model to (1) is

$$X_{rs} = \beta_1(X_{r-1,s}+X_{r+1,s})+\beta_2(X_{r,s-1}+X_{r,s+1})+Y_{rs}, \tag{6}$$

where the Y_{rs} are independent. The spectral function for this simultaneous model is proportional to

$$[1-2(\beta_1\cos\omega_1+\beta_2\cos\omega_2)]^{-2}. \tag{7}$$

The analogue of (3) becomes

$$E\{X_{rs} | X_{r-1,s}, X_{r,s-1}, X_{r+1,s}X_{r,s+1}\}$$
$$= \beta_1(X_{r-1,s}+X_{r+1,s})+\beta_2(X_{r,s-1}+X_{r,s+1}), \tag{8}$$

but it is no longer clear at this stage whether this may be termed a nearest-neighbour model, in the sense that

$$P\{X_{rs} | \text{all other } Xs\} = P\{X_{rs} | X_{r-1,s}, X_{r,s-1}, X_{r+1,s}, X_{r,s+1}\}.$$

Indeed, it has been shown (Besag, 1972) that the latter type of two-dimensional model, sometimes termed a *Markov field*, can only have the linear relation (8) in the case of *normal* processes; such a process has been termed the *autonormal* process.

It is of interest to derive a class of linear spatial–temporal models with the same spectrum as (8). Suppose in *continuous* time we have

$$dX_{rst} = -\lambda G_0 X_{rst}dt + dZ_{rst}, \tag{9}$$

where G_0 is some linear *spatial* operator acting on X_{rst}. The spatial–temporal spectrum (if stationarity ensues) will be proportional to

$$[i\omega_t - \lambda G(\boldsymbol{\omega})]^{-1}[-i\omega_t - \lambda G^*(\boldsymbol{\omega})]^{-1}, \tag{10}$$

where $G(\boldsymbol{\omega})$ is the Fourier factor equivalent to $\mathbf{G}_0$. The marginal spatial spectrum is proportional to

$$\{\tfrac{1}{2}\lambda[G(\boldsymbol{\omega}) + G^*(\boldsymbol{\omega})]\}^{-1}. \tag{11}$$

In particular, for

$$(\mathbf{G}_0 - 1)\,X_{rs} = \beta_1'\,X_{r-1,\,s} + \beta_1''\,X_{r+1,\,s} + \beta_2'\,X_{r,\,s-1} + \beta_2''\,X_{r,\,s+1},$$

this becomes $\qquad [1 - 2(\beta_1\cos\omega_1 + \beta_2\cos\omega_2)]^{-1}, \tag{12}$

where $\beta_1 = \tfrac{1}{2}(\beta_1' + \beta_1'')$, $\beta_2 = \tfrac{1}{2}(\beta_2' + \beta_2'')$. This demonstrates the possibility of processes other than the autonormal process with (12) as spectrum, but it is only the autonormal process which has the conditional nearest-neighbour or Markov field property as defined above. The derivations above are of further interest as the cell size of the lattice shrinks, for it will be found that (7) and (12) then turn into the spectral functions (5) and (6) of § 9·4, thus enabling us to classify the respective processes of that section in terms of the above theory.

A useful general form of spatial nearest-neighbour lattice models has been given by Besag (1974) as

$$\frac{P\{X_i|\text{all other }Xs\}}{P\{X_i = 0|\text{all other }Xs\}} = \exp\{X_i G(X_i) + X_i \Sigma_j \beta_{i,j} X_j\}, \tag{13}$$

where i is a lattice site ($i \equiv r, s$ in the rectangular two-dimensional case), j refers to nearest-neighbours of i, $X = 0$ is assumed a possible value, and P is always assumed non-zero. We must also have $\beta_{i,j} = \beta_{j,i}$. In the autonormal case $G(X_i)$ must be linear. The other important case where $G(X_i)$ must be linear is the *autologistic* model for binary variables 0, 1 (e.g. presence or absence of a disease). In the latter case, $X_i^2 = X_i$, and $X_i G(X_i)$ also necessarily becomes linear in X_i.

It is remarkable that for a physical lattice model of n sites with a binary variable $X_{r,s}$ for which the total energy is of the form

$$E = -\{\alpha\Sigma_{r,s}\,X_{rs} + \beta_1\Sigma_{r,s}\,X_{rs}X_{r-1,\,s} + \beta_2\Sigma_{r,s}\,X_{rs}X_{r,\,s-1}\}, \tag{14}$$

we have a simultaneous probability of the form $C\,\exp\{-E/(kT)\}$ (see § 2·22), and hence a conditional probability

$$P\{X_{rs}|\text{all other }Xs\} = P\{\text{all }X\}/[P\{\text{all }X\}_{X_{rs}=0} + P\{\text{all }X\}_{X_{rs}=1}]$$

which may be identified with the autologistic model. In physics this model is known as the Ising model, and is notoriously intractable in more than one dimension, though some exact results in the two-dimensional case are available (Onsager, 1944; Newell and Montroll, 1953). With this physical model it is usual to consider the possible values ± 1 for X instead of $0, 1$. (It is of course simple to transform from one range to the other by writing $X' = \frac{1}{2}(1 + X)$ if $X = \pm 1$, or $X = 2X' - 1$. Note that now $X^2 = 1$.) With this range in (14), it is the case $\alpha = 0$ which was first solved by Onsager (1944) in the two-dimensional model, and shown (as n increased) to give rise to long-range order in the correlations as the temperature decreased to a critical value T_c. This long-range order may be interpreted as a non-zero mean arising with value $\pm m$ even in the nominally zero mean case when $\alpha = 0$, such that the product moment ρ_{ij} between two sites i and j has the form

$$\rho_{ij} = w_{ij} + m^2, \tag{15}$$

where w_{ij} is the covariance between the values of X at these two sites (cf. Bartlett, 1974b).

To examine the structure of these multi-dimensional conditional nearest-neighbour models, let us define them rather generally, first of all for any finite set of n sites i, with 'neighbours' of i affecting the value x_i. It is convenient to suppose the possible values x to be a finite set, and without loss of generality we may suppose that 0 is one of these possible values. It is supposed that $P\{x_i | \text{all other site values}\}$ is specified and greater than zero for all i, and moreover that no restrictions hold on the simultaneous realizations. Then for two complete realizations denoted by $\mathbf{x}$ and $\mathbf{y}$ respectively, we have

$$p(\mathbf{x}) = p(x_n | x_1, \dots, x_{n-1}) \, p(x_1, \dots, x_{n-1})$$

$$= p(x_n | x_1, \dots, x_{n-1}) \frac{p(x_1, \dots, x_{n-1}, y_n)}{p(y_n | x_1, \dots, x_{n-1})}$$

$$= \dots = p(\mathbf{y}) \prod_{i=1}^{n} \frac{p(x_i | x_1, \dots, x_{i-1}, y_{i+1}, \dots, y_n)}{p(y_i | x_1, \dots, x_{i-1}, y_{i+1}, \dots, y_n)}. \tag{16}$$

Next we define the set of 'neighbours' of site i by the condition: $p(x_i | \text{all other values})$ is dependent on x_j if and only if site $j (\neq i)$ is a 'neighbour' of site i. Any set of sites consisting of a single site i, or in which every site is a neighbour of every other site in the set is called a 'clique'. Now define

$$L(\mathbf{x}) = \log p(\mathbf{x}) - \log p(\mathbf{0}). \tag{17}$$

We may expand $L(\mathbf{x})$ uniquely as

$$\Sigma_i x_i G_i(x_i) + \Sigma_{i,j} x_i x_i G_{i,j}(x_i, x_j) + \ldots$$
$$+ x_1 x_2 \ldots x_n G_{1,2,\ldots,n}(x_1, x_2, \ldots, x_n), \tag{18}$$

where in the summations no i, j are repeated. For example,

$$\begin{aligned}
x_i x_j G_{i,j}(x_i, x_j) &= \log p(0, \ldots, 0, x_i, 0, \ldots, 0, x_j, 0, \ldots, 0) \\
&\quad - \log p(0, \ldots, 0, x_i, 0, \ldots, 0) \\
&\quad - \log p(0, \ldots, 0, x_j, 0, \ldots, 0) + \log p(\mathbf{0}),
\end{aligned}$$

vanishing if either x_i or x_j is 0. Then the functions $G_{i,j}, \ldots$ may be non-null if and only if the sites form a clique (the Hammersley–Clifford theorem).

For let us denote $(x_1, \ldots, x_{i-1}, 0, x_{i+1}, \ldots, x_n)$ by $\mathbf{x}_i$; then

$$\exp\{L(\mathbf{x}) - L(\mathbf{x}_i)\}$$
$$= p(x_i | x_1, \ldots, x_{i-1}, x_{i+1}, \ldots, x_n)/p(0 | x_1, \ldots, x_{i-1}, x_{i+1}, \ldots, x_n)$$

and can only depend on x_i and values at 'neighbouring' sites. Consider in particular site 1. We have

$$L(\mathbf{x}) - L(\mathbf{x}_1) = x_1\{G_1(x_1) + \Sigma_j' x_j G_{1,j}(x_1, x_j) + \ldots\},$$

where the dashed summations do not involve site 1. Suppose site $k(\neq 1)$ is not a neighbour of site 1. Then $L(\mathbf{x}) - L(\mathbf{x}_1)$ must be independent of x_k for all $\mathbf{x}$. Putting $x_i = 0$ for $i \neq 1$ or k, we see that $G_{1,k}(x_1, x_k) = 0$. Similarly the higher-order G functions involving x_1 and x_k must be null. Similar results also follow for any pair of sites that are not neighbours, and the theorem follows. (The above argument is taken from Besag, 1974.)

For more general x-variates, the sum or integral of $\exp L(\mathbf{x})$ must of course remain finite.

When $L(\mathbf{x})$ refers to cliques containing no more than two sites, the expansion for $L(\mathbf{x})$ must have the relatively simple expression $\Sigma_i x_i G_i(x_i) + \Sigma_{ij} x_i x_j G_{i,j}(x_i, x_j). \tag{19}$

If moreover x is binary $(0, 1)$ then $G_i(x_i)$ and $G_{i,j}(x_i, x_j)$ must necessarily be constants for each i, j. If the system is also 'stationary' then G_i is an absolute constant, and $G_{i,j}$ depends only on the (signless) vector displacement of i and j.

Notice that the number n of sites is finite in the above discussion, so that stationarity is not strictly possible except for periodic boundary conditions; moreover, the further problem of ergodicity already considered will arise if n increases indefinitely.

Spatial–temporal processes for discrete lattice variables. The autonormal Markov field could be generated by the linear spatial–temporal process referred to in the last section, but spatial Markov fields for discrete lattice variables require non-linear temporal processes for their generation. Consider the process

$$P\{x_{i,t+dt} = y|\mathbf{x}_t\} = \lambda G_i(y, \mathbf{x}_t)\,dt \quad (y \neq x_{i,t}). \tag{20}$$

A given limiting equilibrium distribution $p(\mathbf{x})$ must satisfy

$$\Sigma_i S_y p(\mathbf{y}_i)\, G_i(x, \mathbf{y}_i) = p(\mathbf{x})\, \Sigma_i S_y G_i(y, \mathbf{x}), \tag{21}$$

where S_y denotes summation over values of $y \neq x$ at site i, and $\mathbf{y}_i = \mathbf{x}$ except that the value at site i is not x but y. A convenient solution of (21) is obtained from the choice

$$p(\mathbf{y}_i)\, G_i(x, \mathbf{y}_i) = p(\mathbf{x})\, G_i(y, \mathbf{x}). \tag{22}$$

For example, for integer-valued variables x_i which only change at any one time by at most one unit, let

$$G_i(y, \mathbf{x}) = \begin{cases} (n - x_i)\,\gamma(\mathbf{x}) & \text{for} \quad y = x_i + 1, \\ x_i\,\delta(\mathbf{x}) & \text{for} \quad y = x_i - 1, \end{cases}$$

where
$$\gamma(\mathbf{x}) = \exp(\alpha + \Sigma_j \beta_j x_j),$$
$$\delta(\mathbf{x}) = \exp(\alpha' + \Sigma_j \beta'_j x_j),$$

x_j being the nearest neighbours of x_i. For

$$\beta(\mathbf{x}) = \gamma(\mathbf{x})/\delta(\mathbf{x}),$$

and $\beta_j - \beta'_j$ assumed to have at least lateral symmetry in relation to x_i and x_j, the stationary conditional distribution at site i is binomial with parameters n and $\theta = \beta(\mathbf{x})/[1 + \beta(\mathbf{x})]$. This reduces to the autologistic model when $n = 1$.

If alternatively

$$G_i(y, \mathbf{x}) = \begin{cases} \gamma(\mathbf{x}) & \text{for} \quad y = x_i + 1, \\ x_i\,\delta(\mathbf{x}) & \text{for} \quad y = x_i - 1, \end{cases}$$

then the conditional distribution at i is Poisson with mean $\beta(\mathbf{x})$. The requirement that $p(\mathbf{x})$ is a valid distribution imposes, however, the rather severe restriction $\beta_j - \beta'_j \leqslant 0$, implying competition rather than positive contagion between neighbouring sites (Besag, 1974).

Chapter 7

PREDICTION AND COMMUNICATION THEORY

7·1 Linear prediction for stationary processes

In Chapter 1 we saw that some stochastic processes may be termed deterministic, and that their states may be predicted exactly at any future time from suitable information about the past. When a process is non-deterministic, the problem still arises as to how to predict its future state with, in some sense, a minimum of error. This is of particular relevance when the past history of the process can be examined in detail, and used as a basis for the prediction. The solution has been developed by Wiener in the important practical case of linear 'least-squares' prediction for stationary processes, and extended to cover associated problems such as the design of linear wave filters to eliminate 'noise'. An introduction to this technique is given in this section; for a more complete treatment reference should be made to Wiener (1949).

Let us first examine the prediction problem for the linear stationary process

$$X(t) = \int_{-\infty}^{t} g(t-u)\,dY(u), \tag{1}$$

where natural damping is counterbalanced by continual further independent disturbances. We at once have

$$X(t+\tau) = \int_{-\infty}^{t+\tau} g(t+\tau-u)\,dY(u)$$
$$= \int_{t}^{t+\tau} g(t+\tau-u)\,dY(u) + \int_{-\infty}^{t} g(t+\tau-u)\,dY(u),$$

where the first term on the right represents a contribution to $X(t+\tau)$ from disturbances arising *after* time t. We therefore have an essential error of prediction with variance

$$\kappa_2 \int_0^{\tau} g(\tau-x)\,g^*(\tau-x)\,dx = \kappa_2 \int_0^{\tau} |g(x)|^2\,dx, \tag{2}$$

where
$$\kappa_2 = \int_0^1 dE\{Y(u)\,Y^*(u)\}.$$

(Notice that even for the non-stationary linear process

$$X(t) = \int_0^t g(t-u)\,dY(u),$$

where $\int_0^\infty g(u)\,g^*(u)\,du$ does not necessarily converge, the above argument would still be applicable.)

More generally, for all stationary processes with a representation (1) up to the second-order moments, i.e. $Y(u)$ orthogonal but not necessarily additive, we should naturally expect the second term above, namely,

$$X_p(t+\tau) = \int_{-\infty}^t g(t+\tau-u)\,dY(u), \tag{3}$$

to provide the best *linear* prediction formula, and its error variance would still be given by (2). In the particular case of *normal* stationary processes it is also evident that the best linear prediction and best prediction will coincide.

In the solution (3) we still require of course to express $Y(u)$ $(u \leqslant t)$, in terms of $X(v)$ $(v \leqslant t)$, and this must be done by solving (1) for $Y(u)$ in terms of $X(u)$.† It may, incidentally, be noted that while the solution so far developed is in general formal, it provides at once the complete and rigorous solution for the important class of linear processes which are themselves the solutions of stochastic equations representing the dynamical behaviour of physical systems, such as the torsional galvanometer, under random independent impulses. The replacement of $dY(u)$ (or any convenient linear expression in $Y(u)$) by a linear expression in $X(u)$ is then immediate. For example, for the Markov process generated by the equation $dX(t) + \mu X(t)\,dt = dY(t)$, for which $g(u) = e^{-\mu u}$ $(u \geqslant 0)$, we obtain

$$X_p(t+\tau) = e^{-\mu\tau} \int_{-\infty}^t e^{-\mu(t-u)}\,dY(u) = e^{-\mu\tau}X(t),$$

this depending, as of course it should for a Markov process, only on the last available value $X(t)$.

To obtain the general solution (equivalent to (3)) we consider the linear autoregressive formula

$$X_p(t+\tau) = \int_0^\infty X(t-u)\,dB(u), \tag{4}$$

---

† For this to be possible, $g(u)$ must satisfy the condition referred to on p. 201, viz. that its Fourier transform, regarded as a function of the complex variable ω, has no zeros in the lower half-plane.

and try to determine $B(u)$ so that

$$E\{|\,X(t+\tau)-X_p(t+\tau)\,|^2\}$$

is a minimum. This problem is quite analogous to the classical least-squares estimation problem for a finite number of regression coefficients (the corresponding problem for stationary sequences is of course even closer to the classical regression problem; see equation (15) of the next section, § 7·11), and we should expect to have as the equation for $dB(u)$

$$\sigma_X^2 \int_0^\infty \rho(v-u)\,dB(u) = \sigma_X^2 \rho(\tau+v) \quad (v>0). \tag{5}$$

That this solution does in fact minimize the least-squares prediction error may be checked by considering any prediction formula

$$X_p'(t+\tau) = \int_0^\infty X(t-u)\,dB'(u).$$

Then

$$E\{|\,X(t+\tau)-X_p'(t+\tau)\,|^2\}$$

$$= E\{|\,X(t+\tau)-X_p(t+\tau)\,|^2\} + E\{|\,X_p(t+\tau)-X_p'(t+\tau)\,|\}^2$$

$$+ \text{real part } 2E\left\{\left[X^*(t+\tau)-\int_0^\infty X^*(t-u)\,dB^*(u)\right]\int_0^\infty X(t-v)\right.$$

$$\left. \times d[B(v)-B'(v)]\right\}.$$

We assume that $\rho(u)$ is continuous and that the integration and expectation symbol in the last term may be commuted. Then the last term becomes the real part of

$$2\sigma_X^2\left\{\int_0^\infty\left[\rho^*(\tau+v)-\int_0^\infty\rho^*(v-u)\,dB^*(u)\right]d[B(v)-B'(v)]\right\},$$

and vanishes in view of (5). Hence X_p' can never give a smaller prediction error than X_p, as the mean-square error for X_p consists of the error for X_p plus a term which is essentially non-negative.

In order to solve the integral equation (5), we shall assume that the linear representation (1) is possible in the wide sense (i.e. for $Y(u)$ orthogonal); that is, we assume (see equation (18), § 6·3) that $X(t)$ has an absolutely continuous spectrum $f(\omega)$ for which (except for cases where perfect linear prediction is possible)

$$\int_{-\infty}^{\infty} \frac{|\log f(\omega)| \, d\omega}{1+\omega^2} < \infty. \tag{6}$$

We may then define $\alpha(\omega)$ as in Chapter 6, with no zeros or singularities in the lower half-plane and so that $\alpha(\omega)\,\alpha^*(\omega) = \sigma_X^2 f(\omega)$, and write correspondingly

$$\sigma_X^2 \rho(u) = \int_0^{\infty} \beta(v+u)\,\beta^*(v)\,dv \quad (\beta(v)=0 \text{ for } v<0),$$

where $\beta(u) = g(u)\sqrt{\kappa_2}$ was formally related to $\alpha(\omega)$ by the Fourier relation

$$\beta(u) = \frac{1}{\sqrt{(2\pi)}} \int_{-\infty}^{\infty} e^{i\omega u} \alpha(\omega)\,d\omega.$$

Now the mean-square error of prediction may be written

$$\sigma_X^2 \left\{ 1 - \int_0^{\infty} \rho(\tau+u)\,dB^*(u) - \int_0^{\infty} \rho^*(\tau+u)\,dB(u) \right.$$
$$\left. + \int_0^{\infty} \int_0^{\infty} \rho(v-u)\,dB(u)\,dB^*(v) \right\}$$
$$= \sigma_X^2 \int_{-\infty}^{\infty} f(\omega)\,|\,e^{i\tau\omega} - b(\omega)\,|^2\,d\omega, \tag{7}$$

where $b(\omega)$ is defined formally by

$$b(\omega) = \int_{-\infty}^{\infty} e^{-i\omega v}\,dB(v) \quad (dB(v)=0 \text{ for } v<0). \tag{8}$$

The problem of prediction may thus from (7) be regarded as approximating 'in the mean' to $\sigma_X \sqrt{f(\omega)}\,e^{i\tau\omega}$ by $\sigma_X \sqrt{f(\omega)}\,b(\omega)$.

Replace $\sigma_X \sqrt{f(\omega)}$ by $\alpha(\omega)$ and write

$$e^{i\tau\omega}\alpha(\omega) = c_1(\omega) + c_2(\omega),$$

where $c_1(\omega)$ is a 'backward' operator, so-called if it is without singularities (and of appropriate order at infinity) below the real axis, and $c_2(\omega)$ is correspondingly 'forward'. Then (7) becomes

$$\sigma_X^2 \left\{ \int_{-\infty}^{\infty} |\,c_2(\omega)\,|^2\,d\omega + \int_{-\infty}^{\infty} |\,c_1(\omega) - b(\omega)\,\alpha(\omega)\,|^2\,d\omega \right\}, \tag{9}$$

this result following (as $b(\omega)\,\alpha(\omega)$ is backward) from the general formula

$$\int_{-\infty}^{\infty} \gamma_1(\omega)\,\gamma_2^*(\omega)\,d\omega = 0$$

for *any* two operators $\gamma_1(\omega)$ and $\gamma_2(\omega)$ of backward and forward type respectively for which the integral exists. Note also that if

$$\gamma(\omega) = \gamma_1(\omega) + \gamma_2(\omega),$$

and

$$\Gamma(t) = \frac{1}{2\pi}\int_{-\infty}^{\infty} e^{i\omega t}\gamma(\omega)\,d\omega,$$

then

$$\gamma_1(\omega) = \int_0^{\infty} e^{-i\omega t}\Gamma(t)\,dt, \quad \gamma_2(\omega) = \int_{-\infty}^0 e^{-i\omega t}\Gamma(t)\,dt.$$

As only $b(\omega)\,\alpha(\omega)$ is at our choice, the expression in (7) is a minimum when

$$b(\omega)\,\alpha(\omega) = c_1(\omega), \tag{10}$$

which must therefore represent the solution of (5). From the above relations for $\gamma_1(\omega)$, $\gamma_2(\omega)$, we may readily see further that

$$c_1(\omega) = \frac{1}{2\pi}\int_0^{\infty} e^{-i\omega u}du \int_{-\infty}^{\infty} e^{iw(u+\tau)}\alpha(w)\,dw. \tag{11}$$

Moreover, the mean-square prediction error is now from (9)

$$\sigma_X^2 \int_{-\infty}^{\infty} |c_2(\omega)|^2\,d\omega, \tag{12}$$

which may be shown without much difficulty to be equivalent to (2).

There are obvious advantages in deferring any attempt to establish conditions to ensure the validity of the above mainly formal arguments. Suppose $\alpha(\omega)$ is of the general ratio of polynomials type referred to in equation (19) of §6·3, and expressed in partial fractions in the form

$$\alpha(\omega) = \Sigma_{p,q}a_{pq}(\omega - \omega_p)^{-q}; \tag{13}$$

then an explicit formal solution for $b(\omega)$ may be shown to be

$$\alpha(\omega)\,b(\omega) = \Sigma_{p,q}a_{pq}e^{i\omega_p\tau}\sum_{r=0}^{q-1}\frac{(i\tau)^{q-1-r}}{(q-1-r)!\,(\omega-\omega_p)^{r+1}}. \tag{14}$$

Now if we consider the particular spectrum

$$f(\omega) = \frac{1}{2\pi}\frac{2\alpha\beta}{(\beta-\omega^2)^2+\omega^2\alpha^2}, \tag{15}$$

we have at once, by making use of the relation of (15) to the second-order process,

$$d\dot{X}(t) + \alpha \dot{X}(t)\,dt + \beta X(t)\,dt = d Y(t), \qquad (16)$$

and the method indicated at the beginning of this section,

$$X_p(t+\tau) = \int_{-\infty}^{t} g(t+\tau - u)\,d Y(u),$$

where

$$g(u) = (e^{\lambda_1 u} - e^{\lambda_2 u})/(\lambda_1 - \lambda_2),$$

$$(x - \lambda_1)(x - \lambda_2) = x^2 + \alpha x + \beta.$$

We thus find that

$$X_p(t+\tau) = \left[\frac{\lambda_1 e^{\lambda_2 \tau} - \lambda_2 e^{\lambda_1 \tau}}{\lambda_1 - \lambda_2}\right] X(t) - \left[\frac{e^{\lambda_2 \tau} - e^{\lambda_1 \tau}}{\lambda_1 - \lambda_2}\right] \dot{X}(t), \qquad (17)$$

involving the differential operator d/dt. The dependence of $X_p(t+\tau)$ on the two quantities $X(t)$, $\dot{X}(t)$, but not on other values of $X(u)$ for $u < t$, would of course be expected from the Markov character of the process (16) when regarded as a joint process in $X(t)$, $\dot{X}(t)$. But the formal Fourier inverse of d/dt is $i\omega$, and it may be verified that (14) gives the solution

$$b(\omega) = \xi(\tau) - i\omega \eta(\tau),$$

where $\xi(\tau)$, $\eta(\tau)$ are the corresponding expressions in square brackets in (17). Thus valid solutions $b(\omega)$ are not restricted to bounded functions, and correspondingly, $B(t)$ need not be of limited total variation. This use of differential operator expansions is considered further in the next section.

7·11 Further associated problems. *Orthogonal expansions.*
If $\rho(\tau)$ is analytic, it has been noted in Chapter 5 that a complete Taylor expansion for $X(t+\tau)$ is possible in terms of its m.s. differential coefficients, $\overset{(n)}{X}(t)$, so that

$$X_p(t+\tau) = \sum_{0}^{\infty} \overset{(n)}{X}(t)\,\tau^n/n! = X(t+\tau), \qquad (1)$$

and the prediction error is zero. Condition (6) of §7·1 is associated with a slightly wider class of 'quasi-analytic' functions (see Paley and Wiener, 1934, Chapter I), though in practical cases the two conditions are usually satisfied, or not satisfied, together.

If the m.s. differential coefficients $\overset{(n)}{X}(t)$ only exist up to a certain order, and it is required to express $X(t+\tau)$ as nearly as possible in terms of these, the appropriate coefficients will, however, no longer be $1, \tau, \frac{1}{2}\tau^2$, etc., but those given by the least-squares solution. The latter is conveniently obtained in terms of an orthogonal set of linear combinations of $\overset{(n)}{X}(t)$. It will be recalled that for a real stationary process

$$\frac{d}{dt}E\{X^2\} = 2E\{X\dot{X}\} = 0, \quad \frac{d}{dt}E\{\dot{X}^2\} = 2E\{\dot{X}\ddot{X}\} = 0,$$

but $$\frac{d}{dt}E\{X\dot{X}\} = E\{X\ddot{X}\} + E\{\dot{X}^2\} = 0,$$

so that $X(t)$ is orthogonal, not to $\ddot{X}(t)$, but to

$$\ddot{X} + X\sigma^2(\dot{X})/\sigma^2(X).$$

In general, let the sth orthogonal function be X_s, and denote $\sigma^2(\overset{(s)}{X})$ by σ_s^2. Evidently

$$E\{\overset{(s)}{X}\overset{(s+2r+1)}{X}\} = \ldots = (-1)^r E\{\overset{(s+r)}{X}\overset{(s+r+1)}{X}\} = 0 \quad (s \geqslant 0, r \geqslant 0),$$

and odd- and even-order derivatives form two separate groups of orthogonal functions, and to every orthogonal function X_{2s} there is another one $X_{2s+1} = \dot{X}_{2s}$. To obtain the recurrence relation for X_{2s} let

$$X_{2s} = \overset{(2s)}{X} + a_{1,s}X_{2s-2} + \ldots + a_{s,s}X.$$

Then $$E\{X_{2s}X_{2s-2r}\} = E\{\overset{(2s)}{X}X_{2s-2r}\} + a_{r,s}E\{X_{2s-2r}^2\} = 0,$$

whence $$a_{r,s} = -E\{\overset{(2s)}{X}X_{2s-2r}\}/E\{X_{2s-2r}^2\}. \tag{2}$$

The least-squares coefficients in the orthogonal expansion of $X_p(t+\tau)$ are then given by $E\{X(t+\tau)X_r(t)\}/E\{X_r^2(t)\}$ and are expressible in terms of $\rho(\tau)$. For example,

$$E\{X(t+\tau)X(t)\}/E\{X^2(t)\} = \rho(\tau),$$

$$E\{X(t+\tau)\dot{X}(t)\}/E\{\dot{X}^2(t)\} = -\rho'(\tau)\sigma_0^2/\sigma_1^2 \quad (\sigma_0 \equiv \sigma_X).$$

In the example $\rho(\tau) = e^{-\mu|\tau|}$, $X(t)$ is not differentiable and prediction by the above series terminates at the first term; in the example (17) of § 7·1 it stops at the second. In these two cases the above method provides the most efficient formula, but this is not necessarily true; for instance, if an extra additive component were superposed on $X(t)$, the resulting process would no longer be differentiable, and no formula of the above type could proceed beyond the first term.

The relation of expansions in terms of $\overset{(n)}{X}(t)$ to the general prediction formula may be seen by noting that the complete Taylor expansion for $X(t)$ corresponds to writing

$$b(\omega) = 1 + i\omega\tau + \tfrac{1}{2}(i\omega\tau)^2 + \dots,$$

and that if $X(t)$ is only differentiable up to order s, the formula for $b(\omega)$ may be written

$$b(\omega) = 1 + i\omega\tau + \tfrac{1}{2}(i\omega\tau)^2 + \dots + (i\omega\tau)^s/s!$$

$$+ \frac{1}{2\pi\alpha(\omega)} \int_0^\infty e^{-i\omega u} du \int_{-\infty}^\infty \alpha(w) e^{iwu} \left[e^{iw\tau} - 1 - iw\tau \dots - \frac{(iw\tau)^s}{s!} \right] dw. \quad (3)$$

But for $\rho(\tau) = e^{-\mu|\tau|}$ and the prediction $X_p(t+\tau) = \rho(\tau) X(t)$, we may verify from (3) that the remainder term is not zero if we put $b(\omega) = 1 + \dots$, though it is for $b(\omega) = \rho(\tau) + \dots$.

The expansion of $b(\omega)$ in powers of ω is replaced in the above orthogonal expansion by an expansion of $b(\omega)$ in polynomials of ω orthogonal with respect to integration over $f(\omega) d\omega$; this is readily seen from the correspondence between the derivatives and powers of ω. Such orthogonal polynomials can of course only be defined when the moments of the spectral density function $f(\omega)$ exist, and this corresponds to the existence of the derivatives of $X(t)$. A more general method of using orthogonal functions in ω has been suggested by Wiener and illustrated by an expansion in terms of Laguerre functions. Let us write

$$b(\omega) = \sum_{n=0}^N a_n b_n(\omega), \quad (4)$$

and approximate to $\alpha(\omega) e^{i\tau\omega}$ by $\alpha(\omega) \Sigma a_n b_n(\omega)$. The admissible functions $b_n(\omega)$ are again to be transforms of functions vanishing for negative argument and to have no singularities below the real axis; this condition applies also to $b_n(\omega)\,\alpha(\omega)$, which are chosen to form an orthogonal set. We may take, for example,

$$b_n(\omega)\,\alpha(\omega) = l_n(\omega) \equiv \frac{(1-i\omega)^n}{\sqrt{\pi}\,(1+i\omega)^{n+1}}. \tag{5}$$

Illustrating this method on the familiar case $\rho(\tau) = e^{-\mu|\tau|}$, or

$$f(\omega) = \frac{\mu}{\pi(\mu^2 + \omega^2)}, \quad \alpha(\omega) = \frac{\sigma_0\sqrt{\mu}}{\sqrt{\pi}\,(\omega - i\mu)},$$

we choose for convenience a time-scale for which $\mu = 1$. Then

$$a_n = \int_{-\infty}^{\infty} \frac{\sigma_0 e^{i\tau\omega} l_n^*(\omega)\,d\omega}{\sqrt{\pi}\,(\omega - i)} = \begin{cases} i\sigma_0 e^{-\tau} & (n = 0), \\ 0 & (n > 0), \end{cases}$$

and

$$a_0 b_0(\omega) = \frac{1}{\sqrt{\pi}\,(1+i\omega)}(i\sigma_0 e^{-\tau})\frac{(\omega - i)\sqrt{\pi}}{\sigma_0},$$

whence

$$X_p(t+\tau) = e^{-\tau} X(t) \quad (\mu = 1).$$

The problem of 'noise'. The problem of a superposed and unwanted component (noise component), important in electrical applications, may be treated by similar methods. A typical problem is to estimate $Y(t+\tau)$ for positive or negative (or zero) τ from the observed process

$$X(t) = Y(t) + Z(t),$$

where $Z(t)$ is the noise component. Thus for the case where $Z(t)$ is uncorrelated with $Y(t)$, Wiener has shown by similar methods that the previous formula (11) of §7·1 for $\alpha(\omega)\,b(\omega)$ should be modified to

$$\alpha(\omega)\,b(\omega) = \frac{1}{2\pi}\int_0^\infty e^{-i\omega u}\,du\int_{-\infty}^\infty e^{i\omega(u+\tau)}\alpha_1(\omega)\,d\omega, \tag{6}$$

where

$$\alpha_1(\omega) = \sigma_Y^2 f_Y(\omega)/\alpha^*(\omega). \tag{7}$$

The error variance in the prediction

$$Y_p(t+\tau) = \int_0^\infty X(t-u)\,dB(u)$$

is given by

$$\int_0^\infty |\beta_Y(u)|^2\,du - \int_\tau^\infty |\beta_1(u)|^2\,du, \tag{8}$$

where $\beta_1(u)$ is the transform of α_1,

$$\beta_1(u) = \frac{1}{\sqrt{(2\pi)}}\int_{-\infty}^\infty e^{i\omega u}\alpha_1(\omega)\,d\omega.$$

In agreement with the well-known efficiency of linear combinations of least-squares estimates, it may also be shown that to approximate to $\dot{Y}(t+\tau)$ we should write

$$\dot{Y}_p(t+\tau) = \int_0^\infty X(t-u)\,dC(u),$$

where $\quad \alpha(\omega)\,c(\omega) = \dfrac{1}{2\pi}\displaystyle\int_0^\infty e^{-i\omega u}du \int_{-\infty}^\infty \dfrac{\partial}{\partial\tau}[e^{i\omega(u+\tau)}\alpha_1(w)\,dw].$ (9)

The error variance is

$$\int_0^\infty |\beta_Y'(u)|^2\,du - \int_\tau^\infty |\beta_1'(u)|^2\,du, \tag{10}$$

where $\beta_1'(u)$ is the transform of $i\omega\alpha_1(\omega)$, etc.

Prediction for stationary sequences. The prediction procedure for sequences is naturally rather similar, and we merely summarize below the corresponding formulae. Formula (1) of §7·1 becomes

$$X_s = \sum_{u=-\infty}^{s} g(s-u)Y_u, \tag{11}$$

and the most efficient prediction formula is

$$[X_{s+\tau}]_p = \sum_{u=-\infty}^{s} g(s+\tau-u)Y_u, \tag{12}$$

with error

$$\sum_{u=s+1}^{s+\tau} g(s+\tau-u)Y_u \tag{13}$$

and error variance

$$\sigma_Y^2 \sum_{v=0}^{\tau-1} |g(v)|^2. \tag{14}$$

Equation (5) of §7·1 is replaced by

$$\sum_{w=0}^{\infty} \rho_{s-w}a_w = \rho_{s+\tau}, \tag{15}$$

with solution for a_w given by

$$\alpha(\omega)\,b(\omega) = \frac{1}{2\pi}\sum_{u=0}^{\infty} e^{-iu\omega}\int_{-\pi}^{\pi}\alpha(w)\,e^{iw(u+\tau)}dw, \tag{16}$$

where $\quad b(\omega) = \displaystyle\sum_{-\infty}^{\infty} e^{-i\omega u}a_u \quad (a_u = 0 \text{ for } u < 0).$

In this solution the function $\alpha(\omega)$, which satisfies the relation $\alpha(\omega)\,\alpha^*(\omega) = \sigma_X^2 f(\omega)$, must be such that

$$\beta_u = \sigma_Y g(u) = \frac{1}{\sqrt{(2\pi)}}\int_{-\pi}^{\pi} e^{iu\omega}\alpha(\omega)\,d\omega$$

is zero for $u < 0$. The general formula for $\alpha(\omega)$ has been given by Wiener as

$$\alpha(\omega) = \frac{1}{\sqrt{(2\pi)}} \sum_{u=0}^{\infty} \beta_u e^{-iu\omega} = \sigma_X \sqrt{f(\omega)}\, e^{\frac{1}{2}i\phi(\omega)}, \tag{17}$$

where

$$i\phi(\omega) = \sum_{u=1}^{\infty} A_u e^{-iu\omega} - \sum_{u=-\infty}^{-1} A_u e^{-iu\omega}; \tag{18}$$

in (18) the coefficients A_u are the Fourier coefficients of $\log f(\omega)$, that is,

$$\log f(\omega) = \sum_{u=-\infty}^{\infty} A_u e^{-iu\omega}. \tag{19}$$

The condition parallel to equation (6) of § 7·1 is

$$\int_{-\pi}^{\pi} |\log f(\omega)|\, d\omega < \infty; \tag{20}$$

otherwise perfect linear prediction is possible.

General autoregressive representations of the sequence type

$$[X_{s+\tau}]_p = \sum_{u=0}^{\infty} a_u X_{s-u}$$

were first considered by Wold (1938); explicit formulae for the coefficients a_u and β_u were obtained, independently of Wiener's work, by Kolmogorov (1941), who also considered the inter-polation problem in the discrete case.

It is also possible to extend all these methods to predicting from multiple time-series, but such extensions are omitted here. In statistical applications at least it seems probable that any cases which arise are best handled directly as autoregressive problems in a finite number of variables (cf. Chapter 9).

7·12   Regulation and control. The theory of prediction is obviously relevant to the problem of the automatic control of industrial or other processes which tend to fluctuate more widely than is desired. Indeed, if a stationary process $X(t)$ is to be corrected by a device which involves a delay τ but is otherwise perfect in its operation, we should merely correct by the predicted value $X_p(t+\tau)$ to have a new value at $t+\tau$ of

$$X_c(t+\tau) = X(t+\tau) - X_p(t+\tau).$$

In practice we should want to correct again and again, or the output process would relax towards the original process $X(t)$, but it should be noticed that the variance σ_c^2 of $X_c(t)$ is the minimum variance attainable, as it represents the unpredictable part of the process.

To examine the effect of repeated or continuous control, it is convenient to introduce operators to represent transformations or operations on a process $X(t)$. We suppose, as in previous sections, that we are considering only linear prediction and hence linear operators. Thus we write

$$X_p(t+\tau) = P_\tau X(t), \quad X(t+\tau) = E_l^\tau X(t).$$

Let the output process after control be still $X_c(t)$, but as control is assumed continuous we no longer have the simple control operator $P_\tau E_l^{-\tau}$. Instead, we have the equation

$$X_c(t) = X(t) - R_\tau E_l^{-\tau} X_c(t), \tag{1}$$

where R_τ is the required operator. As we also require

$$X_c(t) = (1 - P_\tau E_l^{-\tau}) X(t),$$

we obtain $\qquad R_\tau E_l^{-\tau} = (1 - P_\tau E_l^{-\tau})^{-1} - 1. \tag{2}$

$R_\tau E_i^{-\tau}$ in (2) is the ideal control operator under the conditions assumed, but of course the effect of any admissible R_τ can also be investigated from equation (1).

Similar equations obviously hold for sequences. Thus for a linear Markov sequence and $\tau = 1$, so that $P_\tau = \rho$, $P_\tau E_l^{-\tau} = \rho E_l^{-1}$, we obtain

$$R_\tau E_l^{-\tau} = \rho E_l^{-1} + \rho^2 E_l^{-2} + \dots. \tag{3}$$

More general situations can be investigated along similar lines. The control feedback mechanism may not have an instantaneous effect at lag τ, but a distributed effect represented by an operator F. The output process may undergo a transformation G before the control measurements are taken. Thus we assume now the actual correction $FRX_c(t)$ to the initial process $X(t)$ due to the imposed correction $RX_c(t)$ (this correction signal not interfering further with the final output $X_c(t)$, as it might in a purely communication system). Equation (1) is then generalized to

$$G^{-1}X_c(t) = X(t) - FRX_c(t), \tag{4}$$

whence $$X_c(t) = (1 + GFR)^{-1} GX(t),\tag{5}$$

from which the effect of any *admissible* R can be found.

In the Markov sequence example, suppose G is still 1, but F is $pE_t^{-1} + qE_t^{-2}$ instead of E_t^{-1}. If we try writing

$$X_c(t) = Y(t) = (1 - \rho E_t^{-1}) X(t),$$

we require $$(pE_t^{-1} + qE_t^{-2}) R = (1 - \rho E_t^{-1})^{-1} - 1$$

so that formally $$R = \frac{(1 - \rho E_t^{-1})^{-1} - 1}{pE_t^{-1} + qE_t^{-2}}.\tag{6}$$

This provides an admissible expansion in powers of E_t^{-1} if $p > q$, i.e. $p > \frac{1}{2}$. If $p \leqslant \frac{1}{2}$, we might modify the prediction operator to be of the form $a(pE_t^{-1} + qE_t^{-2})$, thereby ensuring the cancellation of the awkward denominator in (6). This easily leads to an optimum choice of

$$a = \frac{\rho(p + q\rho)}{1 - 2pq(1 - \rho)}$$

and a reduced variance

$$\sigma_c^2 / \sigma_x^2 = 1 - \frac{\rho^2 (p + q\rho)^2}{1 - 2pq(1 - \rho)}$$

compared with the value for $p > \frac{1}{2}$ of $1 - \rho^2$.

Further complications will of course arise in practice and also have to be taken into account, for example, the effects of non-stationary fluctuations, non-linear correction devices, multi-variate processes or the relative costs of alternative control systems.

7·2   Theory of information and communication

The theory of prediction, and indeed the theory of stationary processes in general, has a bearing on a topic which has aroused considerable interest in recent years, the general theory of communication. This theory has been systematized by Shannon (1948), and we give later some of his main theorems, treating these in an elementary way in keeping with the physical notions used. First of all, however, it is desirable to summarize the theory of information developed with it, as this is used in, but is not dependent on, the theory of communication; and moreover needs to be distinguished from the information concept introduced by R. A. Fisher in statistical inference (see §4·1, and also Chapter 8).

Suppose a number of possible events A_s have corresponding probabilities p_s defining a discrete probability distribution. Then the uncertainty (or entropy) measure of this distribution can be defined to be

$$I_S = E\{-\log p_S\} = -\Sigma p_S \log p_S, \tag{1}$$

and is the measure used by Shannon. It can be regarded as information in the sense that uncertainty is reduced when the event that has occurred at a trial is known. The convention on sign is rather important; we shall regard information as positive (as in (1)), so that uncertainty should be given a negative sign.

There are some advantages in two modifications in the above definition. First, it is useful to speak of the information, or reduction in uncertainty, associated with the particular event A_S that occurs. Thus if we define this to be $i_S = -\log p_S$, the information measure in (1) is the average or expected value of i_S. In two of the most important fields of application, physics and communication theory, this modification is, however, not usually important. In physics large numbers of similar particles are considered, the uncertainty or entropy measure of which can be obtained either over the entire equilibrium distribution of possible configurations, or from a particular configuration probability provided one of the most probable configurations is chosen. In communication theory rather similarly the information in a long sequence of messages is considered; and the average information in the distribution of possible sequences becomes for an ergodic sequence equivalent to the information in a single typical sequence.

Secondly, it is sometimes useful to extend the definition to 'relative information', in which the probability of A_s does not change from p_S to 1, but to p_S', say. The relative information is then defined by $r_S = -\log(p_S/p_S')$. This definition includes naturally both information and uncertainty, with opposite sign since $-\log p_S = \log(1/p_S)$. The average relative information is now perhaps somewhat less natural, as there are two such quantities, one averaged over p_S and the other over p_S'; however, if the p_S have been modified to conditional probabilities p_S' in the light of some information that has materialized, the p_S representing the absolute probabilities, then averaging over p_S is usually the more useful. This concept is relevant to communication theory when we consider the information in a message received through a 'noisy channel'. Again, however, we shall keep to Shannon's definition in (1), and shall often for convenience refer to it as the entropy measure without always specifying whether it corresponds to information or uncertainty; this will be clear from the context.

These information concepts have no direct connexion with Fisher's information function, but if p_S and p_S' are interpreted as hypothetical probability values changing with the unknown value of a parameter θ, then

$$r_S = -\log\left(p_S/p_S'\right) = (\Delta\theta)\frac{\partial\log p_S}{\partial\theta} + \tfrac{1}{2}(\Delta\theta)^2\frac{\partial^2\log p_S}{\partial\theta^2} + \dots,$$

where $\Delta\theta = \theta - \theta'$. For small $\Delta\theta$ and conditions (cf. §4·1) for which

$$E\left\{\frac{\partial\log p_S}{\partial\theta}\right\} = 0, \quad E\left\{\frac{-\partial^2\log p_S}{\partial\theta^2}\right\} = I_F(\theta),$$

we obtain $\qquad\qquad E\{r_S\} \sim -\tfrac{1}{2}(\Delta\theta)^2 I_F(\theta), \qquad\qquad (2)$

if θ is the true value of the parameter.

7·21   Communication systems. We depict any communication schematically as in diagram (fig. 8); we shall, however, consider first discrete noiseless systems, in which a sequence of choices from a finite set of m elementary symbols is to be transmitted from one point to another via a channel with noise interference absent. For example, as in Morse code, the symbols may consist, say, of a dot, a dash, a letter space and a word space. These different symbols will in general have different durations. There may be in addition certain restrictions on permissible sequences; for example, two spaces will never be contiguous. A suitable definition of the *capacity* C of the channel is

$$C = \lim_{T\to\infty} \log n(T)/T, \qquad\qquad (1)$$

where $n(T)$ denotes the number of permissible sequences of duration T. The base 2 for the logarithm is conventionally used.

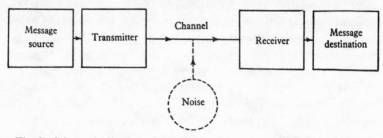

Fig. 8. Schematic diagram of communication system (cf. Shannon (1948)).

Consider next the discrete source of information. This is assumed to consist of a stationary stochastic process (non-deterministic and ergodic) in which symbols (for example, letters or words, and not necessarily the same symbols as in the signal sequence, which represents a *coding* of the message) represent the possible states. Any sequence S of finite duration T (or number of symbols N) will have a probability $p(S)$, and, as above, we define the average information or entropy measure for S by

$$J = -E\{\log p(S)\}. \tag{2}$$

The entropies per symbol or time unit are then defined as

$$H' = \lim_{N \to \infty} J/N, \quad H = \lim_{T \to \infty} J/T \tag{3}$$

(that these limits exist follows from the property

$$J(U) \leqslant J(U, V) \leqslant J(U) + J(V)).$$

The justification of these definitions lies in their relations and use. Thus the fundamental theorem connecting H' and C is that it is possible to encode the information source so as to transmit as near the average rate C/H' as we please, but it is not possible to transmit at a greater rate.

To see this, we note first that the entropy measure in (2) for a sequence of duration T is a maximum when all the permissible sequences S have equal probability, and its value then becomes $\log n(T)$, where $n(T)$ is the number of different possible sequences. Thus the maximum entropy rate for the symbol sequence U, say, of the channel is just its capacity C. Now the entropy of a discrete sequence is not altered by reversible encoding since

$p(S) = p(U)$; and, since (while there may be finite delays) the transmitter will be assumed to cope with the source without allowing any indefinitely increasing time-lag, the entropy *rates* H for source and channel must also be identical. Hence H for the source cannot be greater than C.

One method of showing that it is possible to approach the maximum rate is to find an explicit code that will achieve this. Consider the possible sequences of N symbols from the source, and arrange the probabilities $p(S)$ in order of decreasing probability

$$p_1 \geqslant p_2 \geqslant p_3 \geqslant \ldots \geqslant p_s.$$

Let

$$P_i = \sum_{j=1}^{i-1} p_j.$$

The ith message is encoded by expanding P_i as a binary fraction and using only the first t_i places, where

$$\log_2 1/p_i \leqslant t_i < 1 + \log_2 1/p_i, \tag{4}$$

or

$$1/2^{t_i} \leqslant p_i < 1/2^{t_i - 1}.$$

This ensures a reversible code. Thus P_{i+1} differs by p_i from P_i and its binary expansion will therefore differ in one or more of the first t_i places, and similarly for all others. The average length of the encoded message will be

$$E(t) = \Sigma p_i t_i,$$

where from (4)

$$-\Sigma p_i \log p_i \leqslant \Sigma p_i t_i \leqslant \Sigma p_i (1 - \log p_i)$$

or

$$J_N \leqslant E(t) \leqslant 1 + J_N.$$

The average number of binary digits per message symbol is $E(t)/N$, so as $N \to \infty$ we obtain H' as the limit of this average. (If the permissible signal sequences are not binary digits, we may set up a correspondence between the one set and the other.) It will be noticed that this code employs the intuitive choice of the shortest symbols for the most probable sequences of message symbols.

Noisy channels. When the received signal is distorted by noise, the transmitted signal sequence U and received signal sequence V, say, must be distinguished. Let the joint probability of U and V be $p(U, V)$ and the conditional probability of U for given

V be $p_V(U)$. We define the conditional entropy $J_V(U)$ of U, given V, by the relation

$$J(U, V) = -E\{\log p(U, V)\}$$
$$= -E\{\log p(V)\} - E_V\{E_{U|V} \log p_V(U)\}$$
$$= J(V) + J_V(U),$$

and correspondingly as the lengths of the sequence increase indefinitely

$$H(U, V) = H(V) + H_V(U). \tag{5}$$

The information lost through the noise effect or the *equivocation* is defined as this conditional entropy $H_V(U)$ of the transmitted signals when the received signals are known. The actual rate of transmission (per unit time) is thus defined as

$$R = H(U) - H_V(U), \tag{6}$$

and the capacity C of the noisy channel is defined as max R as U varies.

To establish the relevance of this definition for noisy channels, i.e. that we may still transmit at a rate approaching C with arbitrarily small equivocation, we shall first digress for a moment on formula (2). For long enough sequences the ergodic condition is assumed to imply that 'nearly all' such sequences are 'typical', and do not need further averaging, that is,

$$-\log p(S)/T \to H \tag{7}$$

in probability (for explicit conditions ensuring this, see §§ 2·21 and 8·2). For such a sequence this implies that its probability $p(S)$ tends to be independent of S and given by 2^{-TH}, that is, we may replace the actual *ensemble* of sequences by a set of 2TH equally probable sequences, ignoring the remainder (the rigorization of this argument will be omitted).

Hence for a message source with an information rate R_0, we have in a long time T a set of 2^{TR_0} equiprobable relevant sequences. We may associate these with the $2^{TH(U)}$ signal sequences as we wish. Consider first their *random* association; then the probability of a particular signal sequence being a possible message will be $2^{T[R_0 - H(U)]}$. Each received signal sequence may have come from $2^{TH_V(U)}$ transmitted signal sequences, and the probability that none of these signal sequences is a possible message (apart from the actual message) is P, where

$$\log P = 2^{TH_V(U)} \log \{1 - 2^{T[R_0 - H(U)]}\}.$$

Now if $R_0 < H(U) - H_V(U)$, we may write

$$R_0 - H(U) = -H_V(U) - \eta,$$

where $\eta > 0$. Hence $\log P \sim 2^{-T\eta} \to 0$ as $T \to \infty$, or $P \to 1$. The above result shows that the average probability of error $\to 0$ for the combination of message and signal sequences in all possible ways. It follows that there exists at least one combination for which the probability of error $\to 0$.

For $R_0 > R$, it is only possible to transmit with a finite equivocation $R_0 - R$. This follows since we may transmit the fraction R/R_0 of the message accurately, and the remainder gives an equivocation $R_0 - R$.

The elimination of the noise effect has of course to be achieved by a redundancy in the coded or transmitted signal sequence compared with the original message sequence. This may have to be achieved in practice in a rather complicated way, but redundancy in the original message may reduce the need for it. We define redundancy in a message sequence by $1 - H/H_{\max}$. For the English language, it appears to be well over 50 %.

Use of continuous signal processes. With continuously varying signals the information content still remains effectively finite, owing to the presence of noise and to the finite resolving power of any apparatus. Using for convenience a Fourier representation (corresponding closely to the common use of electromagnetic signals), we assume that the channel is restricted to the frequency band $0-W$ cycles per second. Now if any realized function $x(t)$ is so limited it is completely determined by its values at intervals $1/(2W)$. In fact we may write

$$x(t) = \sum_{-\infty}^{\infty} x_n \frac{\sin \pi(2Wt - n)}{\pi(2Wt - n)} \tag{8}$$

(cf. J. M. Whittaker, 1935). The restriction to the same frequency range of spectral frequencies of a stationary process only defined at values $1/(2W)$ apart is another aspect of this result. Any section S of this function of duration T will thus be represented by $2TW$ coordinates, and will correspond to a point in a space of $2TW$ dimensions. With this representation, it should be noticed that

$$\frac{1}{T}\int_0^T x^2(t)\,dt = \frac{1}{2TW}\Sigma x_n^2,$$

when $x(t) = 0$ outside the interval $0-T$, since

$$2W\int_{-\infty}^{\infty}\frac{\sin\pi(2Wt-m)}{\pi(2Wt-m)}\frac{\sin\pi(2Wt-n)}{\pi(2Wt-n)}\,dt = \delta_{m,n},$$

where $\delta_{m,n} = 1$ $(m = n)$, 0 $(m \neq n)$. The average power P of the signal is defined by the above mean square; thus the square of the distance of the signal point from the origin is $2WTP$.

The particular process defined by

$$X(t) = \sum_{-\infty}^{\infty} A_n \frac{\sin\pi(2Wt-n)}{\pi(2Wt-n)}, \tag{9}$$

where the A_n are independent normal coefficients, all with the same variance N, will be called '(band-limited) white noise' of average power N.

The information (entropy) measure J of a set of continuous variables is defined by $-E\{\log f\}$, where f is the probability density function. In contrast with the measure for discrete variables, it is not defined absolutely, and will change with a change of coordinates affecting the density function. Entropy rates per second or per degree of freedom (one for each coordinate) are also readily defined.

Entropy loss in linear filters. The change in entropy with a change of coordinates from $\mathbf{x}$ to $\mathbf{y}$ is easily seen to be

$$\Delta H' = -E\{\log K\}, \tag{10}$$

where K is the Jacobian $(\partial\mathbf{x}/\partial\mathbf{y})$ of the transformation. For the linear transformation $\mathbf{y} = \mathbf{A}\mathbf{x}$, this gives

$$\Delta H' = \tfrac{1}{2}\log|\mathbf{A}|^2. \tag{11}$$

A linear filter merely changes each frequency ω by a factor $\gamma(\omega)$, and as each frequency has two degrees of freedom (sine and cosine) it is readily shown (e.g. by taking a discrete set of *frequencies* at intervals $2\pi/T$ and proceeding to the limit as $T \to \infty$) that

$$\Delta H' = \frac{1}{2\pi W}\int_0^{2\pi W}\log|\gamma(\omega)|^2\,d\omega. \tag{12}$$

Optimum entropy properties of white noise. It should be noted that the distribution in one variable with maximum entropy for a given standard deviation is not the uniform or rectangular distribution, but the normal or Gaussian one. For if

$$J = -\int f(x) \log f(x)\, dx$$

is maximized subject to

$$\int f(x)\, dx = 1, \quad \int x f(x)\, dx = 0, \quad \int x^2 f(x)\, dx = \sigma^2,$$

we have, by the calculus of variations, to maximize

$$\int f(x)\, [\lambda x^2 + \mu x + \nu - \log f(x)]\, dx,$$

whence

$$-1 - \log f(x) + \lambda x^2 + \mu x + \nu = 0,$$

or

$$f(x) = \frac{1}{\sqrt{(2\pi\sigma^2)}}\, e^{-\frac{1}{2}x^2/\sigma^2}.$$

For this distribution it is easily shown that $J = \log \sqrt{(2\pi e\sigma^2)}$. Similarly in n dimensions, for given second-order moments, the n-dimensional normal distribution is the optimum. Further, given average σ^2, the entropy is obviously a maximum for independent variables and when all the σ's are equal.

Long realizations of an ergodic process will again be typical of the stochastic process, and we assume

$$-\log f(S)/n \to H' \tag{13}$$

in probability as n, the number of coordinates, tends to infinity. We have also, since there are $2W$ coordinates per second, the identity

$$2WH = H'. \tag{14}$$

For white noise, $H' = \log \sqrt{(2\pi e N)}$, $H = W \log (2\pi e N)$. From (13) we may think of the typical density function for a long realization as $2^{-nH'}$, constant over a volume $2^{nH'}$. For example, in the case of white noise, this volume is the sphere of radius $\sqrt{(nN)}$.

The entropy of white noise we have seen depends on its power N. It is convenient to define the *entropy power* N_1 of any other process as the white noise power having the same entropy H', i.e.

$$N_1 = 2^{2H'}/(2\pi e). \tag{15}$$

A useful inequality concerning the sum of two independent series, $Z = X + Y$, is that

$$N_1(X) + N_1(Y) \leqslant N_1(Z) \leqslant N(X) + N(Y), \tag{16}$$

where N_1 and N denote entropy power and average power respectively. The upper limit follows immediately because $N_1(Z) \leqslant N(Z) = N(X) + N(Y)$. To obtain the lower bound, let us investigate the minimum entropy for Z, given the X and Y entropies. We denote the density functions for X, Y and Z by f, g and h respectively, the argument being a vector set of coordinates $\mathbf{u}$. Introducing undetermined multipliers, and remembering also that

$$\int f(\mathbf{u})\, d\mathbf{u} = \int g(\mathbf{u})\, d\mathbf{u} = 1,$$

$$h(\mathbf{u}) = \int f(\mathbf{v})\, g(\mathbf{u} - \mathbf{v})\, d\mathbf{v},$$

we minimize

$$I = -\int \{h \log h + \lambda f \log f + \mu g \log g + \xi f + \eta g\}\, d\mathbf{u},$$

$$\delta I = -\int \{(1 + \log h)\,\delta h + \lambda(1 + \log f)\,\delta f + \mu(1 + \log g)\,\delta g$$
$$+ \xi\,\delta f + \eta\,\delta g\}\, d\mathbf{u}$$

$$= -\int \{(1 + \log h)\,\delta h + \lambda \log f\,\delta f + \mu \log g\,\delta g + \xi'\,\delta f + \eta'\,\delta g\}\, d\mathbf{u} = 0,$$

for all permissible δf, δg. Since

$$\delta h = \int g(\mathbf{u} - \mathbf{v})\,\delta f(\mathbf{v})\, d\mathbf{v} + \int f(\mathbf{u} - \mathbf{v})\,\delta g(\mathbf{v})\, d\mathbf{v},$$

we obtain

$$\left. \begin{aligned} \int \log h(\mathbf{u})\, g(\mathbf{u} - \mathbf{v})\, d\mathbf{u} + \lambda \log f(\mathbf{v}) + \xi'' = 0, \\ \int \log h(\mathbf{u})\, f(\mathbf{u} - \mathbf{v})\, d\mathbf{u} + \mu \log g(\mathbf{v}) + \eta'' = 0. \end{aligned} \right\} \tag{17}$$

Consider the case X and Y normal, i.e.

$$f(\mathbf{u}) = \frac{\exp\{-\tfrac{1}{2}\mathbf{u}'\mathbf{A}^{-1}\mathbf{u}\}}{(2\pi)^{\frac{1}{2}n}\,|\,\mathbf{A}\,|^{\frac{1}{2}}}, \quad g(\mathbf{u}) = \frac{\exp\{-\tfrac{1}{2}\mathbf{u}'\mathbf{B}^{-1}\mathbf{u}\}}{(2\pi)^{\frac{1}{2}n}\,|\,\mathbf{B}\,|^{\frac{1}{2}}}.$$

Then $h(\mathbf{u})$ is normal with $\mathbf{C} = \mathbf{A} + \mathbf{B}$, and

$$\int g(\mathbf{u} - \mathbf{v}) \log h(\mathbf{u})\, d\mathbf{u}$$

$$= \int \{-\tfrac{1}{2}n \log 2\pi - \tfrac{1}{2}\log|\,\mathbf{C}\,| - \tfrac{1}{2}(\mathbf{u}' - \mathbf{v}')\,\mathbf{C}^{-1}(\mathbf{u} - \mathbf{v})\}$$

$$\times \frac{\exp\{-\tfrac{1}{2}\mathbf{u}'\mathbf{B}^{-1}\mathbf{u}\}}{(2\pi)^{\frac{1}{2}n}\,|\,\mathbf{B}\,|^{\frac{1}{2}}}\, d\mathbf{u}$$

$$= -\tfrac{1}{2}n \log 2\pi - \tfrac{1}{2}\log|\,\mathbf{C}\,| - \tfrac{1}{2}\mathbf{v}'\mathbf{C}^{-1}\mathbf{v} - \tfrac{1}{2}\,\text{trace}\,(\mathbf{B}\mathbf{C}^{-1}).$$

From (17) we require the equality of this expression with

$$\lambda\{-\tfrac{1}{2}n\log 2\pi - \tfrac{1}{2}\log|\mathbf{A}| - \tfrac{1}{2}\mathbf{v}'\mathbf{A}^{-1}\mathbf{v}\} + \xi'',$$

whence we must have $\lambda\mathbf{A}^{-1} = \mathbf{C}^{-1}$, or $\mathbf{A} = \lambda\mathbf{C}$. Then $\mathbf{B} = (1-\lambda)\mathbf{C}$, and equations (17) can be satisfied by suitable choice of $\lambda, \mu, \xi'', \eta''$. Now for this normal case

$$J(X) = \tfrac{1}{2}n\log(2\pi e) + \tfrac{1}{2}\log|\mathbf{A}|, \tag{18}$$

etc., or as $n \to \infty$,

$$N_1(X) = \lim_{n\to\infty}|\mathbf{A}|^{1/n} = \lambda N_1(Z), \quad N_1(Y) = (1-\lambda)N_1(Z).$$

Since in this case $N_1(Z) = N_1(X) + N_1(Y)$ is a minimum, we have in general $N_1(Z) \geqslant N_1(X) + N_1(Y)$.

Capacity of a continuous noisy channel. By means of our enumerable system of coordinates (finite for finite T), we are able to define the capacity of our channel (in general affected by noise) as before, namely, $\max R$, where R is the input entropy minus the equivocation. If the coordinates for the transmitted and received signals U, V are $\mathbf{u}$, $\mathbf{v}$, then (in formal notation)

$$R = H(U) - H_V(U)$$
$$= \lim_{T\to\infty}\frac{1}{T}\left\{-\int f(\mathbf{u})\log f(\mathbf{u})\,d\mathbf{u} + \iint f(\mathbf{u},\mathbf{v})\log\frac{f(\mathbf{u},\mathbf{v})}{f(\mathbf{v})}\,d\mathbf{u}\,d\mathbf{v}\right\}$$
$$= \lim_{T\to\infty}\frac{1}{T}\iint f(\mathbf{u},\mathbf{v})\log\left[\frac{f(\mathbf{u},\mathbf{v})}{f(\mathbf{u})f(\mathbf{v})}\right]d\mathbf{u}\,d\mathbf{v}. \tag{19}$$

It should be noticed that this formula is invariant to a change of coordinates in $\mathbf{u}$ or $\mathbf{v}$. Moreover, being defined here as the expected value of the logarithm of a *likelihood ratio* (cf. Chapter 8) R exists for rather general U and V. In the case when the noise $V - U$ is *independent* of the signal U, so that $p_V(U) = p(V-U)$, we have

$$R = H(U) - H_V(U) = H(U) + H(V) - H(U,V) = H(V) - H_U(V)$$
$$= H(V) - H(V-U). \tag{20}$$

Maximizing R for variation of U means maximizing $H(V)$, since the noise entropy is independent of U.

Hence to maximize R for given average signal power P we

must if possible choose V to correspond to 'white noise' with power $P + N$, where N is the average noise power. Since we cannot do better than this, we have

$$R \leqslant \max\{H(V)\} - W \log(2\pi e N_1)$$

$$\leqslant W \log\{2\pi e(P + N)\} - W \log(2\pi e N_1),$$

where $W \log(2\pi e N_1)$ is the noise entropy $H(V - U)$. If the noise is white noise, $N_1 = N$ and V is white noise when U is made white noise.† In this case we have the equality

$$C = W \log\{1 + P/N\}. \tag{21}$$

A comparison of the capacities of some standard systems with this ideal capacity has been made by Shannon (1949).

In other cases, suppose we still make U white noise. Then from the result (16) we have for the received signal an entropy power not less than $P + N_1$; that is,

$$\max\{H(V)\} \geqslant W \log\{2\pi e(P + N_1)\}$$

and $$C \geqslant W \log\{1 + P/N_1\}.$$

We have thus obtained in general

$$W \log\frac{P + N_1}{N_1} \leqslant C \leqslant W \log\frac{P + N}{N_1}. \tag{22}$$

For N/P small, these limits become nearly the same.

Since $N_1 < N$ except for white noise, (22) shows that white noise is the worst type of noise. For other Gaussian noise (with varying power for differing frequencies), it is evident that we can obtain white noise for V if the average power $P + N$ is greater than the maximum noise power at any frequency, by choosing the appropriate power for the signal, made Gaussian, at each frequency. Thus the upper limit in (22) is then reached. It cannot be strictly reached in other cases, for no Gaussian variable can be resolved into two independent non-Gaussian variables.

Continuous messages. In the preceding section we have seen that the capacity of a noisy continuous channel, with band-limited frequency and finite signal and noise power, is finite, in relation to the transmission of a discrete message sequence with

---

† This may ideally be achieved by an asymptotic coding procedure, as in the discrete channel case. (The *messages* are here still assumed discrete.)

finite entropy rate. This may be immediately seen in terms of the geometrical representation of the process. While we have represented our signals in an entire space of $2TW$ dimensions, the restriction on signal power implies for long sequences restriction to signal points within a distance $\sqrt{(2TWP)}$ of the origin. Noise shifts each signal point. For example, with white noise the received signals lie within a distance $\sqrt{[2TW(P+N)]}$ of the origin, and $\sqrt{(2TWN)}$ of the corresponding transmitted signals. The number of distinguishable signals thus remains finite, and for white noise is limited by the volume ratio $(1+P/N)^{WT}$, as we saw in result (21).

If the message source also refers to a continuously varying quantity, it is therefore not possible to transmit messages exactly, for their exact specification would require an infinite number of binary digits at each instant. However, given some permissible *tolerance*, we may replace the original message source X by an equivalent coarser signal. This tolerance v will in general be defined in terms of some measure of the difference between the transmitted and finally received process, e.g. their mean-square discrepancy. It will be assumed that the stochastic process corresponding to the message source can be represented to as close an approximation as desired by coordinate representations of the type considered above. For any such approximating representation, a length T will still have a finite set $\mathbf{x}$ of coordinates. If the final received message Y is represented by $\mathbf{y}$, we shall define the entropy rate of the source, relative to a given fidelity or tolerance v, by $R_0 = \min R$, where R is defined, as in (19), by

$$R = \lim_{T \to \infty} \frac{1}{T} E \left\{ \log \frac{f(\mathbf{x}, \mathbf{y})}{f(\mathbf{x}) f(\mathbf{y})} \right\},$$

$f(\mathbf{x}), f(\mathbf{y})$ denoting the (different) density functions of X and Y and $f(\mathbf{x}, \mathbf{y})$ the simultaneous density function. The minimum is to be obtained for constant v, by varying the conditional distribution of Y for given X.

With this definition, a source with rate R_0 can be transmitted over a channel of capacity C if $R_0 \leqslant C$, but not if $R_0 > C$. The last part of this theorem follows immediately, since R_0 is an R, and no R exceeds C. To establish the first part, suppose $f_0(\mathbf{x}, \mathbf{y})$ corresponds to the system which minimizes R to R_0. Without repeating previous arguments in detail, we can see that, by suitable encoding, the received messages Y could have been transmitted at rate R_0 with arbitrarily small equivocation, and thus be available at the transmitter. The transition from an X to a Y is arranged according to $f_0(\mathbf{x}, \mathbf{y})$, this 'coarsening' of the message closely corresponding to coarsening of a message through noise. We choose at random $2^{R_0 T}$ 'high-probability' Y's out of the total such set of $2^{T H(Y)}$, and associate, with each, $2^{T H_Y(X)} X$'s. As T increases this covers almost all X's, for each of which at least one appropriate Y in the related set will thus be available. From the ergodic assumptions this selection of the transmitted Y's does not affect the fidelity v, which is ensured for each 'high probability' pair X, Y without further averaging.

To illustrate this result, suppose that v is based on the mean-square discrepancy, and that the original message source is white noise. Then

$$R_0 = \min\{H(X) - H_Y(X)\} = H(X) - \max\{H_Y(X)\}.$$

But $\max\{H_Y(X)\}$ occurs when the 'noise' $Y - X$ is white noise. and is $W_1 \log(2\pi e v)$, where W_1 is the band-width of the message source. Thus

$$R_0 = W_1 \log(2\pi e Q) - W_1 \log(2\pi e v) = W_1 \log(Q/v), \qquad (23)$$

where Q is the average message power. More generally, for any message source of band-width W_1, Q average power and Q_1 entropy power,

$$W_1 \log(Q_1/v) \leqslant R_0 \leqslant W_1 \log(Q/v). \qquad (24)$$

The lower bound follows because $H(X) = W_1 \log(2\pi e Q_1)$ and the value above for $\max\{H_Y(X)\}$ can only be reached in the case of white noise. The value for the upper bound may be seen from the geometrical representation. In the message space the message point is confined to a sphere of radius $\sqrt{(2TW_1 Q)}$, and the root-mean-square discrepancy v allows a distance $\sqrt{(2TW_1 v)}$ from message X to the received message Y. The consequent entropy rate, determined by the number of different messages which can be transmitted, is restricted by the ratio of these two volumes.

Further notes. In practice, apart from the technical difficulty of achieving optimum results, the above criteria of efficiency may not of course always be the most relevant. This limitation refers not merely to the use of the root-mean-square discrepancy between original and received message as a measure of fidelity, but to the ignoring of finite delays (however long) between message and reception in comparison with the indefinitely long processes considered. The rapid transmission of short signals would thus require separate consideration.

It is also evident that the theory would not necessarily be relevant to the important problem of detecting slight signals, e.g. the detection of radar signals above the noise background. In fact this will often be more like the inference problems to be considered in Chapter 8, and in particular, the most efficient detection of a signal in cases where the probability or likelihood function of the received data S is known to be $f(S \mid H_0)$ in the presence of noise alone and $f(S \mid H)$ if the signal is also present must be based on the likelihood ratio f/f_0 (cf. Lawson and Uhlenbeck (1950), Chapter 7). Similarly the efficient estimation of, say, a target distance θ by a radar signal must conform to the principles of estimation summarized in Chapter 8.

Chapter 8

THE STATISTICAL ANALYSIS OF
STOCHASTIC PROCESSES

8·1   Principles of statistical inference

The statistical analysis of stochastic processes arising in nature does not differ in principle from the analysis of other types of statistical data, but the existence of some dependence or continuity in the successive observations will often mean that the classical methods become inadequate, and need extension. Moreover, there are repercussions on the practical side, for unless the statistician has a well-defined and realistic model of the actual process he is studying, his analysis is likely to be abortive. It is of course true that a statistician must always be fully cognisant of how his data were collected and of any other relevant information, but in the case of stochastic processes this ancillary information should certainly include as thorough a theoretical knowledge of the mechanism and structure of the process as possible. This is largely because dependence has so many more possibilities *a priori* than independence that these will usually need to be drastically restricted in any particular context. However, before we consider the problem of inference for stochastic processes further it will be as well to summarize the statistical principles to be used (for further details the reader is referred to Cramér (1946) or M. G. Kendall (1946)).

The problems of statistics may broadly be classified into (i) problems of specification, (ii) problems of statistical inference. Theoretically we may discuss the first without the second, as we have been doing for stochastic processes up till now in this book, but, as stressed above, the converse is not true. It is also perhaps true to add that the practical use of theoretical specifications can hardly be separated from inference problems, for one of the latter's functions is to check the adequacy of the specification. The more detailed the specification the narrower is the inference problem, but at the same time such a detailed specification may prove untenable as a representation of the data.

Where possible, the specification or probability model H should specify precisely the probability of the data S (and of any alternative set of data S' which might have arisen under the same conditions). It is sometimes important to distinguish between H and the 'structural model' leading to H, though if two different structural models give rise to the same H, no statistical analysis can of course discriminate between them. The probability $P\{S \mid H\}$ will be denoted by p. If S refers to observations having a continuous range of possible values, p will be strictly zero, but we may consider alternatively the density function $f(S \mid H)$. The function p or f is called the likelihood function, and we shall denote $\log p$ or $\log f$ by L.

Statistical estimation. In many cases, while H is not known exactly, it may be known or provisionally assumed known apart from one or more unknown constants θ_i (each with possible values assumed extending over a continuous range). In the case of one unknown only, we shall refer to it as θ. In the reference to sequential analysis in §4·1 the information function introduced by R. A. Fisher has already been defined as

$$I(\theta) = E\left\{\left(\frac{\partial L}{\partial \theta}\right)^2\right\} \quad \left(E\left\{\frac{\partial L}{\partial \theta}\right\} \text{ assumed zero}\right). \tag{1}$$

In the case of more than one unknown, we define the information matrix

$$\{I_{ij}\} \equiv E\left\{\frac{\partial L}{\partial \theta_i} \frac{\partial L}{\partial \theta_j}\right\}. \tag{2}$$

Then under suitable conditions (the most important practical condition to check is noted in (1), that $E\{\partial L/\partial \theta\}$ is zero, this usually following if the range of the random variables does not depend on θ) we have for any estimating function $T(S)$

$$E\{(T - \theta)^2\} \geqslant b^2 + (1 + \partial b/\partial \theta)^2/I(\theta), \tag{3}$$

where $b \equiv E(T) - \theta$ is the bias in T. For unbiased estimates $b = 0$ and (3) reduces to

$$\sigma^2 \geqslant 1/I(\theta), \tag{4}$$

where σ^2 is the variance of T. The estimating function T may be any appropriate function of the observations; if it is a linear function, we shall call T a linear estimate.

Restricting ourselves for simplicity to unbiased estimates (a wide class of estimates is at least asymptotically unbiased in large samples), we have in the case of more than one unknown that the covariance matrix of the set of estimates T_i of θ_i is 'bounded' below by the inverse $\{I^{ij}\}$ of the information matrix $\{I_{ij}\}$, by which we mean in particular that σ_i^2, the variance of T_i, is not less than the corresponding diagonal element of $\{I^{ij}\}$ (and in general that similar inequalities hold for any linear transformation of the estimates).

The condition for the equality sign in (4) is that

$$\frac{\partial L}{\partial \theta} = I(\theta)\,(T - \theta). \tag{5}$$

The corresponding set of conditions for several unknowns θ_i is

$$\frac{\partial L}{\partial \theta_i} = \Sigma_j I_{ij}(T_j - \theta_j) \quad \text{for all } i. \tag{6}$$

The 'maximum-likelihood' estimates $\hat{\theta}_i$ are defined as any set of values of θ_i that maximize L. In particular when $\partial L/\partial \theta_i$ exist, they satisfy the equations

$$\left[\frac{\partial L}{\partial \theta_i}\right]_{\hat{\theta}_i} = 0. \tag{7}$$

Under some further conditions, which have until recently usually included the assumption of independent observations, the estimates $\hat{\theta}_i$ have the property that as the number n of observations is increased they tend to be normally distributed about θ_i with their covariance matrix the optimum compatible with the above results. Thus while optimum estimates in the above variance sense may not exactly exist (if they do, they are identical with the maximum likelihood estimates) equation (7) will provide estimates which have these optimum properties at least asymptotically. The relation

$$E\left\{\frac{\partial L}{\partial \theta_i}\frac{\partial L}{\partial \theta_j}\right\} = E\left\{-\frac{\partial^2 L}{\partial \theta_i \partial \theta_j}\right\} \tag{8}$$

usually holds, and is then often convenient for evaluating I_{ij}. The asymptotic approximation (cf. equation (7), §7·2)

$$-\frac{\partial^2 L}{\partial \theta_i \partial \theta_j}\Big/ I_{ij} \sim 1 \tag{9}$$

is also useful.

Asymptotic confidence intervals. The classical large-sample estimation procedure is in effect to use the normal approximation, the standard error indicating on this basis the possible interval on either side of the estimate in which the unknown is likely to lie. This procedure may often be improved where worth while by making use directly of quantities such as L or $\partial L/\partial\theta$. For example, under the conditions assumed above, we have the quantity $L' \equiv \partial L/\partial\theta$ with mean zero and variance $I(\theta)$. For independent observations at least, L' tends to normality, and an approximate confidence interval† is obtained by solving the equation

$$T(\theta) \equiv L'/\sqrt{I(\theta)} = \pm\lambda, \qquad (10)$$

where $\pm\lambda$ are the upper and lower limits corresponding to any stipulated probability risk of a standardized normal variate falling outside these limits. Further improvements can be obtained by studying further the skewness or other higher moments of L'. Thus the general formula for the third cumulant of L' is

$$\kappa_3 \equiv E\left\{\left(\frac{\partial L}{\partial\theta}\right)^3\right\} = 2E\left\{\frac{\partial^3 L}{\partial\theta^3}\right\} + 3\frac{\partial I(\theta)}{\partial\theta}, \qquad (11)$$

and in place of (10) we may use the equation

$$T(\theta) - \tfrac{1}{6}\kappa_3(\lambda^2 - 1)/I^{\frac{3}{2}} = \pm\lambda. \qquad (12)$$

The correcting term is $O(1/\sqrt{n})$, and the neglected terms are $O(1/n)$. Similar modifications are possible in the case of more than one unknown, though they may become complicated.

While these asymptotic methods have been replaced where feasible by more precise small-sample methods, they have a wide range of applicability, and as precise small-sample methods are less often available for the stochastic process problems we shall want to study, are particularly relevant here.

In some cases where p cannot be completely specified, estimates can still be chosen with reasonable properties for a wider class of H. An important example is that of 'least-squares' estimates, which in classical problems are, for a particular class of H,

---

† Anyone unfamiliar with the precise probability interpretation of a confidence interval should also consult the references already mentioned (Cramér, Kendall (1946)).

unbiased with minimum variance in the group of linear estimates. As we shall see in Chapter 9, this property may be extended in a certain asymptotic sense to estimation problems for auto-regressive series.

Statistical tests. For testing one hypothesis H_0 against a rival hypothesis H the best criterion to use is the 'likelihood ratio' p/p_0, or f/f_0 if only densities exist, where $p_0 \equiv P\{S \mid H_0\}$, etc., or equivalently $L - L_0$ (this was assumed in the sequential sampling rule of §4·1). If p has the form

$$p \equiv P\{S \mid H\} = P\{U \mid H\} P\{S \mid U\},$$

where the last factor is independent of H (or the aspects of it which are in doubt) and U is a reduced set of statistics obtained from the original observations S, then p/p_0 is a function only of U, which is said to be a set of *sufficient* statistics in regard to H. In particular, where there is only one unknown θ, U may be a single quantity or 'statistic' T.

The manner of using p/p_0 will in general depend on the situation, especially if the rival hypothesis H is merely one of a class. But in the probability space of S a certain region will be favourable to H_0 as against H, and this is defined by some condition $p/p_0 \leqslant \lambda$. When a sufficient statistic T exists, the region will be defined in terms of critical values of T. It may be noted that if equation (5) holds then T is sufficient (but the converse need not hold).

In certain cases when H and H_0 are not completely specified, but depend on further 'nuisance' parameters which are unknown, it is sometimes possible and desirable to remove these completely by considering the conditional probabilities $p\{S \mid H, U\}$, where U is a set of sufficient statistics for the nuisance parameters. When these cannot be removed exactly, they may be removable approximately by the substitution of their maximum-likelihood estimates.

Suppose we now wish to test the 'goodness of fit' of a model specified entirely by the set of parameters θ_i, ϕ_j, where θ_i $(i = 1, ..., r)$ are supposed known and ϕ_j $(j = 1, ..., s)$ unknown. The alternative class of hypotheses for comparison has both θ_i, ϕ_j unknown. For a useful asymptotic test we substitute for ϕ_j in the former case their maximum-likelihood estimates (θ_i given), and for both θ_i, ϕ_j their simultaneous estimates in the latter case. Denote θ_i, ϕ_j jointly by ψ_m. Then for $\hat{\psi}_m - \psi_m$ small,

$$L(\psi_m) \sim L(\hat{\psi}_m) - \tfrac{1}{2}\Sigma_{m,n}(\hat{\psi}_m - \psi_m)\left(-\frac{\partial^2 L}{\partial \psi_m \partial \psi_n}\right)(\hat{\psi}_n - \psi_n),$$

$$\equiv L(\theta_i, \phi_j)$$

$$\sim L(\theta_i, \hat{\phi}_j(\theta_i)) - \tfrac{1}{2}\Sigma_{j,k}(\hat{\phi}_j(\theta_i) - \phi_j)\left(-\frac{\partial^2 L}{\partial \phi_j \partial \phi_k}\right)(\hat{\phi}_k(\theta_i) - \phi_k),$$

whence, by subtraction,

$$2[L(\hat{\psi}_m) - L(\theta_i, \hat{\phi}_j(\theta_i))] \sim \chi_1^2 - \chi_2^2 = \chi_3^2, \tag{13}$$

where, if the standard asymptotic properties for the maximum-likelihood estimates hold, χ_1^2, χ_2^2 and χ_3^2 are on the basic or *null* hypothesis χ^2 quantities (sums of squares of independent normal variables with zero means and unit variances) with degrees of freedom (number of independent variables) $r + s$, s and r respectively. When $s = 0$, (13) becomes $-2[L - L(\hat{\psi}_m)]$ or χ_1^2 with r degrees of freedom. The asymptotic χ^2 form for the expression in (13) depends on the sample size n being large; it is equivalent to a χ^2 form *in distribution* up to and including the $O(1/\sqrt{n})$ terms, but neglecting $O(1/n)$.

All statistical inference problems do not fall under these headings of estimation, discrimination and goodness of fit tests, but we shall see that they cover most problems of analysis arising from stochastic processes.

8·11 Application to stochastic processes.

If we now attempt to survey the various types of data arising from stochastic processes some types may be classed as analysable by classical methods. The simplest type is (a) the purely random sequence of independent events, though here one should add the warning that if the order and independence of the observations is a feature to be tested, as in sequences of random numbers, the relevant tests will be those belonging in general to dependent sequences (see the analysis of probability chains in §8·2). Any non-random component can be tested and separated, so that the classical tests of means and regression coefficients were many years ago extended to test for the significance of strictly periodic components in time-series. But again it will be noticed that the validity of this procedure depends on the class of alternative hypotheses envisaged, and if in the last case the alternative is a non-deterministic time-series with continuous spectrum, the classical procedure will be inadequate. A second general type of problem analysable by classical methods occurs if (b) independent repetitions of the process are available, and some particular feature such as the number of bacteria or number of particles at a given time is observed. Examples are the numbers of mutated bacteria in replicate cultures (§ 4·31) and the numbers of electrons in a cascade shower (mentioned in §§ 3·4 and 3·42). Here the theory of stochastic processes is essential to provide a model for the observed phenomena, but the replication still allows classical methods of analysis and comparison of the data with the theoretical model. In many applications, however, the dependence of successive observations in a single observed sequence, or, if more than one sequence are available, a mutual interdependence of the observations, makes new methods necessary. For continuous time-records, a re-examination of the inference problem is also clearly necessary.

In the case of a random sequence for which a finite number of observations is available no new difficulty in principle arises, for the methods summarized in § 8·1 are largely applicable to dependent as well as independent observations. The known theorems on the asymptotic theory of maximum-likelihood estimates do not in general apply to dependent observations, and have to be extended. These extensions, required also for the asymptotic properties of L', are associated with the extension of the Central Limit Theorem to dependent observations. Another more practical difficulty is that the distribution and dependence of the observations representing stochastic processes may well in many cases be imperfectly known, so that the precise formulation of the likelihood function may not be feasible without excessive and dubious idealization; in such cases methods of broader validity (like the 'least-squares' estimates to be considered in the discussion of autoregressive series) are advisable.

When we consider continuous time-records, the further theoretical problem is to set up an equivalent of the probability function for such a record of given duration; this may be done if, as has been assumed in this book (cf. § 1·3), we can describe it fully by an enumerable sequence of coordinates. It will be sufficient for our purpose to note two of the most useful ways of doing this:

(1) For processes m.s. continuous we may consider the values at n points $t_1, \ldots, t_n$, and then let n increase so that max $(t_r - t_{r-1})$ decreases to zero.

(2) For processes for which $dX(t)$ is zero except at an enumerable set of random times $T_1, T_2, \ldots, T_N$ (N also random with $P\{N < \infty\} = 1$), then the probability may be specified in terms of these times and the corresponding values of $dX(t)$, together with $X(0)$ and N. (An example is the birth-and-death process of §3·4.)

More abstract representations have been considered by Grenander (1950), who has shown that it is legitimate to evaluate the likelihood ratio for the data either as in (2) (when available) or as the limit of (1) as n increases. This likelihood ratio, if evaluated for a fixed hypothesis H_0 in the denominator, can obviously also be used in estimation problems. The exact or 'small-sample' theory of estimation then still applies. The problem of elucidating minimum conditions under which the asymptotic properties of maximum-likelihood estimates hold becomes at first sight even more formidable, though by representations such as (1) or (2) the problem should be reducible to a random-sequence problem, or the limit of one. We should moreover expect, by analogy with the relatively simple case of probability chains discussed in the next section, that in the case of completely stationary and ergodic processes for which the dependence drops off sufficiently rapidly the classical asymptotic properties will still apply. Relevant results for autoregressive and other sequences will be referred to in due course.

8·2   The analysis of probability chains

It is instructive to examine in some detail the inference problem for a simple probability chain, which we shall suppose to be a stochastic process defined for discrete values of both variable and parameter, and whose probability dependence does not extend more than a finite number of intervals (k, say). For example, we may wish to test the adequacy of a Markov chain model to the observational sequences obtained by Svedberg and Westgren (see Chandrasekhar, 1943) from their counts of colloidal particles in a small volume. We assume at present that the total number s of states is finite; in addition to stationarity (or ultimate stationarity) of the series, we shall also require to assume an ergodic property of the type called 'positive regularity' in Chapter 2.

Denote the random sequence by

$$S \equiv X_1, X_2, ..., X_{n-1}, X_n,$$

where the suffices to X refer of course to its serial order, and not to its realized value. The probability of this sequence S is

$$P\{S\} = P\{X_1\}P\{X_2 \mid X_1\}P\{X_3 \mid X_1, X_2\} ... P\{X_k \mid X_1, X_2, ..., X_{k-1}\}$$

$$\times \prod_{i=1}^{n-k} P\{X_{k+i} \mid X_i, ..., X_{k+i-1}\}. \tag{1}$$

The variable X can take s values denoted conventionally by the states $1, 2, ..., s$, and hence a subsequence $X_h, X_{h+1}, ..., X_{h+k-1}, X_{h+k}$ can take s^{k+1} 'values' (specified by the simultaneous values of the $k+1$ X's). Let the frequency of any such specified value $(i, j, ..., q, r)$ for the subsequence of length $k+1$ be $N_{ij...qr}$. For brevity we shall often denote the value $(i, j, ..., q)$ of the subsequence $X_h, ..., X_{h+k-1}$ by u, and correspondingly the frequency $N_{ij...qr}$ by N_{ur}. We denote also the conditional probability of the value r for the last variable of a subsequence of length $k+1$, given the value u for the subsequence of length k, by p_{ur}. Then (1) may be written

$$\log P\{S\} = \sum_{j=1}^{k} \log P\{X_j \mid X_1, X_2, ..., X_{j-1}\} + \Sigma_{u,r} N_{ur} \log p_{ur}. \tag{2}$$

As n increases, the second sum in (2) will become the dominant part of $\log P\{S\}$. The maximum-likelihood estimates of the p_{ur} are given by finding the maximum of $L \equiv \log P\{S\}$ subject to the condition

$$\Sigma_r p_{ur} = 1, \tag{3}$$

whence we easily obtain for n large enough the estimates

$$\hat{p}_{ur} = N_{ur}/N_u, \tag{4}$$

where

$$N_u = \Sigma_r N_{ur}. \tag{5}$$

The goodness of fit criterion $-2[L - L_{\max}]$ or λ, say, for n large thus becomes

$$-2\Sigma_{u,r} N_{ur} \log (p_{ur} N_u / N_{ur})$$
$$= 2[\Sigma_{u,r} N_{ur} \log (N_{ur}/m_{ur}) - \Sigma_u N_u \log (N_u/m_u)], \quad (6)$$

where $m_{ur} = nP_{ur} = nP_u p_{ur}$, $m_u = nP_u$ (P_{ur}, P_u denoting absolute probabilities of the 'values' (u, r) and u). In the case $k = 0$, we have the independent case; (2) becomes exactly

$$L = \Sigma_r N_r \log p_r \quad (7)$$

and (6) becomes $\lambda = 2\Sigma_r N_r \log (N_r/m_r)$.   (8)

It is well known that the expression (8) is the appropriate criterion in this case, and has asymptotically the χ^2 distribution with $s - 1$ degrees of freedom as n increases. If we wish we may replace it by the χ^2 expression

$$\chi^2 = \Sigma_r \frac{(N_r - m_r)^2}{m_r}, \quad (9)$$

to which it is approximately equivalent as the m_r increase. Similarly, we may replace the sums in (6) by sums like (9) if we wish, though there are some advantages in retaining the natural form (6).

One incidental result from (2) is that if the N_{ur}/n are consistent estimates of $P_{ur} = P_u p_{ur}$ (this follows from results established below), then

$$\lim_{n \to \infty} \frac{L}{n} = \Sigma_{u,r} P_u p_{ur} \log p_{ur}. \quad (10)$$

This formula is important in the theory of communication (see §7·2).

From the definition of (6) it might be hoped that its asymptotic distribution will be the χ^2 distribution. Now *if we may assume that the asymptotic distribution of the N_{ur} is normal*, it follows that the criterion λ has such a limiting distribution. This would follow indirectly as a special case of equation (13), §8·1. More directly, for any limiting non-degenerate normal distribution in q variables denoted by the (column) vector $\mathbf{X}$, we have (in vector and matrix notation)

$$P\{\mathbf{X}\} \propto \frac{1}{(2\pi)^{\frac{1}{2}q} |\mathbf{V}|^{\frac{1}{2}}} \exp \{-\tfrac{1}{2}(\mathbf{X} - \mathbf{m})' \mathbf{V}^{-1} (\mathbf{X} - \mathbf{m})\} \quad (11)$$

and $-2[\log P\{X\} - \log P_{\max}\{X\}] = (X - m)' V^{-1}(X - m)$,   (12)

which is a χ^2 with q degrees of freedom. If the distribution is degenerate, with r constraints, we see, by an orthogonal transformation of the variables, that λ has a χ^2 distribution with $q - r$ degrees of freedom.

The asymptotic normality of the N_{ur} will be demonstrated below. First of all, however, we may note the number of degrees of freedom for λ. The N_{ur}, even if normal, are not unrestricted.

By definition in (5) $N_u \equiv N_{ij\ldots q}$, say, is obtained from $N_{ur} \equiv N_{ij\ldots qr}$ by summing over r. Alternatively, if we consider the frequencies $N_{hu} \equiv N_{hij\ldots q}$ and obtain N_u', say, by summing over the first suffix h, we are obtaining a total frequency for subsequences of k consecutive terms, of exactly the same type u as those represented in N_u. We have therefore the linear relation for each u (s^k of them)

$$N_u' \equiv \Sigma_h N_{hij\ldots q} = \Sigma_r N_{ij\ldots qr} \equiv N_u,   (13)$$

except perhaps for an end-effect, in the total sequence, which becomes negligible as n increases. The relations (13) are not algebraically independent, for summing over all u gives the same total frequency n', say, on each side. However, we have also the condition

$$\Sigma_{u,r} N_{ur} = n',   (14)$$

where $n' = n - k \sim n$ is fixed. Hence there are $(s^k - 1) + 1 = s^k$ restrictions, and the number of degrees of freedom for N_{ur} and hence for λ will be (provided all m_{ur} are non-zero)

$$s^{k+1} - s^k = s^k(s - 1).   (15)$$

It may be noted that if we subtract from (6) the similar quantity defined for $(k - 1)$-dependent sequences we have two additive quantities with degrees of freedom $s^{k-1}(s - 1)$ and

$$s^k(s - 1) - s^{k-1}(s - 1) = s^{k-1}(s - 1)^2.$$

However, the two components will only be χ^2's separately if the sequence is not more than $(k - 1)$-dependent. If it is k-dependent, the expectation of each component may be *above or below* its nominal number of degrees of freedom, depending on the transition-probability matrix. For example, anticipating the detailed results given below for the case $k = 1$, $s = 2$, we find for these asymptotic expected values (calculated for the standard quadratic form for χ^2) the analyses:

(i) $p_{11} = p_{21}$ (independence):
$$(3 \cdot 0 - 2 \times 1 \cdot 0) + 1 \cdot 0 = 1 \cdot 0 + 1 \cdot 0 = 2 \cdot 0,$$

(ii) $p_{11} = \frac{1}{3}, \; p_{21} = \frac{2}{3}$:
$$(2 \cdot 5 - 2 \times 0 \cdot 5) + 0 \cdot 5 = 1 \cdot 5 + 0 \cdot 5 = 2 \cdot 0,$$

(iii) $p_{11} = \frac{2}{3}, \; p_{21} = \frac{1}{2}$:
$$(3 \cdot 4 - 2 \times 1 \cdot 4) + 1 \cdot 4 = 0 \cdot 6 + 1 \cdot 4 = 2 \cdot 0.$$

For a Markov chain $(k = 1)$ the simultaneous distribution of the N_{ur}, which we can in this case write N_{ij}, may be investigated by the method described in § 2·22. We replace the transition probability matrix $\mathbf{Q}$ by

$$\mathbf{R}(\mathbf{\theta}) \equiv \{p_{ji} e^{\theta_{ji}}\} \tag{16}$$

and easily find for the cumulant function

$$K_n(\mathbf{\theta}) \equiv \log E\{\exp \Sigma_{ij} \theta_{ij} N_{ij}\}$$

the asymptotic form

$$K_n(\mathbf{\theta}) \sim n \log \mu_1(\mathbf{\theta}). \tag{17}$$

Here $\mu_1(0) = \lambda_1 = 1$ is the dominant root of $\mathbf{Q}$ (*the chain is assumed regular*), and we assume further that $\mu_1(\mathbf{\theta}) \neq 1$ if *some* $\theta_{ij} \neq 0$. This last condition, which is required to ensure the validity of (17), is automatically satisfied if $\mu_1(\mathbf{\theta}) \neq 1$ for *any* $\theta_{ij} \neq 0$, which is required if all the frequencies N_{ij} are to be represented in (17), and follows if all the p_{ij} are non-zero, as already implicitly assumed in the discussion on degrees of freedom. Under this last condition we shall see further that the variances of the N_{ij} are asymptotically of the form $n\sigma_{ij}^2$, where σ_{ij}^2 is non-zero and finite, and it follows from the Central Limit Theorem that the simultaneous distribution of the N_{ij} (appropriately scaled) is asymptotically normal.†

The distribution will be degenerate, owing to the linear restrictions on the N_{ij}. However, all that we need to know to calculate the criterion λ are the *expected* values $m_{ij} \sim n P_i p_{ij}$,

---

† It might be noticed that the essential feature is the unlimited increase with n of the number of independent 'trials' starting from each specified state; in this form the argument may be extended to chains with an enumerable but 'effectively finite' set of states, such as the emigration-immigration process.

where P_i refers to the final distribution of X, and is given by the column latent vector $\mathbf{s}_1$ corresponding to the latent root 1. The expected values m_{ij} are asymptotically correct even if the chain is not initially stationary, and only becomes so during the observed sequence, but it should be an improvement in such cases to substitute more exact expected values based on the observed initial value. Moreover, for an initially stationary process, the *exact* expected values are available, namely, $(n-1)P_i p_{ij}$ (based on the available number of transitions), or, more generally for k-dependent chains, $(n-k)P_u p_{ur}$.

If in (16) we put $p_{ij} = p_j$, we obtain the case of an *independent* sequence ($k = 0$), such as a sequence of random numbers, and we see that the above theory is required for an adequate discussion even of this case, though this has often been overlooked. In the case of $k > 1$, the only further point to make is that any k-dependent chain with s possible states can be regarded as a Markov chain in a variable with s^k possible states, specified by the set of k consecutive values of the original variable. This may be illustrated by considering the case $s = 2$, $k = 2$. We have the following transition probability matrix for the composite variable with four possible states:

	1, 1	1, 2	2, 1	2, 2
1, 1	p_{111}	0	p_{211}	0
1, 2	p_{112}	0	p_{212}	0
2, 1	0	p_{121}	0	p_{221}
2, 2	0	p_{122}	0	p_{222}

Here p_{hij} denotes the conditional probability of $X_{r+2} = j$, given $X_r = h$, $X_{r+1} = i$; and the composite variable at time $r + 2$ is specified by the two states (i, j). Since i also occurs in the composite variable at time $r + 1$, the new composite variable can only take values for which i remains constant; hence the zeros in the above table.

This transformation to an equivalent Markov chain is of course given to enable the asymptotic normality of the N_{hij} (or N_{ur} in the abbreviated notation, with $u \equiv (h, i)$) to be demonstrated by the Markov chain technique, and does not affect the previous argument about degrees of freedom. It may be verified that the occurrence of zeros in the above transition matrix does not affect the appearance of all the p_{hij} in the equation for $\mu_1(\theta)$.

The non-vanishing of these permissible probabilities p_{hij} is, as in the case $s = 2$, a sufficient condition for the required regularity property to hold. This may be seen in general for the transition matrix $\mathbf{R}$ obtained from a k-dependent chain by considering the matrix power $\mathbf{R}^k$. This powering has the effect of freeing the possible transitions between the states from the artificial restrictions imposed by their definition, and all the terms of $\mathbf{R}^k$ become positive if the original probability coefficients p_{ur} are all positive. It now follows from one of the conditions quoted in § 2·21 (see end of second paragraph, p. 34; this particular condition is due to Markov) that the process is positively regular.

The condition $p_{ur} > 0$ is necessary for the degrees of freedom $s^k(s-1)$ to be valid, but is not essential otherwise. If the degrees of freedom are adjusted to allow for zero p_{ur} weaker conditions ensuring positive regularity are sufficient; for example, for processes with no cyclic groups (no root $\mu_r(0) \neq 1$ of modulus unity) but possessing 'paths' from any state to any other, including itself, in a finite number of steps (see § 2·3; the absence of cyclic groups is automatic if the chain is an intermittent sequence obtained from a Markov chain defined for *continuous* time). If this test is used as an approximation for sequences with an enumerable but 'effectively finite' number of states, such as the colloidal particle counts, an appropriate adjustment of the degrees of freedom is of course also required.

Explicit evaluation of $\mu_1(\theta)$. An exact expression for $\mu_1(\theta)$ will in general be impossible, but its expansion in ascending powers of θ_{ij} can be investigated if required from the determinantal equation for $\mu_1(\theta)$, starting from the solution 1 for $\theta = 0$ (we may thus verify that the coefficient of $\frac{1}{2}\theta_{ij}^2$ in the expansion of $\log \mu_1(\theta)$ is non-zero and finite, this giving the variance of the corresponding 'scaled' frequency $N_{ij}/\sqrt{n}$). In the case $k = 1$, $s = 2$, however, we easily obtain (cf. the expression for the root in § 2·22)

$$\mu_1(\theta) = \tfrac{1}{2}(p_{11}e^{\theta_{11}} + p_{22}e^{\theta_{22}})$$
$$+ \tfrac{1}{2}\sqrt{\{(p_{11}e^{\theta_{11}} - p_{22}e^{\theta_{22}})^2 + 4p_{21}p_{12}e^{\theta_{12}+\theta_{21}}\}}.$$

Up to the second degree in θ_{ij}, the expansion of $\log \mu_1(\boldsymbol{\theta})$ is

$$\theta_{11}\left(\frac{p_{11}p_{21}}{p_{12}+p_{21}}\right)+\theta_{22}\left(\frac{p_{12}p_{22}}{p_{12}+p_{21}}\right)+(\theta_{21}+\theta_{12})\left(\frac{p_{21}p_{12}}{p_{12}+p_{21}}\right)$$

$$+\tfrac{1}{2}\theta_{11}^2\left\{\tfrac{1}{2}p_{11}+\frac{p_{11}^2-\tfrac{1}{2}p_{11}p_{22}}{p_{12}+p_{21}}-\frac{p_{11}^2p_{21}^2}{(p_{12}+p_{21})^2}-\frac{\tfrac{1}{2}p_{11}^2(p_{11}-p_{22})^2}{(p_{12}+p_{21})^3}\right\}$$

$$+\tfrac{1}{2}\theta_{22}^2\left\{\tfrac{1}{2}p_{22}+\frac{p_{22}^2-\tfrac{1}{2}p_{11}p_{22}}{p_{12}+p_{21}}-\frac{p_{12}^2p_{22}^2}{(p_{12}+p_{21})^2}-\frac{\tfrac{1}{2}p_{22}^2(p_{11}-p_{22})^2}{(p_{12}+p_{21})^3}\right\}$$

$$+\tfrac{1}{2}(\theta_{12}+\theta_{21})^2\left\{\frac{p_{21}p_{12}}{p_{12}+p_{21}}-\frac{p_{21}^2p_{12}^2}{(p_{12}+p_{21})^2}-\frac{2p_{21}^2p_{12}^2}{(p_{12}+p_{21})^3}\right\}$$

$$+\theta_{11}\theta_{22}\left\{\frac{-\tfrac{1}{2}p_{11}p_{22}}{p_{12}+p_{21}}-\frac{p_{11}p_{22}p_{21}p_{12}}{(p_{12}+p_{21})^2}+\frac{\tfrac{1}{2}p_{11}p_{22}(p_{11}-p_{22})^2}{(p_{12}+p_{21})^3}\right\}$$

$$+\theta_{11}(\theta_{12}+\theta_{21})\left\{-\frac{p_{11}p_{21}^2p_{12}}{(p_{12}+p_{21})^2}-\frac{p_{11}p_{21}p_{12}(p_{11}-p_{22})}{(p_{12}+p_{21})^3}\right\}$$

$$+\theta_{22}(\theta_{12}+\theta_{21})\left\{-\frac{p_{21}p_{12}^2p_{22}}{(p_{12}+p_{21})^2}+\frac{p_{21}p_{12}p_{22}(p_{11}-p_{22})}{(p_{12}+p_{21})^3}\right\}+\ldots.$$

The mean values given by the coefficients of θ_{ij} agree with the values P_ip_{ij}, and it may be verified that the variance-covariance matrix $\mathbf{V}$ given by the quadratic expression $\tfrac{1}{2}\boldsymbol{\theta}'\mathbf{V}\boldsymbol{\theta}$ (in which $\boldsymbol{\theta}'$ stands for the row vector $(\theta_{11}, \theta_{12}, \theta_{21}, \theta_{22})$) has rank 2, in agreement with the degrees of freedom $s(s-1)$ when $k=1$, $s=2$. The variance matrices for the three particular cases referred to earlier are recorded for reference (as $N_{12} \sim N_{21}$, the middle two rows and columns are identical, and for convenience are not separated).

Case (i). $p_{11}=p_{21}=p=1-q$ (independence):

$$
\begin{array}{cccc}
 & N_{11} & N_{12}, N_{21} & N_{22}
\end{array}
$$

$$n^{-1}\mathbf{V} \sim \begin{pmatrix} p^2q(1+3p) & -2p^3q+p^2q^2 & -3p^2q^2 \\ -2p^3q+p^2q^2 & pq(1-3pq) & -2pq^3+p^2q^2 \\ -3p^2q^2 & -2pq^3+p^2q^2 & q^2p(1+3q) \end{pmatrix}.$$

Case (ii). $p_{11}=\tfrac{1}{3}$, $p_{21}=\tfrac{2}{3}$:   *Case* (iii). $p_{11}=\tfrac{2}{3}$, $p_{21}=\tfrac{1}{2}$:

$$72n^{-1}\mathbf{V} \sim \begin{pmatrix} 13 & -4 & -5 \\ -4 & 4 & -4 \\ -5 & -4 & 13 \end{pmatrix}. \quad 125n^{-1}\mathbf{V} \sim \begin{pmatrix} 62 & -14 & -34 \\ -14 & 8 & -2 \\ -34 & -2 & 38 \end{pmatrix}.$$

The effect of estimating unknown parameters. As shown by Fisher in the classical χ^2 frequency theory, the estimation of m unknown parameters reduces the degrees of freedom by m, but asymptotically has no other effect on the χ^2 distribution provided that the estimation is fully efficient. The above results show how these features apply to the extended problem. Suppose (in our general notation) that the probabilities p_{ur} are unknown until m parameters α_ν have been estimated. However, the maximum-likelihood estimates of such p_{ur} individually are based on the N_{ur}, which tend to normality for large n; and, moreover, their joint distribution may be regarded as arising from the convolution of *independent* components. In these circumstances these estimates $\hat{p}_{ur}$ are efficient. The estimates of the α_ν are similarly from (2) given asymptotically by the equations

$$\Sigma_{u,r} N_{ur} \frac{\partial \log p_{ur}}{\partial \alpha_\nu} = 0, \tag{18}$$

which may be regarded both as m linear restrictions on the N_{ur}, and as equations for α_ν in terms of $\hat{p}_{ur}$ or of N_{ur}. It follows from the first of these remarks that m degrees of freedom are lost, as in the classical case; it also follows from the second that equation (18) gives efficient estimates for α_ν, and that the asymptotic χ^2 distribution, apart from this loss of degrees of freedom, is preserved.

The efficiency properties of the maximum-likelihood estimates $\hat{p}_{ur}$ provide an interesting alternative method of investigating the asymptotic fluctuation formulae for the N_{ur}. For it follows from the form of the likelihood function $P\{S\}$ in (2), i.e.

$$L \equiv \log P\{S\} \sim \Sigma_{u,r} N_{ur} \log p_{ur} \quad (\Sigma_r p_{ur} = 1),$$

that the information matrix for the p_{ur}, namely,

$$E\left\{ -\frac{\partial^2 L}{\partial p_{ur} \partial p_{tq}} \right\},$$

(which provides the inverse of the asymptotic variance matrix for the estimates $\hat{p}_{ur}$), is asymptotically equivalent to that for multinomial probabilities p_{ur} (u fixed) from $E\{N_u\} = m_u$ *independent* observations, with, moreover,

$$E\{ -\partial^2 L / \partial p_{ur} \partial p_{tq} \} = 0 \quad (u \neq t).$$

This gives immediately the variances and covariances of the $\hat{p}_{ur} = N_{ur}/N_u$ as standard multinomial formulae, with the extra results $\mathrm{cov}\,(\hat{p}_{ur}, \hat{p}_{lq}) = 0$ $(u \neq t)$. We may now use the further standard asymptotic formulae

$$N^2 P_u P_t \,\mathrm{cov}\,(\hat{p}_{ur}, \hat{p}_{lq}) \sim \mathrm{cov}\,(N_{ur}, N_{lq}) - p_{lq}\,\mathrm{cov}\,(N_{ur}, N_t)$$
$$- p_{ur}\,\mathrm{cov}\,(N_u, N_{lq}) + p_{ur} p_{lq}\,\mathrm{cov}\,(N_u, N_t) \quad (19)$$

to obtain relations between $\mathrm{cov}\,(\hat{p}_{ur}, \hat{p}_{lq})$ and $\mathrm{cov}\,(N_{ur}, N_{lq})$, leading to a set of linear equations to determine the latter in terms of the former. While the number of equations in (19) is nominally equal to the number of unknowns, some of these are not algebraically independent. However, it will be found that the degeneracies among the $\mathrm{cov}\,(N_{ur}, N_{lq})$ counterbalance this so that a unique solution is in fact possible by this method. (It was used, for example, to check the covariance matrix for the N_{ij} in the case of the Markov chain with $s = 2$, $p_{11} = \frac{2}{3}$, $p_{21} = \frac{1}{2}$.) This follows in general from the relation between the amount of degeneracy and the degrees of freedom which for the maximum-likelihood estimates are obviously $s - 1$ for each u, and hence $s^k(s - 1)$ for all u (in agreement with the total number of degrees of freedom deduced earlier). The number of independent equations represented by (19) is thus not $\frac{1}{2}s^{k+1}(s^{k+1} + 1)$, but $\frac{1}{2}s^k(s - 1)\,[s^k(s - 1) + 1]$.

Example. We are indebted to B. J. Prendiville for the following numerical illustration of the above method. An artificial Markov chain was constructed with the aid of Tippett's random numbers to correspond to the transition probability matrix

$$\begin{pmatrix} 0\cdot625 & 0\cdot25 & 0\cdot25 \\ 0\cdot25 & 0\cdot5 & 0\cdot375 \\ 0\cdot125 & 0\cdot25 & 0\cdot375 \end{pmatrix}.$$

The first 150 values of the 'state' variable were:

```
0 1 2 2 1 0 0 0 0 0 0 1 2 2 2 1 0 0 1 0 0 0 0 0 0

1 1 2 0 0 2 1 1 0 0 0 0 0 0 0 0 0 0 0 0 0 0 1 1 1

1 0 0 0 0 2 1 0 0 2 1 0 0 0 0 0 0 1 1 1 2 2 0 0 2

1 1 1 1 2 1 1 1 1 1 1 1 1 1 0 2 0 1 1 0 0 0 1 2 2

0 0 0 0 0 0 2 2 2 1 1 1 1 0 1 1 1 1 0 0 2 1 1 0 0

0 0 0 2 2 1 1 1 1 1 2 1 2 0 0 0 1 2 2 2 0 0 0 1 1
```

The observed and expected frequencies for N_{ur}, given $k = 1$, are shown in the table below:

u / r	0	1	2	Total
0	51 (37·25)	12 (13·83)	6 (8·52)	69 (59·60)
1	11 (14·90)	31 (27·67)	11 (12·77)	53 (55·34)
2	8 (7·45)	9 (13·84)	10 (12·77)	27 (34·05)
Total	70 (59·60)	52 (55·34)	27 (34·06)	149 (149·00)

The log formula gives

$$\chi^2 = 10·02 - 3·50 = 6·52 \quad \text{(6 degrees of freedom)},$$

and the quadratic χ^2 formula gives the approximately equal value

$$\chi^2 = 10·06 - 3·48 = 6·58,$$

indicating that the frequencies of the realized series accord very well with expectation.

B. J. Prendiville examined the adequacy of the emigration-and-immigration Markov chain model (equation (5), §3·41) to counts of colloidal particles by such methods. This model, while a useful first approximation, cannot be strictly correct for counts of particles which move continuously in space, owing to the non-Markovian character of grouped counts (see end of §5·21), but rather extensive data are needed to detect any but large discrepancies. (One or two significant anomalies were detected in the colloidal particle counts, but no marked systematic discrepancy.)

Another example of the use of this particular Markov model, in the study of the movements of spermatozoa, is referred to later.

8·21 Goodness of fit of marginal frequency distributions.
In the preceding theory the relevant frequencies were the transition frequencies N_{ur}, but it often happens that a marginal frequency distribution of, for example, N_u, is obtained and compared with a theoretical law such as a Poisson or normal distribution. The standard χ^2 theory, as was clear from particular cases in §8·2, no longer necessarily applies, and needs re-examination.

From equation (12) of the last section, a modified χ^2 is in principle available from the joint normal distribution of the N_u, but requires the inversion of the covariance matrix $\mathbf{V}$. In view of this complication and the familiar use of the quadratic expression

$$\chi_0^2 = \Sigma_u (N_u - m_u)^2/m_u \tag{1}$$

for measuring goodness of fit, it seems useful to retain the criterion (1) and to determine the effect on its asymptotic distribution due to the stochastic process dependences. An examination of this effect need not be confined to probability chains, though the possible values of a continuous variable X of course require final grouping before any χ^2 theory can be used.

Some theoretical results due to V. N. Patankar (1953) are quoted for reference, though it should be added that they are as yet incomplete, and do not include the modifications due to estimating unknown parameters. For any fully specified process grouped into k classes with frequencies N_u, the mean and variance of χ_0^2 may be evaluated. Thus

$$E\{\chi_0^2\} = \sum_{u=1}^{k} \frac{\sigma_u^2}{m_u}, \tag{2}$$

and if the N_u are asymptotically joint normal variables we easily find also

$$\sigma^2\{\chi_0^2\} \sim 2 \sum_{u,\,v=1}^{k} \frac{w_{uv}^2}{m_u m_v}, \tag{3}$$

where $w_{uv}/\sqrt{(m_u m_v)}$ is the asymptotic covariance of $N_u/\sqrt{m_u}$ and $N_v/\sqrt{m_v}$. For a normal stationary process, with marginal normal distribution, these formulae become approximately (for equal grouping intervals)

$$E\{\chi_0^2\} \sim k - 1 + 2 \sum_{s=1}^{\infty} \frac{\rho_s}{1 - \rho_s}, \tag{4}$$

$$\sigma^2(\chi_0^2) \sim 2(k-1) + 8 \sum_{s=1}^{\infty} \frac{\rho_s}{1 - \rho_s} + 8 \sum_{s,\,t=1}^{\infty} \frac{\rho_s \rho_t}{1 - \rho_s \rho_t}, \tag{5}$$

which become for a normal Markov process

$$E\{\chi_0^2\} \sim k - 1 + 2 \sum_{s=1}^{\infty} \frac{\rho_1^s}{1 - \rho_1^s}, \tag{6}$$

$$\sigma^2(\chi_0^2) \sim 2(k-1) + 8 \sum_{s=1}^{\infty} \frac{\rho_1^s}{(1 - \rho_1^s)^2}. \tag{7}$$

An investigation of the emigration-and-immigration Markov model showed that formulae (6) and (7) apply to this case also, at least to $O(\rho_1^3)$, and suggests that they may be of more general applicability.

In this last case the marginal distribution is Poisson. An observed marginal distribution of colloidal particle counts by Westgren (quoted by Chandrasekhar, 1943), with theoretical mean $1\cdot428$ and with $\rho_1 = 0\cdot606$, is shown in the table below:

	0	1	2	3	4	5	6 or more	Total
Observed	381	568	357	175	67	28	7	1583
Expected	379·6	542·0	387·0	184·2	65·8	18·8	5·6	1583·0

We find from (6) and (7)

$$E\{\chi_0^2\} \sim 11\cdot63, \qquad \sigma^2(\chi_0^2) \sim 56\cdot80.$$

The observed value of χ_0^2 is $8\cdot95$, and as this is less than expectation the fit is satisfactory. However, to illustrate an approximate quantitative method of making use of standard χ^2 theory we note that $A\chi_0^2/B$ has mean $f = A^2/B$ and variance $2f$, where $A \equiv E\{\chi_0^2\}$, $2B = \sigma^2(\chi_0^2)$, and hence should be an approximate χ^2 with f degrees of freedom. Here we obtain $A\chi_0^2/B = 3\cdot66$ with $f = 4\cdot76$.

The formulae quoted above have also been extended to apply to marginal distributions obtained from two-dimensional stochastic processes.

8·22   Analysis of two-dimensional chains. The χ^2 technique of § 8·2 cannot in general be extended to multi-dimensional spatial probability chains, as the dependence on neighbouring values is not for spatially symmetric models such that the likelihood function has any simple factorization of the type in § 8·2. An intermediate class of two-dimensional models, however, is that where the postulated dependence is still one-sided. Consider a rectangular array of states X_{rs}, and suppose

$$P\{X_{rs}|\text{all } X_{ij} \text{ values for } i < r, j \leqslant s\} = P\{X_{rs}|X_{r-1,s}, X_{r,s-1}\}.$$

Then the likelihood function conditional on the first row and column is similar in structure to that considered in § 8·2, the only difference being the nature of the conditioning variables $C_{rs} \equiv X_{r-1,\,s},\, X_{r,\,s-1}$.

Example. A one-sided probability model of the above type for a binary variable $X = 0$ or 1 was specified by

$$P\{X_{rs} = 1\,|\,C_{rs}\} = 0\cdot2 + 0\cdot3(X_{r-1,\,s} + X_{r,\,s-1}). \tag{1}$$

A simulated set of values from this model were as follows (Bartlett, 1967, 1968)

1	0	1	0	1	0	0	0	1	1	0
0	0	0	0	0	1	0	0	1	1	0
0	0	0	1	1	0	0	1	1	1	0
0	0	0	1	0	1	0	1	1	0	0
0	0	0	1	1	1	0	1	1	1	1
1	1	1	0	0	1	1	0	1	0	0
0	1	1	0	0	0	1	1	0	0	1
0	1	1	0	0	1	1	1	0	0	0
1	1	1	0	0	1	1	1	0	0	0
0	1	1	1	1	1	1	1	1	0	0
1	1	1	1	1	0	0	1	1	0	0

The configuration frequencies from this set of results is shown in the table below, together with the expected values obtained from (1) for *given* column totals. Note that it is more convenient to use the observed column totals, as otherwise the marginal probabilities must be known. The value of χ^2 is $2\cdot69$ (4 d.f.).

	0 0 .	1 0 .	0 1 .	1 1 .	Total
0	20 (20·8)	7 (10·0)	12 (11·5)	8 (5·2)	47
1	6 (5·2)	13 (10·0)	11 (11·5)	23 (20·8)	53
Total	26	20	23	31	100

To find the expected marginal frequencies for the model

$$P\{X_{rs} = 1\,|\,C_{rs}\} = p + (\tfrac{1}{2} - p)\,(X_{r-1,\,s} + X_{r,\,s-1}), \tag{2}$$

note that the right-hand side is also the expected value of X_{rs}, given C_{rs}. Averaging further over C_{rs} and assuming that the chain has become 'stationary', we obtain

$$m = E\{X_{rs}\} = \tfrac{1}{2},$$

and by multiplying by $X_{r-i,\,s-j}$ ($i, j > 0$ and $i \leqslant 0, j > 0$) before averaging we obtain also for the covariance w_{ij}

$$w_{ij} = (\tfrac{1}{2} - p)\,(w_{i-1,\,j} + w_{i,\,j-1}). \tag{3}$$

The solution of (3) has been given by Whittle (1954; cf. also Besag, 1972), and we have for the product moment $m_{ij} = w_{ij} + \frac{1}{4}$

$$m_{10} = m_{01} = \tfrac{1}{2}\{q_1 - \surd(p_1 q_1)\}/(q_1 - p_1)$$

$$m_{11} = \tfrac{1}{2}\{1 - \surd(p_1 q_1)\}, \quad m_{-1,1} = \{\tfrac{1}{2} - \surd(p_1 q_1)\}/(q_1 - p_1)^2,$$

giving in particular for $p_1 = 0\cdot 2$,

$$m_{10} = m_{01} = 0\cdot 333, \quad m_{11} = 0\cdot 300, \quad m_{-1,1} = 0\cdot 278.$$

The first and last configuration frequencies in the table thus have expected values 30, and the other two, 20.

It is important to note that $m_{11} \neq m_{-1,1}$, so that complete spatial symmetry is *not* achieved with the above model (for the specification of such models, see § 6·84).

8·3   Estimation problems

In this section we shall indicate the use of the likelihood function in some particular stochastic process estimation problems; the autoregressive estimates used in the correlation analysis of the next chapter are, however, not dependent on a complete specification of the likelihood function, and for the moment are only referred to incidentally.

Normal Markov sequence. For a stationary linear sequence $X_0, X_1, ..., X_n$, where

$$X_r = \beta X_{r-1} + Y_r, \tag{1}$$

the Y_r being independent normal variables with zero mean and variance σ_Y^2, the log likelihood function is, apart from a constant,

$$L = -\tfrac{1}{2}\log \sigma_X^2 - \frac{n}{2}\log \sigma_Y^2 - \frac{1}{2}\frac{X_0^2}{\sigma_X^2} - \frac{1}{2}\sum_{r=1}^{n} \frac{Y_r^2}{\sigma_Y^2} \tag{2}$$

$$\sim -\frac{n}{2}\log \sigma_Y^2 - \frac{1}{2}\sum_{r=1}^{n} \frac{Y_r^2}{\sigma_Y^2}, \tag{3}$$

the neglect of the end-correction in (3) having a relative error only of $O(1/n)$. We have

$$\frac{\partial L}{\partial \beta} = \sum_{r=1}^{n} \frac{(X_r - \beta X_{r-1}) X_{r-1}}{\sigma_Y^2}, \tag{4}$$

whence $$\hat\beta = \sum_{r=1}^{n} X_r X_{r-1} \bigg/ \sum_{r=1}^{n} X_{r-1}^2. \tag{5}$$

We have further $\quad I(\beta) = E\left\{-\dfrac{\partial^2 L}{\partial \beta^2}\right\} = \dfrac{n}{1-\beta^2}.$ $\qquad$ (6)

It is known, as will be referred to again in the next chapter, that the classical asymptotic properties for the expressions in (4) or (5) still hold, so that the asymptotic standard error of $\hat{\beta}$ is $\sqrt{\{(1-\beta^2)/n\}}$. We have also if required

$$\frac{\partial^3 L}{\partial \beta^3} = 0, \quad \frac{\partial I(\beta)}{\partial \beta} = \frac{2n\beta}{(1-\beta^2)^2}, \quad E\left\{\left(\frac{\partial L}{\partial \beta}\right)^3\right\} = \frac{6n\beta}{(1-\beta^2)^2}; \qquad (7)$$

thus a more accurate confidence interval for β, with allowance for the skewness of $\partial L/\partial \beta$, should be determined from the relevant roots of the equation

$$\sum_{r=1}^{n} \frac{X_{r-1}^2}{\sigma_Y^2} (\hat{\beta} - \beta) - \frac{\beta}{1-\beta^2}(\lambda^2 - 1) = \pm \lambda \sqrt{\left(\frac{n}{1-\beta^2}\right)}, \qquad (8)$$

(where $\lambda = \pm 1.96$ for upper and lower 0.025 probability limits). It is interesting to notice the appearance of the skewness term in (8)—in contrast with the classical regression case.

It has been assumed here that σ_Y^2 is known. If it is unknown, the estimation equations must be extended to include $\sigma_Y^2 \equiv \alpha$, say. It will be found that

$$\frac{\partial L}{\partial \alpha} = \frac{n}{2\alpha^2}\left(\sum_{r=1}^{n} \frac{(X_r - \beta X_{r-1})^2}{n} - \alpha\right),$$

whence $\qquad I_{\alpha\alpha} = \dfrac{n}{2\alpha^2}, \quad I_{\alpha\beta} = 0, \quad (I_{\beta\beta} \equiv I(\beta)),$

and $\qquad \hat{\alpha}(\beta) = \displaystyle\sum_{r=1}^{n} \dfrac{(X_r - \beta X_{r-1})^2}{n}.$ $\qquad$ (9)

With $I_{\alpha\beta} = 0$, the confidence interval obtained from (8) may be shown not to be affected to its relative accuracy $O(1/\sqrt{n})$ by the substitution of $\hat{\alpha}(\beta)$ in (8) for σ_Y^2 (if $I_{\alpha\beta} \neq 0$, the variance of $[\partial L/\partial \beta]_{\alpha=\hat{\alpha}(\beta)} \sim I_{\beta\beta} - I_{\alpha\beta}^2/I_{\alpha\alpha}$). If the mean $m = E\{X_r\}$ is unknown and thus not necessarily zero, this must of course also be estimated, as in the example below.

Continuous time case. Consider the analogous problem in continuous time

$$dX(t) + \mu X(t)\, dt = dZ(t), \qquad (10)$$

where $Z(t)$ is a normal additive process. This example has been discussed by Grenander (1950), who, however, considered only the estimation of the mean m of $X(t)$. We shall accordingly suppose more generally that the intermittent observations satisfy an equation similar to (1), but with a non-zero mean m included, i.e.

$$X_r - m = \beta(X_{r-1} - m) + Y_r, \tag{11}$$

where, for observations at intervals Δt, $\beta \equiv \exp(-\mu\Delta t)$. It appears relevant in most practical problems where autoregressive schemes of the type (10) are used to assume that the underlying disturbances $dZ(t)$ or Y_r are of constant variance increment σ^2 per unit time. Then $\sigma_X^2 = \sigma^2/(2\mu)$. For

$$S_n \equiv X_1, X_2, \dots, X_n,$$

the log likelihood function now becomes, apart from a constant,

$$L(S_n \mid m, \mu, \sigma^2) = -\tfrac{1}{2}\log\sigma_X^2 - \frac{n-1}{2}\log\sigma_Y^2 - \frac{1}{2\sigma_X^2}(X_1 - m)^2$$

$$- \frac{1}{2\sigma_Y^2}\sum_{r=2}^{n}[X_r - \beta X_{r-1} - m(1-\beta)]^2,$$

where σ_Y^2 is, as in (1), the variance of Y_r. If we now let $\Delta t \to 0$, we must first stabilize the log likelihood function by the subtraction of $L(S_n \mid H_0)$, where H_0 is an appropriate invariant hypothesis. We defer the question of the estimation of σ^2, and subtract $L(S_n \mid 0, 1, \sigma^2)$. This gives

$$\lim_{\Delta t \to 0} [L(m, \mu, \sigma^2) - L(0, 1, \sigma^2)]$$

$$= -\frac{\mu[X(0) - m]^2}{\sigma^2} + \frac{X^2(0)}{\sigma^2} + \frac{1}{2\sigma^2}\left\{-2\mu\int_0^T [X(t) - m]\,dX(t)\right.$$

$$-\mu^2\int_0^T [X(t) - m]^2\,dt\Big\} + \frac{1}{2\sigma^2}\left\{2\int_0^T X(t)\,dX(t) + \int_0^T X^2(t)\,dt\right\} + \tfrac{1}{2}\log\mu.$$

Thus (for the limit as $\Delta t \to 0$)

$$\frac{\partial L}{\partial m} = \frac{\mu(2 + \mu T)}{\sigma^2}\left\{\frac{X(0) + X(T) + \mu\int_0^T X(t)\,dt}{2 + \mu T} - m\right\}, \tag{12}$$

$$\frac{\partial L}{\partial \mu} = \frac{1}{2\mu} - \frac{[X(0)-m]^2}{\sigma^2} - \frac{\int_0^T [X(t)-m]\,dX(t)}{\sigma^2} - \frac{\mu \int_0^T [X(t)-m]^2\,dt}{\sigma^2}.$$

$$(13)$$

From the form of equation (12) we see that if μ is known, then the optimum unbiased estimate of m is

$$\hat{m} = \frac{X(0)+X(T)+\mu \int_0^T X(t)\,dt}{2+\mu T},$$

$$(14)$$

and that its variance is $\sigma^2/[\mu(2+\mu T)]$. For large T the estimate becomes asymptotically the mean $\int_0^T X(t)\,dt/T$ with asymptotic variance $\sigma^2/(\mu^2 T)$. If μ is not known, we should be obliged to substitute an estimate obtained from (13), and as the pair of equations is not of the form required for minimum variance, no advantage over the asymptotic estimate is necessarily gained. It will be noticed further that the precise estimate of μ from (13) requires a knowledge of σ^2; however, for large T the first two terms are neglected (cf. the first example) and the asymptotic estimate

$$-\int_0^T [X(t)-m]\,dX(t) \Big/ \int_0^T [X(t)-m]^2\,dt \qquad (15)$$

is obtained (in which, if m is unknown, we substitute its asymptotic estimate above). This we shall see is equivalent to the asymptotic least-squares estimate, and its asymptotic variance is $1/I_{\mu\mu} \sim 2\mu/T$.

In the above equations we have not considered the estimation of σ^2, for this leads to difficulties which are strictly due to the above formulation becoming unrealistic in the limit as $\Delta t \to 0$. We should find a term

$$\lim_{\Delta t \to 0} \Sigma \frac{(\Delta X)^2}{\Delta t}$$

arising, which, as $(\Delta X)^2$ is of order Δt, exists but could hardly be evaluated in practice. A similar situation arises in the model for ordinary Brownian motion, where a variance estimate could theoretically be based on such a limit and be determined exactly for any finite period (being based on an infinite number of degrees of freedom). This *impasse* is associated with the assumption of a *normal* disturbance term, and would not arise if the disturbances were assumed to be of more general additive type. However, if they were purely of transition type, they would give rise to a finite number of discontinuities in any finite interval T, and the parameter μ could be measured exactly from the observed decay in $X(t)$ between such jumps! It is not impossible for processes of this kind to arise in some physical situations, but in other contexts the model would be an over-idealized one, and estimates not dependent on such a precise use of the likelihood function would be preferable. In the above Markov linear process the simple estimate $2\mu \int_0^T [X(t) - m]^2 dt/T$ for σ^2 will usually be adequate.

As (14) gives (for μ known) the unbiased estimate of m with minimum variance when $Z(t)$ in (10) is normal, it will also provide a linear estimate with the same mean and variance properties for any stationary process with the same autocorrelation function $\exp(-\mu\tau)$. For other such processes it is not, however, necessarily the optimum estimate, and it may be possible to find a better estimate by making use of the correct likelihood function. An example given by Grenander is the following.

A simple Poisson additive process with events occurring at average rate μ has associated independent normal variables Z_r at each time of occurrence T_r. The process $X(t)$ is defined as Z_0 $(0 \leqslant t < T_1)$, $Z_1(T_1 \leqslant t < T_2)$, ... As the contribution to the autocorrelation for an interval Δt is 1 with probability $\exp(-\mu\Delta t)$, and 0 with probability $1 - \exp(-\mu\Delta t)$, this process has the required correlation function. The distribution is moreover normal at each point t, but is not of course a normal process identical with (10), as we shall see by setting up its likelihood function and obtaining a fully efficient estimate of m. Let the number of occurrences in $(0, T)$ be N, a Poisson variable with mean μT. When N is n, say, the distribution of the times of occurrence $T_1, T_2, ..., T_n$ is uniform in $(0, T)$ and does not provide any information on m or μ. Using the method of representation (2) in §8·11, we have

$$f(S \mid m, \mu) = \frac{(\mu T)^n e^{-\mu T}}{n!} \frac{\exp\left\{ -\frac{1}{2\sigma_z^2} \sum_0^n (Z_r - m)^2 \right\}}{(2\pi)^{\frac{1}{2}(n+1)} \sigma_z^{n+1}},$$

where σ_z^2 is the variance of the Z_r. Hence

$$L(m, \mu)$$
$$= N \log \mu - \mu T - (N+1) \log \sigma_z - \frac{1}{2\sigma_z^2} \sum_0^N (Z_r - m)^2 + \text{constant},$$

$$\frac{\partial L}{\partial m} = \frac{1}{\sigma_z^2} \left\{ \sum_0^N (Z_r - m) \right\}, \tag{16}$$

$$\frac{\partial L}{\partial \mu} = \frac{N}{\mu} - T. \tag{17}$$

Equation (16) is not, with N variable, of the required form to provide an unbiased estimate with minimum variance. However, the maximum-likelihood estimate of m is

$$\hat{m} = \frac{1}{N+1} \sum_0^N Z_r, \tag{18}$$

with asymptotic variance

$$\frac{1}{I_{mm}} = \frac{\sigma_z^2}{E\{N+1\}} = \frac{\sigma_z^2}{1+\mu T} \sim \frac{\sigma_z^2}{\mu T}. \tag{19}$$

As for this process $\sigma_X^2 = \sigma_z^2$, the linear estimate $\int_0^T X(t) \, dt/T$, with asymptotic variance $2\sigma_X^2/(\mu T)$, has a limiting efficiency, measured by the ratio of these variances, of $\frac{1}{2}$. (Formula (18) is not a 'linear estimate' in $X(t)$, as it is an integral over $X(t)$ with weights depending on the realization.)

For μ, (17) gives $\hat{\mu} = N/T$, with variance μ/T. The 'least-squares' estimate (15) is rather irrelevant here, as equation (10) no longer holds. The estimate of σ_X^2 is obviously

$$\sum_0^N (Z_r - \hat{m})^2/(N+1).$$

The 'emigration-immigration' process. This process (defined by equation (5), § 3·41) deserves some consideration in view of its practical applications. We saw in § 6·31 that the regression of $X(t)$ on $X(0)$ was linear, and for large mean $m = \nu/\mu$, the Poisson marginal distribution becomes approximately normal and equation (1) above would approximately hold for observational counts made at unit intervals, with $\beta = e^{-\mu}$. However, the estimate $\hat{\beta}$ in (5) would still not be the maximum-likelihood estimate of β, for we have additional information that $\sigma_X^2 = m$.

The exact likelihood function for the sequence $X_0, X_1, ..., X_n$ is set up as

$$p(X_0) \prod_{r=1}^{n} p(X_r \mid X_{r-1}),$$

and with the usual neglect of the first term, we obtain

$$L(m, \mu) = \sum_{r=1}^{n} \log p(X_r \mid X_{r-1}),$$

where it will be found that $p(X_r = i \mid X_{r-1} = j)$ is given by the rather awkward sum

$$Q_{ij} = e^{-m(1-\beta)} \sum_{s=0}^{\min i, j} \frac{m^{i-s}(1-\beta)^{i+j-2s}\beta^s j!}{(i-s)!\,(j-s)!\,s!}. \tag{20}$$

We thus have

$$\frac{\partial Q_{ij}}{\partial m} = -(1-\beta)\,Q_{ij} + \frac{i}{m}Q_{ij} - \frac{R_{ij}}{m},$$

$$\frac{\partial Q_{ij}}{\partial \beta} = mQ_{ij} - \frac{i+j}{1-\beta}Q_{ij} + \left[\frac{1}{\beta} + \frac{2}{1-\beta}\right]R_{ij},$$

where

$$R_{ij} = \sum_{s=0}^{\min i, j} \frac{m^{i-s}(1-\beta)^{i+j-2s}\beta^s j!}{(i-s)!\,(j-s)!\,(s-1)!}.$$

The maximum-likelihood equations for m and β are thus (cf. Patankar, 1953)

$$\Sigma\left\{-(1-\beta) + \frac{i}{m} - \frac{R_{ij}}{mQ_{ij}}\right\} = 0, \tag{21}$$

$$\Sigma\left\{m - \frac{i+j}{1-\beta} + \frac{1+\beta}{\beta(1-\beta)}\frac{R_{ij}}{Q_{ij}}\right\} = 0, \tag{22}$$

where Σ denotes summation over all pairs i, j in the sample (each pair occurring its appropriate number of times). From (21),

$$\hat{m} = \frac{\Sigma(i - R_{ij}/Q_{ij})}{n(1-\hat{\beta})}, \tag{23}$$

and from (22) $\hat{m}n - \dfrac{2\Sigma i}{1-\hat{\beta}} + \dfrac{(1+\hat{\beta})}{\hat{\beta}(1-\hat{\beta})}\Sigma(R_{ij}/Q_{ij}) = 0,$

as $\Sigma i = \Sigma j$ (apart from a possible end-effect which we are neglecting). Substituting for $\hat{m}$ in this last equation, we obtain

$$\hat{\beta} = \Sigma(R_{ij}/Q_{ij})/\Sigma i, \tag{24}$$

and inserting this value in (23) obtain also

$$\hat{m} = \Sigma i/n. \tag{25}$$

The last estimate for m is simply the sample mean, as might be expected. The estimate of β in (24) is awkward to use, in view of the complicated expressions above for R_{ij} and Q_{ij}, and it is usually convenient to adopt simpler estimates.

By proceeding as in the first example, but with the restriction $\sigma_X^2 = m$, it is possible to obtain estimates efficient at least for large m. In addition to the estimate (25), the equation is obtained (we omit the details of the deduction)

$$\hat{\beta} = \frac{C - 2(C - \hat{\beta}V)/(1 - \hat{\beta}^2)}{\hat{m}}, \tag{26}$$

where C is the observed covariance between consecutive observations and V the observed variance. As in large samples $\hat{m} \to m = \sigma_X^2$, $C - \hat{\beta}V \to 0$ (both in probability), $\hat{\beta}$ is evidently a consistent estimate for any value of m; it could be found by iteration. From the asymptotic theory, the validity of which will not be affected by the restriction $\sigma_X^2 = m$, it is found further that

$$\sigma^2(\hat{m}) \sim \frac{m}{n} \frac{1 + \beta}{1 - \beta}, \tag{27}$$

a result which is easily shown directly to be still true for finite m (cf. §9·1), and

$$\sigma^2(\hat{\beta}) \sim \frac{(1 - \beta^2)^2}{(1 + \beta^2)\,n}, \tag{28}$$

this last result only established for large m. Rothschild (1953), in an interesting use of this model to estimate sperm speeds from counts of consecutive numbers in cinemicrographs, has employed the estimate

$$\check{\beta} = 1 - \tfrac{1}{2}\overline{\delta^2}/\hat{m}, \tag{29}$$

where

$$\overline{\delta^2} \equiv \sum_1^n \frac{(X_r - X_{r-1})^2}{n}.$$

This estimate is also obviously consistent for any m, and Rothschild has given its asymptotic variance (obtainable from the joint moments of X_r up to the fourth order) as

$$\sigma^2(\check{\beta}) \sim \frac{(1 - \beta)^2(3 + \beta)}{(1 + \beta)n} + \frac{\beta(1 - \beta)}{nm}. \tag{30}$$

For variable time interval Δt between observations, we had $\beta = e^{-\mu \Delta t}$. It is the quantity μ rather than β which is required

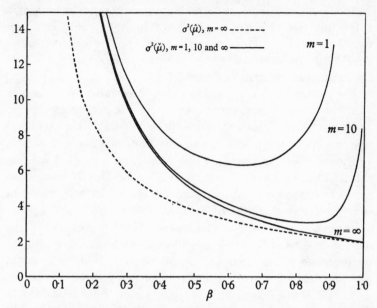

Fig. 9. The variance $\sigma^2(\hat{\mu})$ of the maximum-likelihood estimate $\hat{\mu}$ for the emigration-immigration process compared for large m with $\sigma^2(\check{\mu})$, where $\check{\mu}$ is an estimate based on the mean-square difference. The variance of $\check{\mu}$ for $m = 1$ and $m = 10$ is also shown.

(being directly proportional to the average sperm speed), and as for large n

$$\frac{\sigma^2(\mu_e)}{\mu^2} \sim \frac{\sigma^2(\beta_e)}{(\beta \log \beta)^2}$$

for any consistent estimate β_e, we require first to divide (28) or (30) by $(\beta \log \beta)^2$ when comparing the efficiency of estimating μ for different Δt. It will be seen from the graph (fig. 9) that for large m, $\sigma^2(\check{\mu})$ compares reasonably with $\sigma^2(\hat{\mu})$ for $\beta \geqslant 0.5$. From an examination of $\sigma^2(\check{\mu})$ for varying β when m is small, it appears dangerous to go above about $\beta = 0.9$, so that a useful range appears to be $(0.5, 0.9)$.

The Markov chain model (20) in this particular application is, as in the colloidal particle applications (see § 8·2 at end), somewhat over-simplified,† so that it is advisable to make a test of the adequacy of the fit. The data available were of series of about

† Cf. the discussion by Patil (1957) and Ruben (1962), who consider also the use of a multi-region model.

100 consecutive counts, and were not extensive enough to permit a full analysis of the type discussed in § 8·2. The correlation function could still be tested, however, by the technique to be developed in Chapter 9, where a further reference to this particular application will be found (§ 9·13).

The 'birth-and-death' process. The three previous examples exhibit the common feature that they refer to stationary ergodic processes for which the asymptotic properties of the log likelihood function (or ratio) and maximum-likelihood estimates still apply (cf. § 9·1 for the first two examples, and § 8·2 for the third). The last example, which could equivalently have been called a 'death-and-immigration' process, was a Markov chain of positively regular type owing to the existence of immigration; a stationary 'birth-death-and-immigration' process would be more complicated, but not present any new features. However, for processes which are not of a positively regular type owing to the presence of absorbing states such asymptotic properties may well break down; for example, a simple birth-and-death process starting from one individual will terminate at the first occurrence if this is a death. The distribution problem will, moreover, not only be aggravated by such irrevocable 'stops', but also by the accelerating action of births in a multiplicative process, so that in a given interval of time T the number of births is liable to be a very unstable quantity.

The simple birth-and-death process will be briefly examined in view of these difficulties, as while it seems doubtful whether it will often be used without modification as a realistic model of actual events, it helps to indicate a possible method of attacking the estimation problem in similar cases. We consider the case where the full time record is available, and the representation (2) of § 8·11 employed. To try to avoid the complications noted above, we adopt a sampling rule determined by a fixed number of occurrences n, and not by a predetermined time interval T. (The theory of such 'non-classical' sampling procedures is now better known, the sequential sampling of § 4·1, and 'inverse sampling' in which trials are continued until an event of probability p has occurred a predetermined number of times, being familiar examples.)

The interval ΔT_r between the rth and $(r+1)$th events will depend on the population size $N(t)$ at time T_r, the interval ΔT_r having the exponential distribution with mean $1/[(\lambda + \mu) N(T_r)]$. The $(r+1)$th event will, moreover, have probability $\mu/(\lambda + \mu)$ of being a death. As any actual realization can be generated from these facts, they must determine the likelihood function. Regarding $\lambda + \mu$ and $\mu/(\lambda + \mu)$ as new unknowns θ and ϕ, say, we have:

(a) $2\theta N(T_r) \Delta T_r$ is distributed as a χ^2 random variable independently of $N(T_r)$ with two degrees of freedom (for $\frac{1}{2}\chi^2$ with two degrees of freedom has the exponential distribution with mean unity). Hence

$$2\theta \sum_{r=0}^{n-1} N(T_r) \Delta T_r = 2\theta \int_0^{T_n} N(t)\, dt$$

is a χ^2 quantity with $2n$ degrees of freedom, enabling θ to be estimated, e.g. an exact confidence interval can be assigned to θ;

(b) if there are D deaths, then D is a binomial variable in n independent trials with probability ϕ, whence ϕ may be estimated.

The validity of these estimation equations does not necessarily depend on n being fixed, but does depend on the sampling distributions used being independent of any selection on n. For example, (a) will still hold if the number N of occurrences is the sum of a prescribed number d of deaths and the random number of births in the consequent interval T_N, but not if, alternatively, T_N is restricted to be less than T, say. On the other hand, a switch in the 'stop rule' to a given number of deaths will affect the distribution of D/N. In (b) we must have $n \leqslant N(0)$, to ensure that the rule does not break down through the occurrence of extinction, but more information is obtained if we take the number $d = N(0)$. The sampling distribution in (b) is then changed to one of inverse sampling until there are d occurrences with probability ϕ. The probability of $N = n$ is now the probability ϕ of a death at the last occurrence × the probability of $d-1$ deaths in $n-1$ occurrences, i.e.

$$P\{N = n \mid d\} = \frac{(n-1)!}{(d-1)!\,(n-d)!}\,(1-\phi)^{n-d}\,\phi^d \quad (n \geqslant d), \quad (31)$$

a type of negative binomial distribution. The log likelihood derivative in case (b) and for (31) has the same mathematical form, but a different sampling distribution. It may be written in the standard form in case (b)

$$\frac{\partial L}{\partial \phi} = \frac{n}{\phi(1-\phi)}\left[\frac{D}{n} - \phi\right], \tag{32}$$

showing that D/n is an unbiased optimum estimate of ϕ with variance $\phi(1-\phi)/n$, and in the alternative form

$$\frac{\partial L}{\partial \phi^{-1}} = \frac{\phi^2 d}{1-\phi}\left[\frac{N}{d} - \phi^{-1}\right] \tag{33}$$

for (31), showing that N/d provides an unbiased optimum estimate of ϕ^{-1} with variance $(1-\phi)/(\phi^2 d)$. The information on ϕ from (33) is $I(\phi^{-1})/\phi^4 = d/[\phi^2(1-\phi)]$, which for $d = N(0)$ is greater than the information on ϕ in case (b), with $n = N(0)$. The information on θ by method (a) is n/θ^2, and is of course greatest for n as large as possible, subject to the sampling requirements already indicated.

We may illustrate this method of estimation by reference to the artificial birth-and-death series shown in fig. 4, §3·4. Thus for one of the three series shown the realization up to the 5th death is shown in detail in table 3. The true values of λ and μ in this case were unity, so that $\theta = 2$ and $1/\phi = 2$. Consistently with these values, we have $2\theta \times 3·546 = 14·184(\frac{1}{2}\theta)$ as a χ^2 with 16 degrees of freedom, and

$$\hat{\phi}^{-1} = 8/5 = 1·6, \quad \sigma(\hat{\phi}^{-1}) = [(1-\phi)/5]^{\frac{1}{2}}/\phi = 0·63.$$

Similar calculations for the other two series gave (i) $34·104(\frac{1}{2}\theta)$ with 30 degrees of freedom and $\hat{\phi}^{-1} = 3·0 \pm 0·63$, (ii) $34·220(\frac{1}{2}\theta)$ with 28 degrees of freedom and $\hat{\phi}^{-1} = 2·80 \pm 0·63$. When θ and ϕ are unknown, confidence intervals can alternatively be obtained for them from such data.

If independent replications of transitory or other evolutionary processes are available, then it has already been noted that particular features may be estimated by standard methods based on these replications. As an example of such a procedure, Bailey (1953) has discussed the estimation of the 'recovery'/infectivity ratio from observations of the total eventual number

Table 3

Event	Time interval (ΔT)	Time (t)	Birth (B) or death (D)	Population ($N(t)$)	$N(t)\,\Delta T$
1	0·208	0·208	D	5	1·040
2	0·140	0·348	D	4	0·560
3	0·063	0·411	D	3	0·189
4	0·158	0·569	B	2	0·316
5	0·240	0·809	B	3	0·720
6	0·030	0·839	B	4	0·120
7	0·041	0·880	D	5	0·205
8	0·099	0·979	D	4	0·396
					3·546

of persons infected in initial groups of susceptibles of given sizes, the model considered being of the simplest type with constant infectivity and 'recovery' (including transition to non-infectivity by removal, etc.) rates, with no immigration of new susceptibles (cf. §4·4).

8·31   Further examples of the use of the likelihood function.

The principles made use of in the preceding sections will be further illustrated in this section by two fairly general types of stochastic model. For the first, let us note that for a pure birth (or death) process the time intervals between events are sufficient to specify any realization, as the population size $N(t)$ necessarily increases (or decreases) by one. However, the birth process may be quite general, i.e. the chance of an increment, in place of $\lambda N(t)\,dt$ in time t, $t + dt$ for the simple birth process, is, say, $\lambda(N(t)\,dt$, still giving rise to an exponential time-interval ΔT_r with mean $1/\lambda N(T_r)$ for the time-interval between the observed rth and $(r + 1)$th events.

Example of pure birth process. Consider in particular the experimental data reported in Bartlett, Brennan and Pollock (1971) on the mating of sheep blowflies (*Lucilia sericata*). Males and females were put together in a cage, and each mating pair removed at the time of mating. The times to first, second, etc., matings were recorded, and given below for one of the experiments, in which the initial number of males was 20, and females, 40. The 'birth process' model for the number of matings $N(t)$ was $\lambda(N(t)) = M_t F_t/\theta$, where M_t and F_t are the numbers of males and females at time t, and $1/\theta$ is an unknown rate parameter. In this experiment, $M_t = 20 - N(t)$ and $F_t = 40 - N(t)$, so that $\lambda(N(t)) = [20 - N(t)][40 - N(t)]/\theta$.

r	1	2	3	4	5	6	7	8	9	10	11	12	13	14	15	16	17	18	19
T_r	1	1	2	3	3	5	6	6	8	9	11	14	21	23	36†	36†	36†	36†	79

(in min.)

† These four readings represent matings in the time period 30 to 36 min.

The intervals ΔT_r are standardized to exponential variables τ_r with the same mean θ by multiplying by $m_r f_r$ where $m_r = M(T_r)$, $f_r = F(T_r)$. Then the log likelihood function may be written

$$L = \sum_r \left(-\tau_t/\theta - \log\theta \right),$$

whence

$$dL/d\theta = (n/\theta^2)(\textstyle\sum_r \tau_r/n - \theta),$$

and the unbiased estimator of θ with minimum variance θ^2/n is $\bar\tau = \sum_r \tau_r/n$. This may be obtained by plotting the cumulative sum $\sum_{s=1}^{r} \tau$, against r, and τ is the slope of the line joining the last point to the origin. Moreover, given τ, this sum is a random walk conditioned at the points 0 and n, whence asymptotic confidence limits may be drawn on either side of the central line at a vertical distance $\pm 1\cdot 36\tau\sqrt{n}$ (for significance level $P = 0\cdot 05$; cf. use of this test in another context in §9·22).

In the above example the failure to record mating-times more precisely was largely ignored (e.g. the times $T_1 = 1$, $T_2 = 1$ were replaced by $T_1 = \frac{1}{2}$, $T_2 = 1$ and the times $T_{15}, T_{16}, T_{17}, T_{18} = 36$ replaced by $T_{15} = 30$, $T_{16} = 32$, $T_{17} = 34$, $T_{18} = 36$ (for further details see the paper cited). The estimate of θ from the above data is found to be $11{,}561\cdot 5/19 = 608\cdot 5$, and the fit of the model appeared to be adequate for this particular set of data.

It should be noticed that the distributional solution $P\{N(t) = r\}$ for this birth-process model is known but very complicated (see McQuarrie, 1967). The above simple method completely disregards such distributional problems as irrelevant to the inference problem.

In the second type of example, discussed by Cox (1972), the events, and not their times of occurrence, are restored to the centre of interest. Suppose that the failure (or death) of articles (or individuals) depends on known other variables $\mathbf{z}$ (such as an imposed treatment), so that for each individual the chance of 'death' in t, $t + dt$ is assumed to be of the form (independently for each individual)

$$\mu(t, \mathbf{z}) = \mu_0(t) f(\mathbf{z}, \boldsymbol{\beta}) \tag{1}$$

where f is a known function of $\mathbf{z}$ and unknown parameters $\boldsymbol{\beta}$, and $\mu_0(t)$ is an arbitrary function of time, but the same for every individual. For the record of deaths in this population, the time-intervals between deaths depend on the function $\mu_0(t)$, but the chance of the rth death being of individual i is

$$p_r(i) = f(\mathbf{z}_i, \boldsymbol{\beta})/\Sigma_i f(\mathbf{z}_i, \boldsymbol{\beta}).$$

Hence the complete likelihood function $\exp L$ has two factors,

$$L = L_1(T_r) + L_2(i_r),$$

where $L_2(i_r)$ involves the $\boldsymbol{\beta}$ uncontaminated with $\mu_0(t)$, and hence available for inferences about $\boldsymbol{\beta}$. In the simplest case with two treatments and $f(\mathbf{z}, \boldsymbol{\beta}) = \exp \beta z$, where z is 0 or 1, and $\alpha = e^{\beta}$, we have

$$p_r(z = 0) = \frac{n_r}{n_r + \alpha m_r}, \quad p_r(z = 1) = \frac{\alpha m_r}{n_r + \alpha m_r}, \tag{2}$$

where n_r, m_r are the numbers of individuals at the time of the rth event with $z = 0$ & 1 respectively. Note that the times of death of all individuals may not be reached, the method of analysis being unaffected; individuals entering the experiment after its start may also be considered, if the basic model (1) still applies to all individuals.

Chapter 9

CORRELATION ANALYSIS OF TIME-SERIES

9·1 Correlation and regression analysis of stationary sequences

The theory of stationary processes developed in Chapter 6 throws a powerful light on the possibilities of analysing time-series. Correlation analysis and harmonic analysis, often treated as two distinct methods, are seen to be theoretically equivalent if correctly interpreted. On the other hand, actual time-series of a stationary type will not necessarily possess the simple harmonic structure assumed in classical periodogram analysis; they very often have a continuous, not a discrete, spectrum. Thus the statistical analysis of time-series is logically greatly helped by the general mathematical theory; but the latter also raises many more problems connected with sampling fluctuations.

In some physical processes, the extent of the series available for study may be as much as desired, and the measurement of correlation coefficients accurately carried out to the limits of experimental error by means of the ergodic relation between 'phase' (probability) and time averages. This seems true, for example, in turbulence measurements. In statistical time-series in other fields (especially in economics) the length of series available for study is often severely limited. In all cases, however, it is important to bear in mind the magnitude of the sampling errors. This is perhaps most strikingly illustrated in the case of harmonic analysis, where any direct attempt to estimate the spectral function may give rise to sampling fluctuations that do not diminish with the length of series taken.

In this last chapter we confine our attention mainly to the correlational or equivalent harmonic structure of real stationary time-series. In spite of some exact results, the sampling theory of time-series, like that of stochastic processes in general, is still in a comparatively early stage of development, and in our discussion we shall be concerned mostly with presenting some of the more important 'large-sample' results. We do not consider series which are not stationary; if a series has a systematic trend this can be removed, but if it is essentially of the evolutionary type, then a rather well-specified model is likely to be required for a profitable analysis to be possible. The sampling theory of this chapter is of course relevant to the prediction theory of Chapter 7 if the correlation structure of a series is not known *a priori*, but has to be inferred from observation.

We shall consider first stationary sequences. This may appear somewhat in contrast with Chapters 6 and 7, where the theory of continuous series was developed immediately, but the observational data, in statistical applications at least, are usually of the sequence type, and moreover, as we shall see, a specification for a continuous series may sometimes be conveniently tested by means of the sampling theory for sequences.

Sampling fluctuations of means and correlation coefficients. The sampling properties of means and correlation coefficients obtained from a series of n observations $X_1, X_2, \ldots, X_n$ (corresponding to time values $t = 1, 2, \ldots, n$) may to a considerable extent be investigated by straightforward algebra. Thus for the sample mean defined as

$$\overline{X} = \frac{1}{n} \sum_{t=1}^{n} X_t, \tag{1}$$

we obviously have $E\{\overline{X}\} = E\{X\}$, which will usually be put zero. We have further

$$\sigma^2(\overline{X}) = \frac{\sigma^2}{n} \sum_{s=-n+1}^{n-1} \left(1 - \frac{|s|}{n}\right) \rho_s, \tag{2}$$

so that the variance of $\overline{X}$ is asymptotically $\mu\sigma^2/n$, where

$$\mu = \lim_{n \to \infty} \sum_{s=-n+1}^{n-1} \left(1 - \frac{|s|}{n}\right) \rho_s \tag{3}$$

is assumed finite (from the theory of Chapter 6 this will not be true if the spectral function $F(\omega)$ has a non-zero jump at $\omega = 0$; or in more practical terminology, it is assumed that X_t has no component which, while zero on the average, remains constant for any single realization x_t).

Consider next the covariance and correlation functions. If we know $E\{X\} = 0$, we may define the sample covariances by

$$C_s = \frac{1}{n-s} \sum_{t=1}^{n-s} X_t X_{t+s} \quad (s \geqslant 0). \tag{4}$$

In this case $E\{C_s\} = \rho_s \sigma^2$. If we do not know the value of the true mean $E\{X\}$, we may replace X_t and X_{t+s} in (4) by $X_t - \bar{X}_t$ and $X_{t+s} - \bar{X}_{t+s}$, where $\bar{X}_t$ is the mean of the X_t occurring in (4). In this case we obtain a mean value for C_s of

$$\rho_s \sigma^2 - E\{\bar{X}_t \bar{X}_{t+s}\} \sim \left(\rho_s - \frac{\mu}{n-s}\right) \sigma^2. \tag{5}$$

We shall here record only the dominant term in the formulae for the variances and covariances of C_s $(s = 0, 1, \ldots)$, and to this order of approximation the effect on random sampling errors of measuring X_t from the sample mean can be neglected. The bias indicated by (5) is also of a smaller order of magnitude than sampling fluctuations, which are $O(1/\sqrt{n})$, but being a systematic effect is sometimes worth correcting.

From (4)

$$\text{cov}\,(C_s, C_{s+t}) = \frac{1}{n-s} \sum_{v=-(n-s)+1}^{n-s-t-1} \left[1 - \frac{\eta(v)}{n-s-t}\right] \phi(v) \quad (t \geqslant 0),$$

where

$$\eta(v) = \begin{cases} v & (v \geqslant 0), \\ 0 & (-t \leqslant v \leqslant 0), \\ -v-t & (-(n-s)+1 \leqslant v \leqslant -t), \end{cases}$$

and where

$$\phi(v) = \sigma^4(\rho_v \rho_{v+t} + \rho_{v+s+t} \rho_{v-s}) + \kappa_{v,s,t},$$

$\kappa_{v,s,t-s}$ denoting the fourth-order cumulant between X_u, X_{u+s}, X_{u+v} and X_{u+v+t}. (It should be recalled that

$$E\{X_u X_{u+s} X_{u+v} X_{u+v+t}\} = E\{X_u X_{u+s}\} E\{X_{u+v} X_{u+v+t}\}$$
$$+ E\{X_u X_{u+v+t}\} E\{X_{u+s} X_{u+v}\} + E\{X_u X_{u+v}\} E\{X_{u+s} X_{u+v+t}\}$$
$$+ \kappa_{v,s,t-s}\cdot)$$

For large n, we have approximately

$$\text{cov}\,(C_s, C_{s+t}) \sim \frac{1}{n-s} \sum_{v=-\infty}^{\infty} \phi(v) \tag{6}$$

when this sum converges (the more exact sum involving $\phi(v)$ will always converge under specifiable conditions, as for μ in (3), obtainable by treating C_s as a mean of a new stochastic process $Y_r \equiv X_r X_{r+s}$).

Denoting the sample covariance of the observations X_i, X_{i+1},

$\dots, X_j$ more fully by $C_s(i, j)$, we define the sample correlation coefficient R_s by

$$R_s = C_s(1, n)/[C_0(1, n-s) C_0(s+1, n)]^{\frac{1}{2}}. \tag{7}$$

Noting that $R_s \sim C_s/C_0$, we obtain from the formula

$$\operatorname{cov}\left(\frac{U}{W}, \frac{V}{W}\right) \sim$$
$$\frac{\operatorname{cov}(U, V)}{w^2} - \frac{u \operatorname{cov}(V, W)}{w^3} - \frac{v \operatorname{cov}(U, W)}{w^3} + \frac{uv \operatorname{var}(W)}{w^4},$$

which holds when the variation of U, V and W about u, v and w ($w > 0$) respectively becomes small,

$$\operatorname{cov}(R_s, R_{s+t}) \sim \frac{1}{n-s} \sum_{v=-\infty}^{\infty} (\rho_v \rho_{v+t} + \rho_{v+s+t} \rho_{v-s} + 2\rho_s \rho_{s+t} \rho_v^2$$
$$- 2\rho_s \rho_v \rho_{v-s-t} - 2\rho_{s+t} \rho_v \rho_{v-s}), \quad (8)$$

if the component depending on the cumulant term $\kappa_{v,s,t}$ may be neglected. This component is necessarily zero for normal processes. However, it should also be observed that for any linear process of the type

$$X_t = \sum_{v=0}^{\infty} g_{t-v} Y_v,$$

where the Y_v are independent disturbances with the same distribution, we have (cf. equation (16), §5·2)

$$\kappa_{v,s,t} = \kappa_4(Y) \sum_{u=-\infty}^{\infty} g_u g_{u+s} g_{u+v} g_{u+v+s+t} \quad (g_u = 0, u < 0),$$

whence $\sum_{v=-\infty}^{\infty} \kappa_{v,s,t} = \kappa_4(Y) \sum_{u=-\infty}^{\infty} g_u g_{u+s} \sum_{w=-\infty}^{\infty} g_w g_{w+s+t}$

$$= \gamma(Y) \operatorname{cov}(X_r, X_{r+s}) \operatorname{cov}(X_u, X_{u+s+t}),$$

where $\gamma(Y) \equiv \kappa_4(Y)/\sigma^4(Y)$. This result will be found to ensure that for such processes also the neglected $\kappa_{v,s,t}$ term in (8) vanishes automatically.

As special cases of (8) we have, when $\rho_s \to 0$ as s increases, the formulae

$$\left.\begin{array}{c} \operatorname{var}(R_s) \sim \dfrac{1}{n-s} \sum_{v=-\infty}^{\infty} \rho_v^2, \quad \operatorname{cov}(R_s, R_{s+t}) \sim \dfrac{1}{n-s} \sum_{v=-\infty}^{\infty} \rho_v \rho_{v+t}, \\[2mm] \rho(R_s, R_{s+t}) \sim \dfrac{\sum\limits_{v=-\infty}^{\infty} \rho_v \rho_{v+t}}{\sum\limits_{v=-\infty}^{\infty} \rho_v^2}. \end{array}\right\} \tag{9}$$

(It is assumed in (9) that $\Sigma \rho_{v+s} \rho_{v-s}$ is small when ρ_s is small.) The formulae (8) and (9) are useful in indicating the magnitude and correlations of sampling fluctuations in the observed correlogram (graph of the observed correlation coefficients), for processes with true correlograms of the 'damped' type ($\rho_s \to 0$ as s increases). While, however, the sampling fluctuations in R_s are of $O(1/\sqrt{n})$ and decrease with increase of n, they have the disadvantage, when the true values ρ_s are not known, of depending on these unknown values. This stresses the difficulty of purely empirical correlation studies with time-series. It seems desirable to have some theoretical model in mind for the structure of the series, depending on only a few unknown parameters; these can then be estimated, and the theoretical correlogram corresponding to the fitted model then compared with the observed correlogram in the light of the above sampling errors.

Time-series composed of exact harmonic oscillations are more naturally dealt with by harmonic analysis; although if the correlogram is constructed, it will also exhibit corresponding undamped oscillations. The correlogram analysis is more useful for oscillatory series which do not consist of simple undisturbed harmonic oscillations and exhibit a damped correlogram. An important method of analysis of such series is the use of the linear autoregressive model first introduced by Yule. This is a very general model for non-deterministic series if unlimited in extent, and for uncorrelated residuals (cf. §6·3); a much narrower practical hypothesis consists in assuming (i) that a few terms (p, say) of the series

$$X_t - Y_t = a_1 X_{t-1} + a_2 X_{t-2} + \ldots \tag{10}$$

are sufficient, where (ii) the Y_t are independent residuals. The unknown a coefficients in such a model may be estimated by least squares, i.e. from the equations obtained by minimizing

$$\sum_{t=1}^{n-p} Y_{t+p}^2 = \sum_{t=1}^{n-p} (X_{t+p} - a_1 X_{t+p-1} - \ldots - a_p X_t)^2.$$

The simplest model of type (10) which provides a quasi-periodic oscillatory series is the case $p = 2$, i.e.

$$X_t = a_1 X_{t-1} + a_2 X_{t-2} + Y_t. \tag{11}$$

Multiplying (11) by X_{t-s} and averaging, we obtain (cf. equation (9), §5·2)

$$\rho_s = a_1\rho_{s-1} + a_2\rho_{s-2} \quad (s > 0), \tag{12}$$

with in particular ($s = 1$)

$$\rho_1 = a_1/(1 - a_2). \tag{13}$$

The solution of (12) is, for $-1 \leqslant a_2 < 0$ and $a_1^2 + 4a_2 < 0$,

$$\rho_s = \frac{(-a_2)^{\frac{1}{2}s}\sin(s\theta + \psi)}{\sin\psi}, \tag{14}$$

where
$$\cos\theta = \frac{\frac{1}{2}a_1}{\sqrt{-a_2}}, \quad \tan\psi = \frac{1 - a_2}{1 + a_2}\tan\theta.$$

Formula (14), or more simply formulae (12) and (13), may be used to calculate ρ_s when a_1 and a_2 are known. The least-squares estimates $\hat{a}_1$ and $\hat{a}_2$ are given by

$$\hat{a}_1\Sigma X_{t+1}^2 + \hat{a}_2\Sigma X_{t+1}X_t = \Sigma X_{t+2}X_{t+1},$$

$$\hat{a}_1\Sigma X_{t+1}X_t + \hat{a}_2\Sigma X_t^2 = \Sigma X_{t+2}X_t,$$

or approximately, by substituting R_s for ρ_s in (12) for $s = 1$ and 2,

$$\hat{a}_1 \sim \frac{R_1(1 - R_2)}{1 - R_1^2}, \quad \hat{a}_2 \sim \frac{R_2 - R_1^2}{1 - R_1^2}. \tag{15}$$

We cannot quote the usual standard error formulae for least-squares estimates without further justification, since the X's play simultaneously the roles of dependent and independent variables (in the regression sense). Mann and Wald (1943) have, however, shown that these least-squares error formulae are still asymptotically valid for large samples. A simple (unrigorized) demonstration in the case of (11) is given below.

Writing $\Sigma X_{t+1}Y_{t+2} = A$, $\Sigma X_t Y_{t+2} = B$, and then adding these equations, after substituting for Y_{t+2} in terms of X_t from (11), to the exact least-squares equations given above, we obtain

$$\delta a_1\Sigma X_{t+1}^2 + \delta a_2\Sigma X_t X_{t+1} = A,$$

$$\delta a_1\Sigma X_t X_{t+1} + \delta a_2\Sigma X_t^2 = B,$$

where $\delta a_1 \equiv \hat{a}_1 - a_1$, $\delta a_2 \equiv \hat{a}_2 - a_2$. Using such results as

$$E\{Y_{t+1}X_t Y_{u+1}X_u\} = 0$$

unless $u = t$ (e.g. if $u > t$,

$$E\{Y_{t+1}X_tY_{u+1}X_u\} = E\{Y_{u+1}\}E\{Y_{t+1}X_tX_u\} = 0),$$

we have

$$E\{A^2\} \sim E\{B^2\} \sim n\sigma^2(Y)\,\sigma^2(X), \quad E\{AB\} \sim n\rho_1\sigma^2(Y)\,\sigma^2(X),$$

whence

$$\sigma^2(\hat{a}_1) \sim \sigma^2(\hat{a}_2) \sim \frac{\sigma^2(Y)}{n(1-\rho_1^2)\,\sigma^2(X)}.$$

Squaring and averaging both sides of (11), we obtain

$$\frac{\sigma^2(Y)}{\sigma^2(X)} = \frac{(1+a_2)([1-a_2]^2 - a_1^2)}{1 - a_2}, \tag{16}$$

whence

$$\sigma^2(\hat{a}_1) \sim \sigma^2(\hat{a}_2) \sim (1 - a_2^2)/n. \tag{17}$$

Least-squares estimation has been considered here as providing efficient 'linear' methods of estimation, using the term 'linear' in the sense that the deviations of the estimates from the true coefficients are linear in the residuals Y_t, and while the variables X_t are not independent of Y_t as in the classical regression or least-squares model, the above results indicate that asymptotically this does not affect the error formulae. If exact information were available on the distribution of the residuals Y_t, this could be used to provide fully efficient estimates based on the maximum-likelihood method (see § 8·3); in the important case when the Y_t are not only independent but normal (if they are uncorrelated and normal they are necessarily independent), the methods are asymptotically equivalent, owing to the form of the log likelihood function

$$L \sim -\frac{n-p}{2}\log\sigma^2 - \frac{1}{2\sigma^2}\sum_{t=p+1}^{n}Y_t^2 + \text{constant}, \tag{18}$$

where $\sigma^2 \equiv \sigma^2(Y)$.

9·11 Goodness of fit tests.

In conformity with the principles of § 8·1, an asymptotic goodness of fit test of an autoregressive series of low order p could be made within the class of autoregressive series of some higher order q, say, if a full specification of the form of the likelihood function is available. Thus on the assumption of normal residuals we readily obtain from (18) (with neglect of end corrections of relative order $1/n$)

$$\chi^2 \sim (\Sigma_1' - \Sigma_2)/\sigma^2, \tag{1}$$

where Σ_2 denotes the sum of squares of residuals after fitting a series of order q, and Σ_1' after fitting one of order p (the dash denotes that Σ_1 has been made comparable with Σ_2, for example, by taking only the last $n-q$ residuals). From the asymptotic properties of the fitted coefficients, which include their asymptotic normality (Mann and Wald, 1943), the classical asymptotic likelihood theory applies, and the approximate χ^2 quantity in (1) has $q-p$ degrees of freedom.

If σ^2 is also not known, a logarithmic form is obtained, asymptotically equivalent to

$$\chi^2 \sim -(n-q)\log(\Sigma_2/\Sigma_1'). \tag{2}$$

The approximate form (2) has the advantage that it depends only on variance ratios, that is, on correlations, and preserves its asymptotic distributional properties under the wider assumption of independent but not necessarily normal residuals. This follows from the asymptotic joint normal properties of the serial correlations R_s, established by Mann and Wald, whose assumptions were of this wider form (finite moments of all orders for Y_t are assumed).

The use of χ^2 tests of the above type based on the likelihood criterion has been advocated by Whittle. While derived on the basis of normal residuals, the insensitiveness of such a criterion as (2) to this assumption allows their more general use. One practical disadvantage appears to be the explicit formulation of the alternative class of hypothesis, for although we have seen that this class for fairly large q includes a wide class of non-deterministic time-series, it implies that the autoregressive coefficients have in effect to be estimated not only for the model of low order p, but also for the alternative of higher order q, the latter giving rise to simultaneous linear equations with associated determinants of order q.

In view of this, it will sometimes be more convenient to make use of an alternative goodness of fit test due to Quenouille and formulated purely in terms of the assumed autoregressive model of order p. A simple method of deriving this test is as follows. Denote the general linear operator equation connecting X_t with Y_t by

$$H_t X_t = Y_t, \tag{3}$$

where this equation is of autoregressive type (i.e. it can be put in the form $X_t = G_t X_t + Y_t$, where G_t is a 'backward' operator). We consider the properties of $H_t C_t$, where this notation means that the operator H_t is to act on the X's occurring in the covariance C_t. We have

$$E\{H_t C_t\} = E\left\{\frac{1}{n-t} \sum_{u=1}^{n-t} X_u Y_{u+t}\right\} = 0 \quad (t > 0),$$

$$E\{H_t C_t H_\tau C_\tau\} = E\left\{\frac{1}{(n-t)(n-\tau)} \sum_{u=1}^{n-t} \sum_{v=1}^{n-\tau} X_u Y_{u+t} X_v Y_{v+\tau}\right\}$$

$$= \frac{1}{n-\tau} E\{X_v X_{v+t-\tau}\} E\{Y^2\} \quad (0 < \tau \leqslant t),$$

$$= \frac{1}{n-\tau} \sigma^2(X) \sigma^2(Y) \rho_{t-\tau}. \tag{4}$$

Now from (3), multiplying by $H_\tau X_\tau$ and averaging, we have $H_\tau H_t \rho_{t-\tau} \sigma^2(X) = \delta_{t,\tau} \sigma^2(Y)$. This result, together with (4), enables us to choose a linear combination of C_t which is uncorrelated with the corresponding function of C_τ ($\tau \neq t$). Since

$$H_\tau H_t \rho_{t-\tau} \equiv H_{-t} H_{-\tau} \rho_{t-\tau},$$

we have, in fact, a choice between $H_t^2 C_t (t > p)$ and $H_{-t} H_t C_t (t > 0)$. The latter choice has the attraction of a simple correspondence with the covariances of the Y's, since

$$\frac{1}{n-t} \sum_{u=1}^{n-t} Y_u Y_{u+t} = \frac{1}{n-t} \sum_{u=1}^{n-t} H_\tau X_{u+\tau} H_t X_{t+u} \quad (\tau = 0),$$

$$\sim H_\tau H_t C_{t-\tau} \quad (\tau = 0),$$

$$= H_{-t} H_t C_t.$$

However, the alternative choice of $H_t^2 C_t$, which was the form proposed by Quenouille, appears somewhat more satisfactory, at least in the usual case when unknown parameters are estimated. Thus if we consider the asymptotic distribution of $(n-t)^{\frac{1}{2}} h_t^2 C_t$ $(t > p)$, where h_t denotes the operator H_t after substitution of our estimates of the unknown parameters, we find for the difference

$$(n-t)^{\frac{1}{2}} (h_t^2 - H_t^2) C_t = (n-t)^{\frac{1}{2}} (h_t - H_t)(h_t + H_t) C_t$$

$$\sim (n-t)^{\frac{1}{2}} (h_t - H_t)(2H_t C_t)$$

$$\sim (n-t)^{\frac{1}{2}} (h_t - H_t)(2H_t \rho_t) \sigma^2(X)$$

$$= 0 \quad (t > p),$$

where $\sim$ denotes asymptotic equivalence, and it is assumed that the errors in the coefficients are $O\{(n-t)^{-\frac{1}{2}}\}$, the identity $H_t \rho_{t-\tau} = 0 \ (t > \tau)$ following from (3) after multiplication by X_τ. On the other hand,

$$
\begin{aligned}
(n-t)^{\frac{1}{2}} & (h_{-t} h_t - H_{-t} H_t) C_t \\
&= (n-t)^{\frac{1}{2}} (h_{-t} h_t - H_{-t} h_t + H_{-t} h_t - H_{-t} H_t) C_t \\
&= (n-t)^{\frac{1}{2}} (h_{-t} - H_{-t}) h_t C_t + (n-t)^{\frac{1}{2}} H_{-t} (h_t - H_t) C_t \\
&\sim (n-t)^{\frac{1}{2}} (h_{-t} - H_{-t}) H_t \rho_t \sigma^2(X) + (n-t)^{\frac{1}{2}} H_{-t} (h_t - H_t) C_t \\
&= 0 + (n-t)^{\frac{1}{2}} H_{-t} (h_t - H_t) C_t \quad (t > 0),
\end{aligned}
$$

which is *not* zero asymptotically. This establishes the asymptotic equivalence of $(n-t)^{\frac{1}{2}} h_t^2 C_t$ with $(n-t)^{\frac{1}{2}} H_t^2 C_t$, but not of $(n-t)^{\frac{1}{2}} h_{-t} h_t C_t$ with $(n-t)^{\frac{1}{2}} H_{-t} H_t C_t$.

We have also from (4), if we denote $H_t C_t$ by D_t,

$$
E\{H_t D_t D_\tau\} = \frac{1}{n-\tau} \sigma^2(X) \sigma^2(Y) H_t \rho_{t-\tau} = 0 \quad (t > \tau).
$$

Thus the functions $H_t D_t \equiv H_t^2 C_t$ are also uncorrelated with D_τ $(\tau < t)$. Dividing through by $C_0 \sim \sigma^2(X)$, we obtain finally the approximate result

$$
E\{H_t^2 R_t H_\tau^2 R_\tau\} \sim \frac{\delta_{t, \tau}}{n-\tau} \frac{\sigma^4(Y)}{\sigma^4(X)}; \tag{5}
$$

and since the functions $H_t^2 R_t$ tend to normality as n increases, they should be approximately independent normal variables.

In the case of the process (11) of §9·1 we obtain $H_t^2 R_t$ as

$$
R_t - 2a_1 R_{t-1} + (a_1^2 - 2a_2) R_{t-2} + 2a_1 a_2 R_{t-3} + a_2^2 R_{t-4},
$$

to be calculated, when R_1 and R_2 are used for estimating a_1 and a_2, for $t = 3, 4, \ldots$.

In the case of processes for which H_t does not represent a finite autoregressive form, use of the above method leads to some difficulties. For example, in the case of 'moving averages' where we have

$$
X_{t+u} = J_t Y_{t+u}, \tag{6}
$$

where J_t is of finite length, the formal inversion of this equation leads to an operator $H_t \equiv J_t^{-1}$ of infinite length. This case can be dealt with by a method developed by Wold (1949), or alternatively by extending the above method to *correlated* residuals. This latter method has the advantage of covering also those cases of time-series specified for continuous time (see §9·12) which are expressible in this way.

To summarize this latter method, we assume now that in equation (3) Y_{t+u} is 'm-dependent' (i.e. correlated to a length m). It may be shown by similar methods that $H_t D_{t+p}$ and $H_{-t} D_t$ then have zero expectation for $t > m$, and also

$$E\{H_t^2 R_{t+p} H_\tau^2 R_{\tau+p}\} \sim E\{H_{-t} H_t R_t H_{-\tau} H_\tau R_\tau\}$$
$$\sim \frac{1}{n-\tau} \frac{\sigma^4(Y)}{\sigma^4(X)} \sum_{i=-m}^{m} \delta_i \delta_{i+t-\tau}, \quad (7)$$

where δ_i denotes the autocorrelation of lag i of Y_u. Equation (7) holds in general for $t \geqslant \tau > 2m$, and may often be assumed true for $\tau > m$ (e.g. if the process X_u is a discrete 'linear process' or is normal). From (7) it will be seen that we arrive at $2m+1$ sequences such that each term has zero mean and any two terms in each sequence are uncorrelated. This leads to a total χ^2 for each sequence. In practice there is some disadvantage in this multiplicity of partial tests; a simple though rather rough method of combining them is to sum all the χ^2's as if they were independent, and then reduce the total χ^2 and its apparent degrees of freedom (the total number of individual items) by the factor

$$1 \bigg/ \sum_{i=-\infty}^{\infty} \eta_i^2, \quad (8)$$

where the η_i are the correlations between the individual components before squaring. A more exact test would involve the inversion of the matrix with (t, τ)th element

$$\sum_{i=-m}^{m} \delta_i \delta_{i+t-\tau}.$$

This is of Laurent type, and an iterative procedure used by Wold (1949) for finding the inverse of such matrices is available if required.

The above extension of the χ^2 test to m-dependent residuals depends on the correlation forms still being asymptotically normal, but this is covered by theorems due to Diananda (1953) and Walker (1954).

Fitting autoregressive constants successively. It is instructive to examine the effect of fitting one further coefficient to an autoregressive scheme (cf. Lee, 1951). This may be done quite simply, and indicates the relevance of the expressions $h_t^2 r_{t+p}$ $(t = 1, 2, ...)$ for testing significance.

The least-squares equations for the scheme

$$H_t X_t \equiv X_t - a_1 X_{t-1} - ... - a_p X_{t-p} = Y_t \tag{9}$$

are
$$h_t r_t = 0 \quad (t = 1, ..., p). \tag{10}$$

We note also that

$$h_{-t} h_t r_t = [\sigma^2(Y)/\sigma^2(X)]_{\text{est.}} \quad (t = 0). \tag{11}$$

Equation (11), together with (10), may equivalently be written

$$h_t r_t = [\sigma^2(Y)/\sigma^2(X)]_{\text{est.}} \quad (t = 0). \tag{12}$$

Under the new hypothesis

$$H_t' X_t \equiv X_t - a_1' X_{t-1} - ... - a_{p+1}' X_{t-p-1} = Y_t, \tag{13}$$

the equations (10) are replaced by

$$h_t' r_t = 0 \quad (t = 1, ..., p+1). \tag{14}$$

Operating on the set (14) by h_t, and making use of (10), we obtain

$$(h_t r_t)_{t=p+1} - (h_t r_t)_{t=0} a_{p+1}' = 0,$$

or
$$\hat{a}_{p+1}' = \frac{(h_t r_t)_{t=p+1}}{(h_t r_t)_{t=0}}. \tag{15}$$

From (10), equation (15) is equivalently written

$$\hat{a}_{p+1}' = \frac{(h_t^2 r_t)_{t=p+1}}{(h_t r_t)_{t=0}}. \tag{16}$$

From the previous results we have

$$E\{(h_t^2 r_t)_{t=p+1}\} \sim 0,$$

$$E\{(h_t^2 r_t)_{t=p+1}^2\} \sim \frac{1}{n}[\sigma^2(Y)/\sigma^2(X)]^2,$$

whence, under the hypothesis (9), we have

$$E\{\hat{a}_{p+1}'\} \sim 0, \quad \sigma^2\{\hat{a}_{p+1}'\} \sim \frac{1}{n}. \tag{17}$$

a result which permits a simple asymptotic test of the significance of a_{p+1}'. If it is significant, the adjusted values $\hat{a}_i'$ $(i = 1, ..., p)$

may be calculated from $\hat{a}_i$ by a standard regression formula, which becomes in this context

$$\hat{a}_i' = \hat{a}_i - \hat{a}_{p+1}' \hat{a}_{p-i+1} \quad (i = 1, 2, ..., p). \tag{18}$$

More generally, it has been shown that the χ^2 test using the $q - p$ forms $h_t^2 r_{t+p}$ $(t = 1, 2, ..., q-p)$ is asymptotically equivalent to the likelihood criterion developed at the beginning of this section for comparing the hypothesis of an autoregressive model of order p with the wider hypothesis of one of higher order q (for a more detailed discussion, see Walker (1952)).

Superposed error. Complications in fitting autoregressive schemes may arise in practice from additional errors; for example, from a random superposed error depressing the observed correlations. Thus if

$$U_t = X_t + Z_t, \tag{19}$$

where Z_t is an independent sequence independent also of X_t, we have

$$\sigma^2(U) = \sigma^2(X) + \sigma^2(Z), \tag{20}$$

$$\operatorname{cov}(U_t, U_{t+\tau}) = \operatorname{cov}(X_t, X_{t+\tau}) \quad (\tau > 0). \tag{21}$$

Writing $\lambda = \sigma^2(Z)/\sigma^2(X)$, we have a depression of the correlations by a factor $1/(1 + \lambda)$. If this is suspected, an estimate of the quantity λ will also be necessary, and correspondingly for other error effects. It might be noticed that the series U_t is of 'p-dependent residual' type if X_t is autoregressive of order p. The equations $H_t \rho_t = 0$ $(t > 0)$ for X_t are thus satisfied for U_t for $t > p$; this is sometimes useful in providing simple consistent (though not fully efficient) equations of estimation for the original coefficients a_i even in the presence of superposed error.

9·12 Time-series specified for continuous time.

Many time-series are not of the type considered in the last section, but are recorded continuously (e.g. physical processes recorded by electrical or optical means; see, for example, fig. 6, § 6·1). Even if a continuous record is not available, it is sometimes more reasonable to envisage a process of continuous type from which discrete observations are made than a process only definable for discrete time.

Two general remarks are relevant. First, if a discrete series of observations is used in place of a continuous record, the general sampling error formulae for the correlations given in the last section will still be valid. Secondly, if the continuous record is considered directly, the adaptation of the formulae is immediate. For example, the sampling variance and covariances of

$$C(\tau) = \frac{1}{T-\tau} \int_0^{T-\tau} X(t)\,X(t+\tau)\,dt \tag{1}$$

are exactly as in the case of discrete sums, with integrals replacing sums and T replacing n; thus the formula corresponding to equation (6) of §9·1 is

$$\mathrm{cov}\,(C(s), C(s+t))$$
$$\sim \frac{1}{T-s} \int_{-\infty}^{\infty} \{\sigma^4[\rho(v)\,\rho(v+t) + \rho(v-s)\,\rho(v+s+t)] + \kappa_{v,s,t}\}\,dv, \tag{2}$$

from which other formulae may be deduced. The $\kappa_{v,s,t}$ term will be zero for normal processes, and its effect on the leading term of the corresponding formulae for the correlation coefficients $R(\tau)$ will also vanish for linear processes of the type

$$X(t) = \int_{-\infty}^{\infty} g(t-v)\,dY(v) \quad (g(u) = 0 \text{ for } u < 0), \tag{3}$$

where $Y(v)$ is an additive process. Apart from the $\kappa_{v,s,t}$ term, an equivalent formula to (2) (which assumes that the correlogram is damped and hence that the spectrum is continuous) is obtained with the use of Parseval's theorem and is

$$\mathrm{cov}\,(C(s), C(s+t)) \sim \frac{\sigma^4}{2\pi(T-s)} \int_{-\infty}^{\infty} f^2(\omega)\,[e^{i\omega t} + e^{i\omega(2s+t)}]\,d\omega. \tag{4}$$

Regression analysis of these continuous processes, however, raises complications, for such processes are in practice likely to arise from differential equations, in contrast with the difference equations with discrete time, so that simple autoregressive models of the discrete type are not in general appropriate except perhaps as approximations. Methods of fitting models more consistent with the character of these processes can in principle be worked out, and will be illustrated by consideration of the model analogous to equation (11) of §9·1 (cf. §5·2, equation (19))

$$d\dot{X}(t) + \alpha \dot{X}(t)\,dt + \beta X(t)\,dt = dY(t). \tag{5}$$

To estimate the unknown coefficients α and β, we minimize the sum of squares

$$\int_0^T [dY(t)]^2 \equiv \int_0^T [d\dot{X}(t) + \alpha \dot{X}(t)\, dt + \beta X(t)\, dt]^2,$$

obtaining the formal solution

$$\left. \begin{aligned} \int_0^T \dot{X}(t)\,[d\dot{X}(t) + \alpha \dot{X}(t)\, dt + \beta X(t)\, dt] = 0, \\ \int_0^T X(t)\,[d\dot{X}(t) + \alpha \dot{X}(t)\, dt + \beta X(t)\, dt] = 0. \end{aligned} \right\} \tag{6}$$

Since
$$\int_0^T X(t)\,\dot{X}(t)\, dt = [\tfrac{1}{2} X^2(t)]_0^T,$$

$$\int_0^T X(t)\, d\dot{X}(t) = [X(t)\,\dot{X}(t)]_0^T - \int_0^T \dot{X}^2(t)\, dt,$$

the solutions of (6) are asymptotically equivalent to the estimates

$$\left. \begin{aligned} \hat{\alpha} &= -\int_0^T \dot{X}(t)\, d\dot{X}(t) \Big/ \int_0^T \dot{X}^2(t)\, dt, \\ \hat{\beta} &= \int_0^T \dot{X}^2(t)\, dt \Big/ \int_0^T X^2(t)\, dt, \end{aligned} \right\} \tag{7}$$

and their asymptotic errors are

$$\left. \begin{aligned} \sigma^2(\hat{\alpha}) \sim \frac{\kappa_2}{\displaystyle\int_0^T \dot{X}^2(t)\, dt} \sim \frac{2\alpha}{T}, \quad \sigma^2(\hat{\beta}) \sim \frac{\kappa_2}{\displaystyle\int_0^T X^2(t)\, dt} \sim \frac{2\alpha\beta}{T}, \\ \mathrm{cov}\,(\hat{\alpha}, \hat{\beta}) \sim 0, \end{aligned} \right\} \tag{8}$$

where κ_2 is the rate of increase of the variance of $Y(t)$.

The above estimation equations are simplified by the orthogonality of $X(t)$ and $\dot{X}(t)$, implying an asymptotic orthogonality for these quantities in the sample. The estimate $\hat{\alpha}$ remains the same if $\beta = 0$, when (5) reduces to a linear Markov equation in $U(t) \equiv \dot{X}(t)$. For such a process we thus have the least-squares estimate

$$\hat{\alpha} = -\int_0^T U(t)\, dU(t) \Big/ \int_0^T U^2(t)\, dt,$$

with asymptotic standard error $\sqrt{(2\alpha/T)}$ (cf. equation (15), § 8·3).

Returning to the second-order process (5), we recall that the least-squares equations for a and b in the discrete time case were asymptotically equivalent to the first two equations ($s=1$ and 2) obtained from the difference relation (12) of § 9·1 by substituting R for ρ. Similarly, the above estimates correspond to the differential equation

$$\rho''(\tau) + \alpha\rho'(\tau) + \beta\rho(\tau) = 0 \quad (\tau \geqslant 0) \tag{9}$$

(obtained by multiplying equation (5) by $X(t-\tau)$ and averaging), evaluated at $\tau = 0$, and to its derivative, also evaluated at $\tau = 0$.

If the direct evaluation of these estimates is not practically convenient, alternative methods making use of the observed autocorrelations may be used, but the above limiting formulae are still useful in suggesting methods approximating to optimum (least-squares) efficiency, and their validity is incidentally checked by such further investigation. Thus the most direct alternative is to obtain α and β by interpolatory methods from the correlations $R(h)$ and $R(2h)$, using the known relation (equivalent to equation (28) § 5·2)

$$\rho(\tau) = \frac{e^{-\frac{1}{2}\alpha|\tau|}\cos(\lambda|\tau|-\theta)}{\cos\theta}, \tag{10}$$

where $\tan\theta = \frac{1}{2}\alpha/\lambda$, $\lambda = \sqrt{(\beta - \frac{1}{4}\alpha^2)}$ $(\beta \geqslant \frac{1}{4}\alpha^2)$. It is useful to note from (9), or from the solution of (5) given in equation (21), § 5·2, that $\rho(\tau)$ also satisfies the difference equation

$$\rho(\tau + 2h) + a\rho(\tau + h) + b\rho(\tau) = 0 \quad (\tau \geqslant 0), \tag{11}$$

where $\qquad a = -2e^{-\frac{1}{2}\alpha h}\cos\lambda h, \quad b = e^{-\alpha h}. \tag{12}$

This relation is convenient for computing the theoretical correlogram when α and β have been estimated; and may sometimes be useful in providing more rapid (though somewhat less efficient) estimates than from $R(h)$ and $R(2h)$ in conjunction with equation (10). Moreover, such difference equations, *for* $\tau > 0$, are still valid even in the presence of superposed random error (cf. end of § 9·11).

As noted in the preceding section, a method of testing the goodness of fit of a theoretical correlogram of the above type makes use of the difference equation for $X(t)$. For example, from the solution for the process (5) it is easily verified that

$$X_{t+2} + aX_{t+1} + bX_t = Y_{t+2}, \tag{13}$$

where a and b are given by (12), but Y_t, while independent of $Y_{t+\tau}$ ($\tau > 1$) is correlated with Y_{t+1}. Thus in the notation of §9·11 for dependent residuals $m = 1$ (in general, for a pth order linear differential equation, $m = p - 1$). We have further

$$\sigma^2(Y) = E\{(X_t + aX_{t-1} + bX_{t-2}) Y_t\}$$
$$= E\{(X_t + aX_{t-1} + bX_{t-2}) (X_t + aX_{t-1})\}$$
$$= \sigma^2(X) \{1 + a^2 + 2a\rho_1 + b\rho_2 + ab\rho_1\}$$
$$= \sigma^2(X) \{1 + a^2 - b^2 + 2a\rho_1\},$$

$$E\{Y_t Y_{t+1}\} = E\{(X_t + aX_{t-1} + bX_{t-2}) Y_{t+1}\}$$
$$= E\{X_t(X_{t+1} + aX_t + bX_{t-1})\}$$
$$= \sigma^2(X) \{\rho_1 + b\rho_1 + a\}.$$

With these results we find

$$\frac{\sigma^4(Y)}{\sigma^4(X)} \sum_{i=-m}^{m} \delta_i \delta_{i+t-\tau} = \begin{cases} (1 + a^2 - b^2 + 2a\rho_1)^2 + 2(a + \rho_1 + b\rho_1)^2 \\ \hspace{7cm} (t = \tau), \\ 2(a + \rho_1 + b\rho_1) (1 + a^2 - b^2 + 2a\rho_1) \\ \hspace{7cm} (t = \tau + 1), \\ (a + \rho_1 + b\rho_1)^2 \hspace{3.5cm} (t = \tau + 2), \\ 0 \hspace{6.5cm} (t > \tau + 2). \end{cases}$$

The first of these expressions μ, say, leads to the asymptotic variances of the functions $(n-t)^{\frac{1}{2}} H_t^2 R_{t+2}$ (or $(n-t)^{\frac{1}{2}} H_{-t} H_t R_t$) and correspondingly the conversion factor from $(H_t^2 R_{t+2})^2$ to a χ^2 component is $(n-t)/\mu$. The other quantities are needed either for testing roughly the significance of the total χ^2 by use of the factor (8) of §9·11, or if a more exact matrix inversion is to be carried out. The process (13) is not a discrete linear process in the precise sense previously defined, and the last results are only true in general for $t \geqslant \tau > 2$.

9·13 Numerical examples. *An artificial autoregressive series.* It is useful to illustrate these methods first on a series of known type, so that the agreement with theory may be indicated. We consider an artificial series of 480 terms of the form

$$X_t = 1 \cdot 1 X_{t-1} - 0 \cdot 5 X_{t-2} + Y_t, \tag{1}$$

constructed by M. G. Kendall (1946); the Y_t were taken directly from a table of random numbers, and were thus not normal. The first fifteen correlations, theoretical and observed, are given in table 4 and fig. 10. From R_1 and R_2 we calculate $\hat{a}_1 = 1 \cdot 132$ (S.E. $0 \cdot 039$), $\hat{a}_2 = -0 \cdot 485$ (S.E. $0 \cdot 039$), consistent with the values $1 \cdot 1$ and $-0 \cdot 5$.

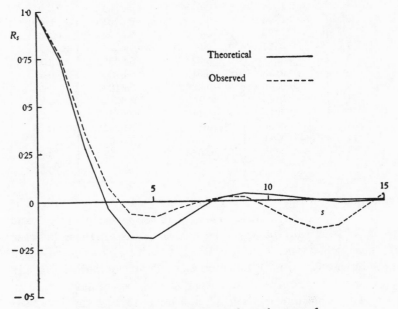

Fig. 10. Theoretical and observed correlograms of M. G. Kendall's artificial series I.

The goodness of fit of this series was demonstrated by Quenouille (1947) using the known values of a_1 and a_2. In practice we should use $\hat{a}_1$ and $\hat{a}_2$, but this would not appreciably affect the results, so we quote the values of $H_s^2 R_{s+2}$ ($s = 1, \ldots, 15$), and the corresponding cumulative sum of squares of the derived standardized variates, which form a χ^2 with degrees of freedom equal to the number of values taken (for statistically significant values of χ^2, see Fisher and Yates, *Statistical Tables*, 1938).

Wolfer's sunspot numbers. This series was the first to be analysed by the autoregressive method in Yule's pioneering paper (1927), and for comparative purposes we have used the

Table 4.   *Correlogram and goodness of fit*
(*M. G. Kendall's artificial series I*)

s	ρ_s	R_s	$H_s^2 R_{s+2}$	χ^2
1	0·7333	0·762	0·02412	2·32
2	0·3067	0·377	0·00417	0·07
3	− 0·0293	0·079	0·01979	1·56
4	− 0·1856	− 0·067	− 0·00812	0·26
5	− 0·1895	− 0·078	− 0·00013	0·00
6	− 0·1156	− 0·039	0·02026	1·62
7	− 0·0325	− 0·007	− 0·02247	1·98
8	0·0221	0·022	− 0·02903	3·31
9	0·0406	0·018	− 0·01047	0·43
10	0·0336	− 0·036	− 0·01226	0·59
11	0·0166	− 0·103	0·00747	0·22
12	0·0015	− 0·145	0·01345	0·70
13	− 0·0067	− 0·128	− 0·00573	0·13
14	− 0·0081	− 0·052	− 0·00517	0·10
15	− 0·0056	0·029	0·00249	0·03
				13·32

graduated series[†] for the years 1749–1924 given by Yule. The observed values of R_s for $s = 1, ..., 23$ are given in table 5 and fig. 11 (there are some unimportant small discrepancies between these values and Yule's for $s = 1, ..., 5$, possibly due to slight differences in the formulae or methods of computation used). Yule fitted a simple second-order autoregressive model from R_1 and R_2, obtaining $\hat{a}_1 = 1·5153$, $\hat{a}_2 = − 0·8025$, and the first 20 χ^2 values obtained from $h_s^2 R_{s+2}$ for this model are also given in the table. The total χ^2, with 20 degrees of freedom, is obviously significant (due mainly to the group of high χ^2 values for low s). It may be argued that a more consistent model of autoregressive type for these numbers, which represent continuous sunspot activity, would be the continuous process of equation (5), §9·12. This yields estimates from R_1 and R_2 of $\hat{\alpha} = 0·3186$, $\hat{\beta} = 0·3631$. From equation (12), §9·12, the corresponding values of a and b are − 1·4255 and 0·7272. This second model appears

---

† The graduated series was introduced by Yule in an attempt to reduce the effect of any superposed error. Its use here in preference to the primary series may be criticized, but the R_s in table 5 were originally worked out for this series in connection with the study of the continuous model, for which they showed more promise for $s \leqslant 5$ (see Bartlett, 1946). They were thereafter made use of for these various further illustrative calculations.

to give slightly better agreement over the first five correlations, but it seems clear that any gain in the goodness of fit near the beginning of the correlogram is counterbalanced by a loss for the later correlations. It must be remembered that an observed correlogram always exhibits less damping than the theoretical, as the observed correlations are inflated by sampling fluctuations

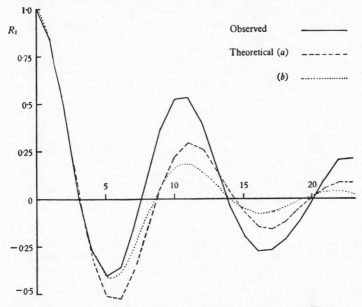

Fig. 11. Observed correlogram of Wolfer's sunspot numbers (Yule's graduated series, 1749–1924), compared with theoretical correlogram of second-order autoregressive model (a) discrete time, (b) continuous time.

(in this case, for the continuous model fitted, the standard error of a correlation has an asymptotic value for large lag of about 0·16); nevertheless, the damping for this second theoretical correlogram appears excessive without detailed test. However, to illustrate the application of the goodness of fit test for the continuous model, the quantities $h_s^2 R_{s+2}$ appropriate to this model are also shown in table 5 with the corresponding χ^2 quantities. (No χ^2 quantity is given for $s = 1$, as we have seen that this would require separate consideration. We have, however, included $s = 2$, assuming that the effect of any non-zero fourth cumulant of the distribution of Y_s is negligible.) Making use of the rough evaluation of significance suggested for this type of application to dependent residuals, we obtain a reduced total χ^2 of 42·85 with $20/1·40 = 14·3 \sim 14$ degrees of freedom, again highly significant.

Table 5. *Correlogram and goodness of fit for Wolfer's sunspot numbers (Yule's graduated series, 1749–1924)*

s	R_s	Discrete autoregressive model			Continuous model		
		ρ_s	$h_s^2 R_{s+2}$	χ^2	ρ_s	$h_s^2 R_{s+2}$	χ^2
1	0·8408	(0·8407)	0·00765	0·92	(0·8407)	0·00554	—
2	0·4715	(0·4714)	0·03138	15·26	(0·4713)	0·03084	12·60
3	0·0470	0·0397	− 0·02571	10·18	0·0605	− 0·02026	5·40
4	− 0·2647	− 0·3181	0·02704	11·19	− 0·2565	0·02613	8·90
5	− 0·4059	− 0·5139	0·01791	4·90	− 0·4096	0·02151	6·02
6	− 0·3601	− 0·5234	0·01679	4·27	− 0·3974	0·02127	5·85
7	− 0·1636	− 0·3807	0·01085	1·77	− 0·2687	0·01594	3·21
8	0·1088	− 0·1569	0·00538	0·43	− 0·0941	0·00832	0·89
9	0·3646	0·0678	0·00450	0·31	0·0612	0·00475	0·25
10	0·5197	0·2286	0·00643	0·60	0·1557	0·00394	0·20
11	0·5273	0·2920	0·01402	2·89	0·1775	0·01041	1·36
12	0·3937	0·2590	0·00625	0·57	0·1398	0·00465	0·25
13	0·1791	0·1582	0·00425	0·26	0·0702	0·00354	0·16
14	− 0·0390	0·0318	− 0·01467	3·10	− 0·0016	− 0·01460	2·58
15	− 0·1947	− 0·0787	0·00241	0·09	− 0·0533	− 0·00008	0·00
16	− 0·2706	− 0·1448	− 0·01200	2·05	− 0·0748	− 0·01340	2·19
17	− 0·2683	− 0·1563	0·02077	6·09	− 0·0679	0·01825	3·98
18	− 0·2179	− 0·1206	− 0·00068	0·01	− 0·0424	0·00193	0·05
19	− 0·1256	− 0·0573	0·01051	1·55	− 0·0111	0·01364	2·26
20	− 0·0096	0·0099	− 0·01519	3·18	0·0150	− 0·01068	1·30
21	0·1143	0·0610			0·0295	− 0·01499	2·59
22	0·2035	0·0845		69·62	0·0311		
23	0·2132	0·0791			0·0229		

These two examples are discussed further in § 9·2 from the point of view of harmonic (periodogram) analysis.

The emigration-immigration process. For this process we had the linear regression equation

$$X_t - m = e^{-\mu t}(X_0 - m) + Y_t,$$

where the successive Y_t were necessarily uncorrelated, but only normal and hence independent for large m. The series referred to below, for which we are indebted to Lord Rothschild (cf. § 8·3), all had means of over 12, and the normal approximation should for such series be sufficient, but a specific theoretical investigation for this process by Miss V. T. Patil has shown that the required asymptotic moment properties of the Quenouille forms are approximately preserved even for finite m. The asymptotic normality of the correlations should also be preserved in general owing to their relation to the transition frequencies, the asymptotic normality of which would still hold for probability chains of this type, even although the number of states is not strictly finite.

Table 6. *Means and correlograms* (*Rothschild's data*)

	Series 1	Series 2	Series 3	Series 4	Series 5
n	112	112	113	117	114
$\hat{m}$	24·99	16·05	15·89	12·89	20·77
R_1	0·5426	0·3755	0·5606	0·4038	0·4313
R_2	0·3109	0·1013	0·3280	0·2055	0·1976
R_3	0·1976	0·0784	0·2026	− 0·0041	0·1617
R_4	0·0588	0·1050	0·1806	− 0·1248	− 0·0033
R_5	0·0560	0·0258	0·1235	− 0·1107	− 0·0520
R_6	− 0·0608	0·0206	0·1096	− 0·0043	0·0024

The relevant numerical results, taken as far as R_6, are shown in table 6. The total χ^2's for the various series, using R_1 as an estimate of ρ_1 in each case, are 4·09, 1·56, 1·09, 4·13, 3·27, each with 5 degrees of freedom, or 14·14 in all with 25 degrees of freedom. In particular, the R_2's for the five series are satisfactory, and thus no appreciable evidence of any inadequacy of the model is indicated. (The five series have been tested separately. In fact nos. 1 and 5 refer to two series of counts in the same size frame, and nos. 2, 3 and 4 to a second size. A test of homogeneity of R_1 and $\hat{m}$ within these two groups, based on the asymptotic sampling variances of these quantities, reveals no heterogeneity for ρ_1, but some for m. For further tests on these data, see also, however, Patil, 1957.)

9·2   Harmonic (periodogram) analysis

It has already been pointed out at the beginning of this chapter that the classical approach to harmonic analysis and the corresponding search for periodicities in observed series by means of 'periodogram analysis' become merely one aspect of the analysis of stationary processes, and that only when regarded from this broader basis is a sound interpretation of observational results likely.

The classical analysis of a discrete (equally-spaced) series of observations consists of computing, usually for integral p, the quantity

$$J_p \equiv A_p + iB_p = \sqrt{\frac{2}{n}} \sum_{r=1}^{n} X_r e^{2\pi ipr/n} \tag{1}$$

(by computing A_p and B_p separately), and hence the intensity $I_p \equiv J_p J_p^*$. The factor $\sqrt{(2/n)}$ has been inserted to make the mean value of I_p $2\sigma^2$ for a completely random series.

It is easily seen that

$$I_p = \frac{2}{n} \sum_{r=1}^{n} \sum_{u=1}^{n} X_r X_u \cos\left[(r-u)\,\omega_p\right], \tag{2}$$

where $\omega_p = 2\pi p/n$, or if we write as before

$$C_s = \frac{1}{n-s} \sum_{r=1}^{n-s} X_r X_{r+s} \quad (s > 0,\ C_{-s} = C_s),$$

then

$$I_p = 2 \sum_{s=-n+1}^{n-1} \left(1 - \frac{|s|}{n}\right) C_s \cos(s\omega_p). \tag{3}$$

For convenience we still assume $E\{X\} = 0$, but since formula (1) is not altered when $E\{X\} \neq 0$, a remark on the effect of this will be inserted directly. With this assumption we have on averaging (3)

$$E\{I_p\} = 2\sigma^2 \sum_{s=-n+1}^{n-1} \left(1 - \frac{|s|}{n}\right) \rho_s \cos(s\omega_p). \tag{4}$$

As n increases, this tends for most values of ω_p to $2\pi\sigma^2 f_+(\omega_p)$, where $f_+(\omega)$ is the (continuous) spectrum of X_r.

The usual periodogram argument considers what happens when X_r contains a harmonic component of frequency λ, so that ρ_s contains a component $\cos \lambda s$. Since in this case the spectrum contains a discrete component at $\omega = \lambda$, it is evident that $E\{I_p\}$ will tend to infinity at this point. The precise effect is readily investigated. Putting $\rho_s = \cos \lambda s$ in (4) we find, noting the identities of the sums

$$\sum_{r=1}^{n} \sum_{u=1}^{n} g(r-u) = \sum_{s=-n+1}^{n-1} (n - |s|)\,g(s) = \sum_{v=0}^{n-1} \sum_{s=-v}^{v} g(s),$$

and also the formulae

$$\sum_{s=-v}^{v} \cos \mu s = \frac{\sin\left[(v+\tfrac{1}{2})\mu\right]}{\sin\tfrac{1}{2}\mu}, \quad \sum_{v=0}^{n-1} \sin(v+\tfrac{1}{2})\mu = \frac{\sin^2 \tfrac{1}{2}n\mu}{\sin\tfrac{1}{2}\mu},$$

$$E\{I_p\} = \frac{\sigma^2}{n} \left\{ \frac{\sin^2[\tfrac{1}{2}n(\omega_p+\lambda)]}{\sin^2[\tfrac{1}{2}(\omega_p+\lambda)]} + \frac{\sin^2[\tfrac{1}{2}n(\omega_p-\lambda)]}{\sin^2[\tfrac{1}{2}(\omega_p-\lambda)]} \right\}. \tag{5}$$

In general, this will be $O(1/n)$, but as $\omega_p \to \lambda$, the second term $\to n\sigma^2$, and $E\{I_p\}$ becomes large, a result which is used to detect the presence of harmonic components in the original series. From the form of the sine terms, the analysis can resolve frequencies which differ by more than $O(1/n)$. The effect of not measuring X from $E\{X\}$ is now apparent; it will merely introduce a spurious discrete component in the spectrum at $\omega = 0$.

However, for finite n, $E\{I_p\}$ from (5) is still finite and it becomes necessary to consider the significance of I_p in comparison with its probable value on other hypotheses. It has been recognized that as the intensity I_p fluctuates about its mean value $2\sigma^2$ for a completely random series, it is advisable to allow for such fluctuations, which are known to follow the probability law

$$P\{I_p \geqslant z\} = \exp\left(-\tfrac{1}{2}z/\sigma^2\right), \tag{6}$$

when X_r is normal (cf. M. G. Kendall, 1946a, § 30·49).

Once we admit the possibility of continuous spectra beyond the uniform 'noise' spectrum of the completely random series the situation is more complex, for a non-uniform spectrum will have peaks which may be comparable to $E\{I_p\}$ in (5) when n is finite. As a simple example, consider the linear Markov process with correlogram $\rho_s = \rho^{|s|}$. This process has a continuous spectrum given by

$$f_+(\omega) = \frac{1}{\pi} \sum_{s=-\infty}^{\infty} \rho^{|s|} \cos s\omega$$

$$= \frac{1-\rho^2}{\pi(1+\rho^2 - 2\rho\cos\omega)} \quad (0 < \omega \leqslant \pi). \tag{7}$$

This function has a maximum at $\omega = 0$, given by

$$f_+(0) = \frac{1+\rho}{\pi(1-\rho)},$$

so that for ρ near 1 this function may be quite large. An example of a series with a more genuine quasi-period is the autoregressive process discussed in § 9·1. By the methods of § 6·3 we have for the series $X_t = a_1 X_{t-1} + a_2 X_{t-2} + Y_t$, with solution $(-1 < a_2 < -\tfrac{1}{4}a_1^2)$,

$$X_t = \sum_{v=-\infty}^{t} \left(\frac{\mu_1^{t+1-v} - \mu_2^{t+1-v}}{\mu_1 - \mu_2} \right) Y_v, \tag{8}$$

where μ_1 and μ_2 are the roots of $\mu^2 - a_1\mu - a_2 = 0$, the spectrum

$$f_+(\omega) = \frac{\sigma^2(Y)}{\pi\sigma^2(X)} h(\omega)\, h^*(\omega) \quad (0 < \omega \leqslant \pi),$$

where
$$h(\omega) = \sum_{v=0}^{\infty} e^{-iv\omega} \left(\frac{\mu_1^{v+1} - \mu_2^{v+1}}{\mu_1 - \mu_2} \right)$$

$$= \frac{1}{\mu_1 - \mu_2} \left[\frac{\mu_1}{1 - \mu_1 e^{-i\omega}} - \frac{\mu_2}{1 - \mu_2 e^{-i\omega}} \right],$$

whence
$$f_+(\omega) = \frac{(1+a_2)\,([1-a_2]^2 - a_1^2)}{\pi(1-a_2)\{[1+a_2]^2 + a_1^2 - 2a_1(1-a_2)\cos\omega - 4a_2\cos^2\omega\}}, \tag{9}$$

with a peak at $\cos^{-1}[-\tfrac{1}{4}a_1(1-a_2)/a_2]$.

For a fixed length of series available it will evidently be more difficult to distinguish strictly harmonic components from peaks of a continuous spectrum than from a uniform 'noise' spectrum, especially in view of sampling fluctuations, the effect of which is considered below. In the case of one or more discrete harmonic components it is in many situations correct to assume that the amplitude of such a component is constant from one series to another. In such a case the estimate of the amplitude has a sampling error which diminishes with n, or equivalently $I_p/E\{I_p\} \to 1$ as n increases (the phase of the component may be random from series to series, and in fact for strict validity of the stationarity assumption this is necessary, but this does not affect I_p, which is a measure of amplitude and not of phase). But in problems for which the spectrum of the process is essentially continuous, as in the autoregressive series of equation (11), § 9·1, or in the more general linear processes of § 6·3, the randomness of the spectral components cannot be isolated into the phase. It might seem possible to imagine physical processes with continuous spectra for which the phases only, and not the amplitudes, of the components are random from series to series. But this implies that the orthogonal function $Z(\omega)$ in the theoretical harmonic analysis of X_t is additive, with, moreover, a fixed spectral contribution at each frequency ω; X_t is then necessarily normal (this is effectively the theorem quoted in § 6·4; cf. also Lévy, 1948, pp. 99–100). The process X_t is therefore stochastically identical with any other normal process with the same autocorrelation function.

For the completely random normal sequence with uniform spectrum we have from (6)

$$\text{var}\{I_p\} = 4\sigma^4 = [E\{I_p\}]^2, \tag{10}$$

so that fluctuations in I_p tend to be of the same order as the mean value itself. It is shown below that this result remains approximately true in general for linear processes. Consider first the quantity

$$J_Y(\omega) = \sqrt{\frac{2}{n}} \sum_{r=1}^{n} e^{ir\omega} Y_r, \tag{11}$$

so that $J_Y(\omega)$ is defined as J_p was for X_t in (1), but for the residuals Y_t. $J_Y(\omega)$ is defined for any ω, including values $2\pi p/n$ for p integral. Then for the expectations

$$E\{J_Y(\omega_1) J_Y(\omega_2)\}, \quad E\{J_Y(\omega_1) J_Y^*(\omega_2)\}$$

we have the single formula (the minus sign in the $\pm$ corresponding to the second expectation)

$$E\left\{\frac{2}{n} \sum_{r=1}^{n} \sum_{u=1}^{n} Y_r Y_u e^{i(\omega_1 r \pm \omega_2 u)}\right\} = \frac{2\sigma^2(Y)}{n} \frac{[1 - e^{in(\omega_1 \pm \omega_2)}] e^{i(\omega_1 \pm \omega_2)}}{1 - e^{i(\omega_1 \pm \omega_2)}}. \tag{12}$$

In particular, when ω_1 and ω_2 are of the form $2\pi p/n$ this vanishes $(\omega_1 \neq \omega_2)$, showing that $J_Y(\omega_1)$ is then uncorrelated with $J_Y(\omega_2)$ and $J_Y^*(\omega_2)$, or equivalently $A_Y(\omega_1)$, $B_Y(\omega_1)$ are uncorrelated with $A_Y(\omega_2)$, $B_Y(\omega_2)$ (similarly also for $J_Y(\omega_1)$ and $J_Y(\omega_2)$ when $\omega_2 = \omega_1$, or equivalently $A_Y(\omega_1)$, $B_Y(\omega_1)$ are uncorrelated and have the same mean square). Further, for the second-order quantities $I_Y(\omega) = J_Y(\omega) J_Y^*(\omega)$, we easily find, either by making use of (12), or by straightforward algebra from the identity (cf. equation (3))

$$I_Y(\omega) = 2 \sum_{s=-n+1}^{n-1} \left\{ \cos s\omega \left[\sum_{r=1}^{n-|s|} \frac{Y_r Y_{r+|s|}}{n} \right] \right\}, \tag{13}$$

that

$$E\{I_Y(\omega_1) I_Y(\omega_2)\} - 4\sigma^4(Y)$$
$$= \frac{4\kappa_4}{n} + \frac{4\sigma^4(Y)}{n^2} \left[\frac{1 - \cos n(\omega_1 + \omega_2)}{1 - \cos(\omega_1 + \omega_2)} + \frac{1 - \cos n(\omega_1 - \omega_2)}{1 - \cos(\omega_1 - \omega_2)} \right]. \tag{14}$$

From (14)

(i) var $I_Y(\omega) \sim 4\sigma^4(Y)$, this result being more general than the corresponding result implicit in (6), as it holds even for Y_t not normal,

(ii) cov $(I_Y(\omega_1), I_Y(\omega_2))$ $(n \mid \omega_1 - \omega_2 \mid \gg 1)$

$$= \begin{cases} O\!\left(\dfrac{1}{n^2}\right) & \text{for } Y_t \text{ normal (zero if } \omega \text{ of form } 2\pi p/n), \\[2ex] O\!\left(\dfrac{1}{n}\right) & \text{if } \kappa_4 \neq 0. \end{cases}$$

If we now examine the corresponding quantity $J_X(\omega)$ $(= J_p$ for $\omega = 2\pi p/n)$, we note when

$$X_r = \sum_{u=-\infty}^{\infty} g_u Y_{r-u} \quad (g_u = 0 \text{ for } u < 0),$$

that

$$J_X(\omega) = \sqrt{\frac{2}{n}} \sum_{u=-\infty}^{\infty} \sum_{r=1}^{n} e^{ir\omega} g_u Y_{r-u}$$

$$= \sqrt{\frac{2}{n}} \sum_{u=-\infty}^{\infty} \sum_{r=1}^{n} e^{i(r-u)\omega} Y_{r-u} g_u e^{iu\omega}$$

$$= h^*(\omega) J_Y(\omega) \left[1 + O\!\left(\frac{1}{\sqrt{n}}\right) \right], \tag{15}$$

where $h(\omega)$ is defined (cf. §6·3) as $\Sigma_u e^{-iu\omega} g_u$, provided $g_u \to 0$ exponentially, say, as u increases, the difference between $J_X(\omega)$ and $h^*(\omega) J_Y(\omega)$ being due to 'end-effects'. The result

$$J_X(\omega) \sim h^*(\omega) J_Y(\omega) \tag{16}$$

for the sample is of considerable importance. It provides at once

$$J_X(\omega) J_X^*(\omega) \sim h(\omega) h^*(\omega) J_Y(\omega) J_Y^*(\omega). \tag{17}$$

Equation (17) gives further

$$E\{J_X(\omega) J_X^*(\omega)\} \sim h(\omega) h^*(\omega) E\{J_Y(\omega) J_Y^*(\omega)\},$$

and, as $n \to \infty$,

$$2\pi\sigma^2(X) f_+(\omega) = 2\sigma^2(Y) h(\omega) h^*(\omega), \tag{18}$$

consistently with the results of §6·3; but (17) indicates also the asymptotic stochastic relation between the periodogram of a linear process and the periodogram of its residuals with uniform spectrum. It shows that the distribution (6) is still asymptotically true, if we replace $2\sigma^2$ by $\lim_{n \to \infty} E\{I_p\}$ (it is assumed $h(\omega) \neq 0$).

Smoothing devices. These results show that while fluctuations in I_p do not diminish as the length of series analysed is increased, the correlation between neighbouring I_p and I_q does, so that the observed periodogram will exhibit a wildly fluctuating appearance. This phenomenon has often been observed in practice (see fig. 12), and raises the question of how the spectrum is to be estimated. A suggestion due to P. J. Daniell is to use

$$\frac{1}{2\epsilon} \int_{\omega-\epsilon}^{\omega+\epsilon} I(\omega')\, d\omega'. \tag{19}$$

Its variance relative to the square of $E\{I(\omega)\}$, that is, the square of its coefficient of variation, is from the above results of order $1/(\epsilon n)$ for ϵ small, ϵn large; its bias will be unimportant for ϵ small if the true spectrum $f_+(\omega)$ is a reasonably smooth function at ω. Such a smoothing device may be especially useful in the automatic electrical or optical analysis of continuous records, for which a similar fluctuation theory may be shown to apply. The averaging would also be arranged to be automatic (e.g. electrically by use of a filter of appropriate band-width). For arithmetical analysis of discrete series, as in (1) and (3), the procedure derived below (a similar procedure is of course also available for continuous processes) has been found useful.

An average taken from m independent series 'lengths', if available, would possess the usual sampling property of its error being proportional to $1/\sqrt{m}$. If the lengths were taken consecutively as contiguous portions of one total length of series, the correlation between quantities like J_p derived from different portions will be negligible as the length n of each portion increases, provided ρ_s decreases fast enough as s increases. For example, even between adjacent portions, we find

$$E\{J_p(J_p^*)'\} = 2\sigma^2 \sum_{s=-n+1}^{n-1} \left(1 - \frac{|s|}{n}\right) e^{is\omega} \rho_{s+n},$$

where J and J' are obtained from the two portions, and this expression tends to zero as n increases provided ρ_{s+n} is $O(1/n)$ for fixed s. Now averaging I_p for the m subseries gives the formula

$$\bar{I}_p = 2 \sum_{s=-n+1}^{n-1} \left(1 - \frac{|s|}{n}\right) \bar{C}_s \cos (s\omega_p),$$

where
$$\bar{C}_s = \frac{1}{m(n-s)} \sum_{u=0}^{m-1} \sum_{r=1}^{n-s} X_{r+nu} X_{r+s+nu} \quad (s \geqslant 0).$$

This latter formula ignores, however, the information available in the interval s at the end of each subseries; replacing this by estimating the covariance from the entire data we finally obtain

$$I_p(\text{smoothed}) = 2 \sum_{s=-n+1}^{n-1} \left(1 - \frac{|s|}{n}\right) C_s \cos(s\omega_p), \quad (20)$$

where C_s is obtained from the entire series. The choice of n in this formula (for given nm) is a compromise between reducing the fluctuations by the approximate factor $1/\sqrt{m}$ and not reducing too much the resolving power which depends on the value of n. Formula (20) is related to the harmonic analysis of the correlogram, which would usually have been calculated already, and it is often convenient to use R_s (calculated using deviations from the mean) in place of C_s in (20); this has the advantage of largely eliminating the spurious component at $\omega = 0$ when $E\{X\} \neq 0$, the effect of which will now persist over the frequency interval $2\pi/n$ rather than $2\pi/(nm)$.

From the sampling results for I_p (unsmoothed), the distribution of mI_p (smoothed) would asymptotically be the convolution or resultant of m such distributions, and I_p (smoothed) will hence be proportional to a χ^2 with $2m$ degrees of freedom. For finite lengths of series this asymptotic theory is of course only an approximation, which will still fail even for large n if the autocorrelations are not sufficiently damped, that is, as the series becomes more like a classical harmonic series (see examples below). (For a more general approach to the smoothing problem for continuous spectra—which includes the·above methods as special cases—see Grenander (1951); cf. also § 9·21.)

Example 1. *Artificial series.* The theory of this section warns us that a straightforward periodogram analysis of the artificial autoregressive series previously discussed in § 9·13 will give violently fluctuating results, as is indicated in fig. 12 (these values for the unsmoothed periodogram have been adapted from table 4·1, M. G. Kendall (1946*b*)). If we knew the series was of second-order autoregressive type, we could of course from the estimated coefficients compute the spectrum by means of formula (9), but if we wish to estimate it more empirically, we may use formula (20), using the observed correlations R_s. Smoothed periodograms, calculated for $n = 15$ $(m = 32)$ and $n = 30$ $(m = 16)$, are also shown in

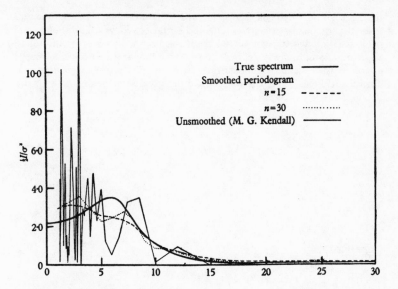

Fig. 12. Periodogram of artificial series I (M. G. Kendall), compared with smoothed periodograms ($n = 15$ and 30) and with true spectrum. (Reprinted from *Biometrika*, **37** (1950), 6.)

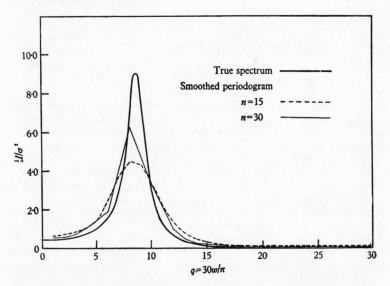

Fig. 13. Smoothed periodograms ($n = 15$ and 30) for M. G. Kendall's artificial series III, compared with true spectrum. (Reprinted from *Biometrika*, **37** (1950), 7.)

fig. 12, together with the true spectrum. It will be seen that there is still some divergence between the true and empirical curves. The theoretical peak is not markedly separated from the origin, this corresponding to the heavy damping in the correlogram, and its separation does not emerge at all in the estimated spectrum. However, the divergence does not significantly exceed its theoretical sampling value, which is of the order $\sqrt{(1/m)}$ (coefficient of variation).

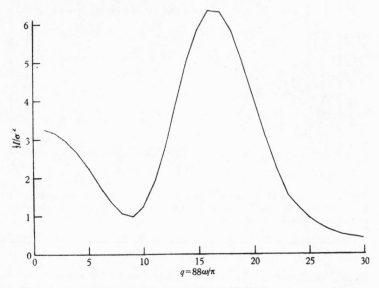

Fig. 14. Smoothed periodogram ($n = 24$) of Wolfer's sunspot numbers.
(Reprinted from *Biometrika*, **37** (1950), 10.)

Smoothed periodograms ($n = 15$ and 30) are also shown in fig. 13 for a second series of 240 terms (M. G. Kendall's series III), for which $a = -1\cdot1$, $b = 0\cdot8$. In this case the true spectrum has a much more pronounced peak, and it will be noticed that the amplitude of the peak of the smoothed periodogram fails to reach the true value, especially for $n = 15$. This effect must be expected for small n in the case of series with lightly damped oscillations. It will be remembered that in the case of classical harmonic analysis with a discrete spectrum the true amplitude (with the standardizing factor being used) would be infinite, whereas that of the periodogram would still be finite. It may be concluded that excessive curtailing of the correlogram is inadvisable if the amplitude of a pronounced peak is to be estimated as well as its location. The smoothed spectrum is only unbiased asymptotically, approaching the true spectrum if the series is long enough for adequate smoothing, and yet n is also large enough for adequate resolving power.

Example 2. *Wolfer sunspot numbers.* In the case of the Wolfer sunspot numbers, no simple autoregressive model was adequate. Nevertheless, the series is one, as Yule pointed out, for which classical harmonic analysis is of doubtful value, although periodogram analysis of the series has at times been made. In order to investigate its spectrum without a knowledge of the underlying mechanism we may proceed by the methods of this section. The smoothed periodogram ($n = 24$) is shown in fig. 14. There is a rise towards $\omega = 0$, which may be partly due to a spurious component at $\omega = 0$, but possibly also indicates a further component of the Markov process type or the effect of finitely dependent residuals (the graduation of the original series by Yule might contribute to this). This suggests that a somewhat better empirical fit might be obtained for the correlogram by including an additional effect of this kind in the model, but such an extension of the empirical model would not greatly contribute to the interpretation of the observed data, and is not considered here in detail. Apart from its behaviour at the origin, the spectrum over the interesting range consists of a smooth hump-backed curve with a peak at about 11·0 years, in conformity with the usually accepted value of the 'period'.

9·21 Smoothing techniques in relation to the mean square error.
A considerable literature has been devoted to the accuracy and efficiency of various smoothing techniques for estimating continuous spectra, of which the two methods referred to in the last section are particular examples.

A more general estimator of the spectral density $f_+(\omega)$, or, rather more conveniently, of the related function $2\pi\sigma_X^2 f_+(\omega)$, can be written in the form

$$T(\omega) = \int_0^\pi I_X(u)\, w(u, \omega)\, du. \tag{1}$$

Thus for the Daniell estimator

$$w(u, \omega) = \begin{cases} 1/(2h) & (\omega - h \leqslant u \leqslant \omega + h), \\ 0 & \text{otherwise;} \end{cases} \tag{2}$$

and for the estimator used in the examples, given by equation (20) of the previous section, it may be shown that effectively

$$w(u, \omega) = \frac{1}{2\pi n} \frac{\sin^2\left[\frac{1}{2}n(u - \omega)\right]}{\sin^2\left[\frac{1}{2}(u - \omega)\right]}. \tag{3}$$

Now from the sampling properties of $I_X(\omega)$ developed in §9·2, it follows, at least for normal series, that

$$E\{T\} \sim \int_0^\pi \gamma(u)\, w(u, \omega)\, du, \tag{4}$$

$$\text{var } T \sim \frac{2\pi}{N} \int_0^\pi \gamma^2(u)\, w^2(u, \omega)\, du, \tag{5}$$

where $\gamma(\omega) \equiv 2\pi\sigma_X^2 f_+(\omega)$, and N is used to denote the total sample length of series available (n being already used in (3) so that $N = nm$). We saw in § 9·2 that a compromise is necessary between accuracy and what was termed resolving power, which in terms of (4) and (5) would be related to the bias $b = E(T) - \gamma(\omega)$ for neighbouring ω. One possible single measure of these two sources of error, variance and bias, is the mean square error,

$$S = \text{var } T + b^2 \tag{6}$$

so that it is instructive to see what choice of the weighting function would minimize S. Unfortunately, it is easily seen that this must depend on the unknown function $\gamma(\omega)$. However, while there is no completely satisfactory way out from this impasse, further progress can be made if $\gamma(\omega)$ is assumed reasonably regular and 'local' weighting functions are considered which effectively vanish outside a small interval round ω. Let us more specifically assume

$$\gamma(u) = \gamma(\omega) + (u - \omega)\gamma'(\omega) + \tfrac{1}{2}(u - \omega)^2 \gamma''(\omega) + \dots$$

near ω, and consider also $w(u, \omega)$ of the form $w(u - \omega)$, where $w(u)$ is an even function of u. We suppose further that

$$\int w(u)\, du = 1.$$

If $w(u)$ is effectively non-zero within $|u| = h$, the variance is $O(1/hN)$. The dominant terms in (6), for $v \neq 0$, are

$$S \sim \frac{2\pi}{N} W\gamma^2(\omega) + \tfrac{1}{4}v^2[\gamma''(\omega)]^2, \tag{7}$$

where
$$\int u^2 w(u)\, du = v, \quad \int w^2(u)\, du = W.$$

For example, when $w(u) = 1/(2h)$ between $-h$ and h, we have $W = 1/(2h)$, $v = \tfrac{1}{3}h^2$ and

$$\frac{S}{\gamma^2(\omega)} \sim \frac{\pi}{hN} + \frac{h^4}{36}\left[\frac{\gamma''(\omega)}{\gamma(\omega)}\right]^2. \tag{8}$$

An optimum choice of h still depends in (8) on the unknown quantity $\beta(\omega) \equiv \gamma''(\omega)/\gamma(\omega)$, but nevertheless we can see that if we put $h = cN^{-\alpha}$, α should for large N be taken to be $\frac{1}{5}$, and S is then minimized if we have chosen c to be $(9\pi/\beta^2)^{\frac{1}{5}}$. Its minimum value is then given by

$$\frac{S}{\gamma^2} \sim \frac{\frac{5}{4}\pi^{\frac{4}{5}}\beta^{\frac{2}{5}}}{9^{\frac{1}{5}}N^{\frac{4}{5}}}, \tag{9}$$

and varies as $N^{-\frac{4}{5}}$. By contrast, it has been shown (see, for example, Grenander and Rosenblatt, 1957) that the weighting function (3) provides a minimum value of S varying ultimately for large N as $N^{-\frac{2}{3}}$. For this reason, when the correlogram is being used with a weighting function that should be simple on this transformed scale other forms of weighting function have been proposed. For example, the function $1 - |s|/n$ for $|s| \leqslant n$ used in the previous section may be replaced by $\frac{1}{2}(1 + \cos[\pi s/n])$ for $|s| \leqslant n$, a function suggested by Tukey that gives a mean square error varying ultimately for large N under appropriate conditions as $N^{-\frac{4}{5}}$, like (9) but with a slightly smaller proportionality factor.

However, if (7) is our criterion we can do better than (9). If, in the frequency domain, we regard $w(u)$ as an unknown function occurring in (7) through W and v, we find by allowing $w(u)$ to vary by some arbitrary small amount that it should be quadratic in u, say

$$w(u) = C(1 - u^2/a^2). \tag{10}$$

Now by suitable choice of a in (10) we can actually make v zero (provided we do not insist on $w(u)$ being always positive). The appropriate value of a is $h\sqrt{\frac{3}{5}}$, whence

$$C = W = 9/(8h),$$

and

$$\frac{S}{\gamma^2} \sim \frac{9\pi}{4hN}. \tag{11}$$

Whilst this is larger than the first term in the expression on the right-hand side of equation (8), it is for constant h of $O(N^{-1})$, and thus decreases more rapidly than S/γ^2 in (9). This is of course not precisely the value of S/γ^2 when v is zero, as we should now include the next-order term in the bias, viz.

$$\frac{1}{24}\gamma^{(iv)}(\omega) \int u^4 w(u) \, du.$$

The 'fourth moment' of $w(u)$, v_4, say, is from (10) $9h^4/110$, whence (11) becomes more accurately

$$\frac{S}{\gamma^2} \sim \frac{9\pi}{4hN} + \left\{ \frac{3h^4}{880} \left[\frac{\gamma^{(iv)}(\omega)}{\gamma(\omega)} \right] \right\}^2. \tag{12}$$

After inclusion of this second term, however, there is nothing to prevent our going back and modifying our optimum function $w(u)$ to arrange for v_4 to be zero as well as v; and so on. In this way the result $S = O(N^{-1})$ can be approached. In practice it seems doubtful whether such a complicated correction would be superior over the first correction, which we have found to be equivalent to the weighting function

$$w(u) = \frac{9}{8h} \left(1 - \frac{5u^2}{3h^2} \right). \tag{13}$$

Indeed, while the quadratic function in (13) does not explicitly depend on $\gamma''(\omega)$ and is thus available for estimating $\gamma(\omega)$, it is by no means preferable to the original Daniell uniform weighting function in all cases. The latter function has the advantage of simplicity, remaining positive (thus ensuring that our estimate remains positive) and providing rather simpler chi-square type sampling distributions for our estimate in the null case of a uniform spectrum. Only if the bias term is considered of probable importance would the quadratic method be preferred.

Both these weighting functions are for use directly with the periodogram, for which there is much to be said when large-scale computing facilities are available. Note that as $I_X(\omega)$ is usually available for integral values of p, where $\omega_p = 2\pi p/N$, the weighting function (13) is still only approximately correct, and the exact values appropriate to the number of discrete points in the interval $-h$ to h may be substituted. These are readily found from the analogous conditions that the sum of the weights should be unity, and the 'second moment' zero. For example, for 8 points in the interval the weights are:

$$-\frac{3}{32} \quad \frac{3}{32} \quad \frac{7}{32} \quad \frac{9}{32} \quad \cdots$$

and for 16 points:

$$\frac{-91}{1344} \quad \frac{-21}{1344} \quad \frac{39}{1344} \quad \frac{89}{1344} \quad \frac{129}{1344} \quad \frac{159}{1344} \quad \frac{179}{1344} \quad \frac{189}{1344} \quad \cdots$$

The value of h must of course be small enough for the remainder term in S to be legitimately neglected.

To compare the performance of the uniform and quadratic weighting functions, let us consider a spectral density with a peak approximately normal, so that

$$\gamma(\omega) \sim C\, e^{-\frac{1}{2}(\omega-m)^2/\sigma^2}$$

near $\omega = m$. We have

$$\gamma''(m)/\gamma(m) = -1/\sigma^2, \quad \gamma^{(\mathrm{iv})}(m)/\gamma(m) = 3/\sigma^4.$$

With the Daniell estimate, we would choose $h^5 = 9\pi\sigma^4/N$ and

$$S/\gamma^2(m) \sim \tfrac{5}{4}\pi/(hN). \tag{14}$$

Suppose for example, we required this to be $0\cdot01$. We then obtain $N = 758$. We then have for comparison from (11), if we choose the same value of h, a value of $\sqrt{S}/\gamma(m) = 0\cdot134$. Of course within limits h may in (11) be given any value we please; and to illustrate this we now take N 32 times larger in (14), with appropriate modification of h, giving $\sqrt{S}/\gamma(m) = 0\cdot025$, whereas from (11), with the *original* value of h, we have the smaller value $0\cdot0237$. In this last case we must make sure that the neglected contribution in (12) is in fact negligible, but this may be easily checked.

The above 'smoothing' techniques may be compared with similar problems for other density estimates, especially for probability densities, where the analogue of the Daniell estimator is the histogram (see Rosenblatt, 1956; Priestley, 1962; Bartlett, 1963). For such problems particular emphasis on *smoothing* has led to the consideration of alternative techniques associated with the use of 'splines' (see Boneva, Kendall and Stefanov, 1971). There is then no need to tie these techniques to a fixed discrete grid, for example, when exact observations from a distribution with probability density are available; for periodogram intensities, however, a basic grid of frequencies seems still natural.

9·22 Further notes and problems related to the spectrum. The coefficients in the transformation

$$A_p = \sqrt{\frac{2}{n}} \sum_{r=1}^{n} X_r \cos \frac{2\pi pr}{n}, \quad B_q = \sqrt{\frac{2}{n}} \sum_{r=1}^{n} X_r \sin \frac{2\pi qr}{n},$$

for $p = 0, 1, \ldots, \tfrac{1}{2}(n-1)$, $q = 1, 2, \ldots, \tfrac{1}{2}(n-1)$ (n odd), or $p = 0, 1,$

$\ldots, \frac{1}{2}n$, $q = 1, 2, \ldots, \frac{1}{2}(n-2)$ (n even) are well known to form an orthogonal set, so that we have the identity (provided the factor 2 in $\sqrt{(2/n)}$ is omitted for $p = 0$ and $p = \frac{1}{2}n$)

$$\sum_{r=1}^{n} X_r^2 \equiv \Sigma_p A_p^2 + \Sigma_q B_q^2, \tag{1}$$

$$\sim \sum_{p=0}^{\frac{1}{2}n} I_p \quad (\tfrac{1}{2}(n-1) \text{ instead of } \tfrac{1}{2}n \text{ if } n \text{ odd}), \tag{2}$$

where the I_p are the periodogram intensities defined in §9·2. Now reference to the sampling theory of distribution functions outlined in §4·1 (see p. 92) will recall that for X_r independent and normal (zero mean and variance σ_X^2), so that I_p has the exponential distribution, the 'random walk' quantity

$$\sum_{p'=0}^{p} I_{p'} \quad \left(\sum_{p'=0}^{\frac{1}{2}n} I_{p'} \text{ given equal to its expectation} \right),$$

or equivalently (from the sampling properties of I_p for normal X_r)

$$T_p = \sum_{p'=0}^{p} I_{p'} \Big/ \sum_{p'=0}^{\frac{1}{2}n} I_{p'} = \frac{S_p}{S_{\frac{1}{2}n}}, \tag{3}$$

say, may be tested against its expectation ($\sim 2p/n$) by checking that the deviation of T_p does not exceed the $\pm \lambda/\sqrt{(\frac{1}{2}n)}$ boundary, the probability of remaining within the boundary being

$$\sum_{s=-\infty}^{\infty} (-1)^s e^{-2\lambda^2 s^2}, \tag{4}$$

(critical values are 0.95 if $\lambda = 1.36$, 0.99 if $\lambda = 1.63$).

This test is asymptotically valid for X_r independent and normal, but as the test criterion is a ratio and a function of the serial correlations R_s of X_r defined in §9·1, it will be insensitive to non-normality, in particular the asymptotic variance of T_p will be independent of the fourth cumulant κ_4 of X_r. It may in fact easily be verified from formula (14), §9·2, that

$$\operatorname{var} T_p \sim \frac{\operatorname{var} S_p}{[E\{S_{\frac{1}{2}n}\}]^2} - \frac{2E\{S_p\} \operatorname{cov}\{S_p, S_{\frac{1}{2}n}\}}{[E\{S_{\frac{1}{2}n}\}]^3} + \frac{[E\{S_p\}]^2 \operatorname{var} S_{\frac{1}{2}n}}{[E\{S_{\frac{1}{2}n}\}]^4}$$

$$\sim 4p(1 - 2p/n)/n^2, \tag{5}$$

independent of κ_4, and in conformity with a random walk restricted to $T_{\frac{1}{2}n} = 1$.

To make use of the above result, we recall further the relation (equation (15), § 9·2)

$$J_X(\omega) = h^*(\omega) \, J_Y(\omega) \, (1 + O(1/\sqrt{n})),$$

so that if the X_r are not independent but constitute a linear process with continuous spectrum $f_+(\omega)$, we have

$$I_X(\omega)/[h(\omega) \, h^*(\omega)] = I_Y(\omega) \, (1 + O(1/\sqrt{n})) \quad (f(\omega) \neq 0),$$

and† $\sum\limits_{p'=0}^{p} \{I_X(\omega')/[h(\omega') \, h^*(\omega')]\} = \left\{ \sum\limits_{p'=0}^{p} I_Y(\omega') \right\} (1 + O(1/\sqrt{n})),$ (6)

as $I_Y(\omega) > 0$, and has the same expectation for different ω. As

$$h(\omega_1) \, h^*(\omega_1)/f_+(\omega_1) = h(\omega_2) \, h^*(\omega_2)/f_+(\omega_2),$$

we obtain finally

$$T_p \sim \sum_{p'=0}^{p} \left[\frac{I_X(\omega')}{f_+(\omega')} \right] \Big/ \sum_{p'=0}^{\frac{1}{2}n} \left[\frac{I_X(\omega')}{f_+(\omega')} \right]. \tag{7}$$

The statistic T_p is available for constructing a confidence band for the entire spectral function $f_+(\omega)$, though as it involves $f_+(\omega)$ in a rather complicated way, would be more readily available for testing the goodness of fit of a function $f_+(\omega)$ given *a priori*.

To avoid the individual weighting of each item $I_X(\omega)$ in (7), one would as an alternative consider the statistic

$$U_p = \sum_{p'=0}^{p} I_X(\omega') \Big/ \sum_{p'=0}^{\frac{1}{2}n} I_X(\omega'). \tag{8}$$

This implies a weighting of each $I_Y(\omega)$ by the factor $f_+(\omega)$. An appropriate test procedure using U_p needs further investigation, for U_p is no longer, even for normal processes, equivalent to a random walk quantity (unless its denominator is fixed in value). Its variance is, however, still asymptotically independent of κ_4 for linear processes, being obtained by similar methods as

$$\frac{4\pi^2}{n^2} \left[(1 - F_+(\omega))^2 \sum_{p'=0}^{p} f_+^2(\omega') + F_+^2(\omega) \sum_{p'=p+1}^{\frac{1}{2}n} f_+^2(\omega') \right], \tag{9}$$

where we have written

$$\sum_{p'=0}^{p} f_+(\omega') \Big/ \sum_{p'=0}^{\frac{1}{2}n} f_+(\omega') \sim F_+(\omega). \tag{10}$$

† In these sums over p', ω' stands of course for $2\pi p'/n$ (and ω for $2\pi p/n$).

If we further standardize U_p by dividing its deviation from $F_+(\omega)$ by

$$2\pi \sqrt{\left[\frac{1}{2n} \sum_{p'=0}^{\frac{1}{2}n} f_+^2(\omega')\right]} \sim \sqrt{\left[\pi \int_0^\pi f_+^2(\omega')\,d\omega'\right]} = \sqrt{v}, \qquad (11)$$

it is suggested that the quantity $[U_p - F_+(\omega)]/\sqrt{v}$ could (as an interim approximate method) be tested in the same way as $T_p - 2p/n$. Such an alternative criterion to T_p still only involves the spectral function $f_+(\omega)$, but it might be convenient to insert a consistent estimate of v from the sample, as this would not affect the test of U_p for large n. Such an estimate is, for example, obtainable from

$$\sum_{p'=0}^{\frac{1}{2}n} I_X^2(\omega') \bigg/ \left[\sum_{p'=0}^{\frac{1}{2}n} I_X(\omega')\right]^2, \qquad (12)$$

provided we recall that the mean value of $I_X^2(\omega)$ is asymptotically *twice* the square of $E\{I_X(\omega)\}$, so that the expression in (12) has asymptotic value

$$\frac{n}{2\pi} \int_0^\pi 2f_+^2(\omega')\,d\omega' \bigg/ \left(\frac{n}{2\pi}\right)^2 = \frac{4v}{n}.$$

Grenander and Rosenblatt (1952) suggested the use of the unweighted and unstandardized quantity

$$V_p = \int_0^\omega I_X(\omega')\,d\omega' \sim \frac{2\pi}{n} \sum_{p'=0}^{p} I_X(\omega'); \qquad (13)$$

even U_p, however, has the advantage, in addition to being insensitive to κ_4, of being an estimate of the spectral function $F_+(\omega)$ corresponding to the correlations, whereas V_p is an estimate of $\sigma_X^2 F_+(\omega)$, corresponding to the covariances, and this will usually be of less interest when σ_X^2 is unknown. In principle, the first statistic T_p appears superior to V_p or U_p.

Mixed spectra. In many problems it is reasonable to suppose it is known *a priori* when the spectrum is discrete, as was always assumed classically, or continuous, as in the case of non-deterministic autoregressive series and linear processes. It might be noticed that any additional discrete components with *known* frequency, such, for example, as annual or other seasonal variation, can be removed as a first step; any other spectrum present complicates the error of estimation of the amplitude and phase of such components, but not much more so than in the elimination of a simple mean.

In a second type of problem, it may be required to discriminate between two hypotheses H_1: a continuous spectrum, H_0: a continuous uniform spectrum plus, say, one discrete component with unknown frequency. Here, as noted in §9·2, the difficulty is that for any finite sample the periodogram intensity for the discrete component is still only finite, and may be confused with the peak of a continuous spectrum (or vice versa). An appropriate criterion in such a case, if we may assume the Y_r normal, would be the likelihood ratio p_0/p_1, where the probabilities p_0 and p_1 for the sample on each hypothesis would require maximizing for any unknown parameters (these being restricted to a few autoregressive coefficients, say, in case H_1). We have (cf. equation (2), §9·11)

$$\log p_0 - \log p_1 \sim -\tfrac{1}{2}(n-2)(\log \Sigma_0 - \log \Sigma_1'), \qquad (14)$$

where Σ_1' denotes an adjusted sum of squares of the residuals after fitting an autoregressive scheme of order 2, say, and Σ_0 the sum of squares of residuals after fitting a discrete component (two degrees of freedom are allowed in Σ_0 for the amplitude and phase, the additional error in locating the frequency being neglected). The significance of this ratio could be based on sequential analysis theory (cf. §4·1, pp. 94–6). We may regard the entire set of n observations as the first of a number of such sets, so that for equal maximum risks of error $\epsilon_0 = \epsilon_1 = \epsilon$, H_1 or H_0 would be adopted according to whether the right-hand side of (14) was lower than $\log \epsilon - \log(1-\epsilon)$ or higher than $\log(1-\epsilon) - \log \epsilon$; if it lay between these two values, no decision could be reached. (Such maximum risks can be rather over-cautious when the probability of a decision with the single sample available is not small.)

In a third class of problem, both discrete and continuous spectra may have to be considered simultaneously. This is not too easy, as it should be noticed that for a finite sample of length N, a discrete component cannot be distinguished from a peak in the continuous spectrum of frequency band-width $O(1/N)$. With this reservation, we may, however, consider possible extensions of tests of a discrete component (cf. §9·2) to the case when the continuous spectrum is non-uniform.

Roughly speaking, we would expect to detect and separate a discrete component in the periodogram either by confining our attention to a frequency band-width containing it for which the continuous component does not effectively vary, or by standardizing the continuous spectrum to the uniform case. The latter approach was suggested by Whittle (1952), but it appears important to use an estimate of the continuous component which does not include the possible harmonic terms. A method of doing this has in effect been indicated by Priestley (1962), who makes use of the correlogram property that a discrete spectral component gives rise to an undamped harmonic term, and a continuous spectrum to a damped function; thus if a weighting function is used with the correlogram to estimate the spectrum, as in §§ 9·2 and 9·21, but of a 'double' form, we have the possibility of separating the two parts of the spectrum. Of course, there will also be such double weighting functions available directly with the periodogram, and we merely cite one of the simplest. In the notation of § 9·21, let

$$w(u) = \begin{cases} 0, & (|u| \leqslant k), \\ 1/[2(h-k)], & (k < |u| \leqslant h). \end{cases}$$

Then we have an estimate of the continuous spectrum at $\omega = \lambda$, say, by use of the weighting function $w(u - \lambda)$, which is asymptotically unaffected by a discrete component at $\omega = \lambda$ (or indeed anywhere within $|\lambda - \omega| = \frac{1}{2}k$, say, provided $k \gg 1/\sqrt{N}$). For asymptotic unbiasedness we require $h \rightarrow 0$ as $N \rightarrow \infty$, and for reasonable efficiency we would take if possible $h \gg k$. In practice we should usually replace the continuous weighting function $w(u)$ on the frequency scale by an effectively equivalent series of weights with the periodogram intensities at $\omega_p = 2\pi p/N$. With this procedure the standardization of the intensities should not appreciably affect statistical tests, provided, as above, $h \gg 1/N$. It would be necessary to arrange that no suspected discrete components coincided with the positive parts of the double weighting function used to estimate the continuous spectrum; and in parts of the frequency range where no such components occurred the double function could be dispensed with.

An apparent criticism of the above method is that there is some selection and positioning of any anomalous peaks, with an associated undervaluation of the continuous spectrum; but if the analysis is merely intended to separate peaks of narrow frequency bandwidth, rather than discrete components only, this criticism is no longer so relevant. Any bias due to the selection of peaks may in any case be corrected for.

Thus if the peak value of $I_X(\omega_p)$ is M, the conditional expected value of a neighbouring periodogram intensity I with unconditional mean m is

$$
E\{I\} = \frac{\int_0^M I\,e^{-I/m}dI}{m(1-e^{-M/m})}
$$

$$
= m\left(1 - \frac{M\,e^{-M/m}}{m(1-e^{-M/m})}\right).
$$

For large enough M/m, the correction is negligible, but if necessary this equation may be used to solve for m by iteration, with $E\{I\}$ as the first iterate.

In cases where only a single discrete spectral component is at most involved in a given range of ω, a more efficient separation of this component from the continuous spectrum may be achieved with the use of the phase of $J(\omega)$. If

$$
X'_r = X_r + \alpha \cos(rv + \tau\Phi),
$$

$$
J_X = \sqrt{\frac{2}{n}} \sum_{r=1}^n X_r \exp(ir\omega_p),
$$

$$
\omega_p = 2\pi p/n, \quad v = \omega_{p-s} + 2\pi\epsilon/n \quad (-\tfrac{1}{2} < \epsilon < \tfrac{1}{2}),
$$

Φ being a random phase angle in the discrete component $\alpha \cos(rv + \Phi)$, then it may be shown (Bartlett, 1967a, b) that asymptotically for n large,

$$
J'_X \sim J_X + \alpha\sqrt{(\tfrac{1}{2}n)}\,e^{-i\Phi}\,\frac{\exp(2\pi i\epsilon)-1}{2\pi i(\epsilon+s)}. \tag{15}
$$

A rapid way of obtaining further information on the spectral density is to measure deviations orthogonal to the fitted line corresponding to the relation (15), when the individual A_p and B_p values, where $J(\omega_p) = A_p + iB_p$, are plotted. The information in the deviations along the fitted line is, however, also recoverable. We minimize

$$
S = \Sigma_s\{(A_s - \beta a_s)^2 + (B_s - \gamma a_s)^2\},
$$

where $a_s = 1/(\epsilon+s)$, β and γ are also functions of ϵ, and it is assumed that the axes have been rotated so that the estimate

$\hat{\gamma} = 0$. Then

$$\partial S/\partial\beta = \Sigma_s(A_s - \beta a_s)a_s = 0, \quad \partial S/\partial\gamma = \Sigma_s(A_s - \gamma a_s)a_s = 0,$$

$$\partial S/\partial\epsilon = \Sigma_s(A_s - \beta a_s)(\beta\,\partial a_s/\partial\epsilon + a_s\,\partial\beta/\partial\epsilon)$$

$$+ \Sigma_s(B_s - \gamma a_s)(\gamma\,\partial a_s/\partial\epsilon + a_s\,\partial\gamma/\partial\epsilon)$$

$$= \Sigma_s(A_s - \beta a_s)\beta\,\partial a_s/\partial\epsilon = 0.$$

From the relation $a_s = 1/(\epsilon + s)$, the last equation may be written

$$\Sigma_s(A_s - \beta a_s)a_s^2 = 0,$$

an equation which, in conjunction with the equation for β, may be used to estimate ϵ.

Numerical illustrations of some of the techniques proposed in this section are for convenience deferred until the next section.

9·23 The spectral analysis of point processes.

The statistical analysis of data represented by a succession of events in time or points in space presents some special problems which are often best considered in the particular context of the example. Just as, however, the variety of stationary processes in quantitative variables can to a large extent be examined on the basis of first and second moment properties, so an analogous technique can be used for stationary point processes. This amounts to a study of the second-order density function, $f_2 = E\{dN(t)\,dN(t')\}$; or equivalently, of the conditional probability-density of an event or point occurring at t', given an event at t. While obviously such a two-point distribution can only (with the mean or first-order density) describe the process very partially, it does represent the simplest specification of any dependence; and, indeed, it is interesting to compare its use in statistical physics in investigating geometrical configurations of molecules or atoms in a gas or liquid, for example, when using X-rays as a probe (cf. Green, 1952, Ch. III).

Suppose we have n occurrences in a sample period of T at times $T_1, T_2, \ldots, T_n$. Then, analogously to the periodogram analysis of a stationary process (in continuous time) for a quantitative variable, we define

$$J(\omega) = \sqrt{\left(\frac{2}{T}\right)}\int_0^T e^{it\omega}\,dN(t) = \sqrt{\left(\frac{2}{T}\right)}\sum_{s=1}^n e^{iT_s\omega}, \tag{1}$$

and

$$I(\omega) = J(\omega)\,J^*(\omega).$$

Note that, for a stationary point process with covariance density $w(\tau)$,

$$E\{I(\omega)\} = 2\left\{\int_{-T}^{T} \left(1 - \frac{|\tau|}{T}\right) e^{-i\tau\omega} w(\tau)\, d\tau + \lambda\right\}$$

$$\sim g_+(\omega) \tag{2}$$

for large T, as defined in § 6·13. However, as in the case of a quantitative variable, it is found that $I(\omega)$ cannot be used to estimate $g(\omega)$ without an appropriate smoothing technique. It will in fact be shown below that for a wide class of stationary point processes the same asymptotic sampling theory for $I(\omega)$ holds as developed in § 9·2, so that the same smoothing techniques are available.

Consider first a class of point processes introduced by Cox (1955), in which a Poisson process with mean density λ is generalized to have a stochastic mean density $\Lambda(t)$, which in the present context will be assumed stationary. For such point processes, which will be termed doubly stochastic Poisson processes, note that the characteristic functional, *for a given realization $\lambda(t)$ of $\Lambda(t)$*, is

$$E\left\{\exp\int_{0}^{T} i\theta(t)\, dN(t)\right\} = \exp\int_{0}^{T} \lambda(t)\,[e^{i\theta(t)} - 1]\, dt. \tag{3}$$

Hence, unconditionally, the characteristic functional is the expression in (3) averaged over all $\lambda(t)$, i.e. the characteristic functional of $N(t)$ with function $i\theta(t)$ is the characteristic functional of $\Lambda(t)$ with function $i\psi(t) = e^{i\theta(t)} - 1$.

To investigate the sampling properties of $J(\omega)$ for such point processes, write

$$\theta(t) = \sqrt{\left(\frac{2}{T}\right)}\{\theta_1 e^{il\omega} + \theta_2 e^{-il\omega}\}. \tag{4}$$

Then

$$M(\theta_1, \theta_2) \equiv E\{e^{\theta_1 J + \theta_2 J^*}\} = E_\Lambda\left\{\exp\int_{0}^{T} \Lambda(t)\,[e^{i\theta(t)} - 1]\, dt\right\},$$

where E_Λ denotes averaging with respect to $\Lambda(t)$ and $\theta(t)$ is given in (4). Expanding in powers of $1/\sqrt{T}$, we find (for $\omega \neq 0$)

$$M(\theta_1, \theta_2) = E_\Lambda\{\exp[\theta_1 J_\Lambda(\omega) + \theta_2 J_\Lambda^*(\omega) + 2\lambda\theta_1\theta_2]\} + O(1/\sqrt{T}),$$

where

$$J_\Lambda(\omega) = \sqrt{\left(\frac{2}{T}\right)}\int_{0}^{T} e^{il\omega}\Lambda(t)\, dt;$$

or, taking logarithms, we obtain

$$K(\theta_1, \theta_2) \sim K_\Lambda(\theta_1, \theta_2) + 2\lambda\theta_1\theta_2 \tag{5}$$

for large T, where the cumulant function on the left is for J, J^* and on the right for J_Λ, J_Λ^*.

The use of complex quantities may obscure the interpretation of the extra term $2\lambda\theta_1\theta_2$ in (5), but the result merely implies that $J(\omega)$ is asymptotically equivalent to $J_\Lambda(\omega)$ apart from the addition of an independent (complex) component with real and imaginary parts each uncorrelated normal with zero mean and variance λ. It then readily follows that the sampling fluctuation of $I(\omega)$ is analogous to that established in § 9·2, provided $\Lambda(t)$ is a linear process (in the sense used in that section). In particular, this is true for the simple Poisson process (for which $\Lambda(t) = \lambda$). The above asymptotic results readily extend to the joint distribution of $J(\omega)$, $J(\omega')$, or to any finite set $J(\omega)$, $J(\omega')$, $J(\omega'')$, ... for $\omega \neq \omega' \neq \omega'' ... \neq 0$.

The application of these results is so far limited by the assumption of the existence and 'linearity' of $\Lambda(t)$, but the form of the results is suggestive. The identity of the covariance density $w(\tau)$ with the autocovariance function of $\Lambda(t)$ in fact indicates a serious limitation in the range of application of this assumption, for it implies that the spectral density component corresponding to $w(\tau)$ should be positive, whereas we have already met examples (e.g. Example 1 in § 10·3) to the contrary. However, the sampling results will now at least be extended to a second important class of point processes, related to but rather more general than the Poisson clustering processes introduced in Example 2 of § 10·3. (It would, as with a quantitative process, be valuable to know the *minimum* conditions necessary for these sampling results to hold.)

Let us associate a 'cluster' with each 'parent' point τ_r, the points τ_r ($r = 1, ..., m$) being a sample from a simple Poisson process. Then the quantity $J(\omega)$ for the entire set of points in T may, if we assume that for large enough T end effects due to boundary clusters being partly outside the interval may be neglected, be written in the form

$$J(\omega) \sim \sqrt{\left(\frac{2}{T}\right)} \sum_{r=1}^m e^{i\tau_r\omega}\phi_r(\omega), \tag{6}$$

where each $\phi_r(\omega)$ is random, but independent of τ_r. $J(\omega)$ in (6) is the sum of m independent (complex) quantities Z_r, say, with the same distribution and hence tends to normality. Each quantity Z_r (for fixed $\omega \neq 0$) has a mean effectively zero from the factor $e^{i\tau_r\omega}$, as $E\{e^{i\tau_r\omega}\} = \sin(\omega T)/(\omega T)$ and hence $E\{Z_r\} = O(T^{-\frac{1}{2}})$. Similarly, the square of Z_r will also have effectively zero mean, for we may write $\phi_r = |\phi_r|\,e^{i\alpha_r}$, where α_r merely shifts the origin of τ_r by a random amount. This implies that Z_r has real and imaginary components with equal variance, and indicates that $I(\omega)$ is asymptotically of the form $I_0(\omega)\,\overline{|\phi|^2}$, where the bar denotes averaging over the $|\phi_r|^2$ in the sample, and $I_0(\omega)$ refers to the Poisson parent points only. For large m, it will be sufficient to write

$$I(\omega) \sim I_0(\omega)\,E\{|\phi|^2\}, \tag{7}$$

a result which implies that the asymptotic sampling results established for doubly stochastic Poisson processes (and hence for simple Poisson processes) also hold for this class of Poisson clustering processes. The result (7) also implies for such processes the result

$$g_+(\omega) = 2\lambda_0\,E\{|\phi|^2\}, \tag{8}$$

where λ_0 is the mean density of clusters.

Example. As an example some data requoted from Bartlett (1963) in table 7 representing the times at which 129 successive vehicles passed a point on a road are analysed below.

The average interval of 15·81 seconds was reduced to the order of unity by taking 16 seconds as a unit, and the values of ω taken were $\omega_p = 2\pi p/n$ with $n = 128$ (intervals being measured from the first vehicle), $p = 1$, $2, \ldots, 256$. Values of A_p, B_p, where $J(\omega_p) = A_p + iB_p$, and $I_p = A_p^2 + B_p^2$ were tabulated by an electronic computer.

Two cautionary remarks about these calculations should be made:

(i) The unit of 16 seconds was chosen for convenience, but it is a safer routine to standardize the calculations more precisely by putting $T' = n$, $T'_s = nT_s/T$. This keeps the orthogonality relations precise for integral values of p, and in particular any bias as $\omega \to 0$ due to the discrete spectral component at $\omega = 0$, with the value

$$2T\lambda^2\left(\frac{\sin\left[\frac{1}{2}T\omega_p\right]}{\frac{1}{2}T\omega_p}\right)^2,$$

vanishes. With the actual scaling adopted, this bias was, however, negligible.

(ii) A sufficient range of ω values was taken to cover the interesting part of the spectrum, though when (as here) the number of p values exceeds $\frac{1}{2}n$ more care is needed with any tests of significance making use of the entire range.

Table 7. *Times to 1/10th sec. at which successive vehicles passed a point*

6,067	6,095	6,129	6,143	6,288	6,307	6,335	6,358
6,511	6,529	6,624	6,649	6,743	6,754	7,640	7,656
7,675	7,690	8,027	8,053	8,182	8,344	8,363	8,566
8,934	9,335	10,040	10,060	10,140	10,161	10,193	10,210
10,775	11,012	11,036	11,250	11,301	11,380	11,581	11,730
11,786	12,303	13,174	13,186	13,213	13,223	13,238	13,251
13,498	14,224	15,422	15,434	15,503	15,542	15,558	15,588
15,606	16,054	16,104	16,143	17,396	17,624	17,643	17,802
17,862	18,068	18,197	18,236	18,366	18,435	18,460	18,583
18,640	18,753	18,778	18,794	18,870	18,893	18,954	18,975
19,322	19,476	19,522	20,079	20,101	20,161	20,179	20,198
20,216	20,636	20,729	21,646	21,670	21,976	21,988	22,076
22,142	22,640	23,221	23,240	23,269	23,274	23,286	23,596
23,715	23,723	23,735	23,743	23,790	23,873	23,946	24,034
24,052	24,083	24,091	24,432	24,462	24,488	24,525	24,938
25,235	25,411	25,430	25,568	25,970	26,071	26,190	26,300
26,302	—	—	—	—	—	—	—

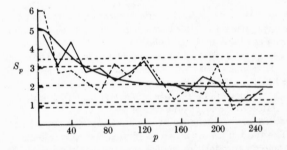

Fig. 15. Smoothed spectra for data in table 7 (uniform weighting: continuous line; quadratic weighting: dotted line). The $P = 0·05$ and $P = 0·01$ significance bands (for individual values in the uniform weighting case) are shown; also the theoretical spectrum for a fitted clustering model. (Reprinted from *J.R. Statist. Soc.* B (1963), **25**.)

Smoothed spectral values are shown in fig. 15 for 16-point averaging, both with uniform weighting of the I_p values and with quadratic weighting (see §9·21). The $P = 0·05$ and $0·01$ significance levels shown are based on fluctuations for individual smoothed values from $2N(T)/T = 2·024$; they are obtained from critical values of χ^2 (with appropriate scaling) with 32 degrees of freedom, and apply to the uniformly weighted estimates.

The rise for small ω_p suggests in the context of the data that some clustering is present. The Poisson clustering process described as Example 2

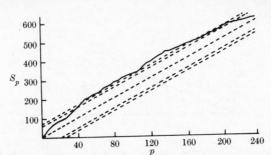

Fig. 16. Cumulative spectral function for data in table 7, with $P = 0.05$ and $P = 0.01$ significance bands (based on 256 degrees of freedom). (Reprinted from *J.R. Statist. Soc.* B, (1963), **25**.)

in § 10·3 was used to obtain a rough fit. This model is not ideal, as traffic clusters would (at least for single-lane traffic with occasional overtaking) be interpreted as separate queues behind slow vehicles, and thus nòt overlap. However, even a good graduation could not show more than compatibility of the model with the data; and no exhaustive search for the best fit, or a better type of model, was made.

The particular model used assumed a modified geometric distribution for cluster size,

$$p(r) = \begin{cases} 1 - c & (r = 0), \\ c\alpha^{-1}(1 - \alpha) & (r = 1, 2, \ldots), \end{cases}$$

and the distribution of intervals between successive vehicles in the same cluster of χ^2 type with characteristic function

$$C(\phi) = (1 - \tfrac{1}{2}i\phi/\mu)^{-2}.$$

This gives, with the choice $c = \tfrac{1}{9}$, $\alpha = \tfrac{2}{3}$, and $\lambda_0 = \tfrac{3}{4}\lambda$,

$$g_+(\omega) = 2\lambda \left\{ \frac{\dfrac{5}{18} + \dfrac{17\omega^2}{24\mu^2} + \dfrac{\omega^4}{16\mu^4}}{\dfrac{1}{9} + \dfrac{5\omega^2}{6\mu^2} + \dfrac{\omega^4}{16\mu^4}} \right\}.$$

The scale for ω is reasonably fitted with $\mu = 15\pi/8$; with $2\lambda = 2.024$ and $0.4\omega/\pi = \Omega$, this gives

$$g_+(\omega) = 2.024 \left\{ \frac{\dfrac{5}{2} + \dfrac{34\Omega^2}{3} + \dfrac{16\Omega^4}{9}}{1 + \dfrac{40\Omega^2}{3} + \dfrac{16\Omega^4}{9}} \right\}.$$

The plot of this theoretical spectrum is also shown in fig. 15. As a check on the adequacy of the fit, the values of I_p for this example were rescaled by dividing by $g_+(\omega)$. The cumulative sums

$$S_p = \sum_{q=1}^{p} I_q$$

are shown in figs. 16 and 17 for the original I_p, and the rescaled values,

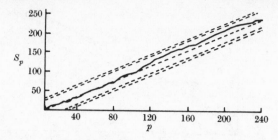

Fig. 17. Cumulative spectral function for data in table 7 after scaling by clustering model spectrum, with $P=0·05$ and $P=0·01$ significance bands (based on 256 degrees of freedom). (Reprinted from *J.R. Statist. Soc.* B, (1963), **25**.)

respectively. By joining the origin to the final sum S_{256} in each case, and drawing the $P=0·05$ and $0·01$ boundaries (for the theory of this test, see § 9·22) we may check whether the spectrum is or has been made uniform. It will be seen that the fit has been made satisfactory by the use of the model described above.

Example with mixed spectrum. In another set of traffic data (referred to by A. J. Miller in the paper already quoted) the times were recorded 'downstream' from a traffic light with a period of 83 secs. A peak in the periodogram at the equivalent frequency was thus not unexpected, and in fact for 340 successive intervals recorded over a total time of 1729 sec. the values of I_{20} and I_{21} were 40·64 and 14·56, compared with an average over I_1 to I_{340} of $475·97/340 = 1·40$. The scale for ω_p was $2\pi p/1700$.

In view of the information available on the nature of the data, the reality of the harmonic term is hardly in question; but to demonstrate the numerical procedure outlined at the end of § 9·22 a test of significance of the peak was made. The I_p values were grouped into 10 sets of 34. For the first set containing the peak, the continuous spectrum was estimated by the weight function $w(u)=0$ for the central 16 values and uniform over the outlying 18 values. This double weight function was unnecessary for the other sets, for which a uniform weight over the 34 values was used. The mean I_p obtained are shown in table 8.

Table 8. *Smoothed estimates for the spectral density* $(p' = 17\frac{1}{2}p)$

p'	Estimate	p'	Estimate
1	1·849	6	0·877
2	1·723	7	0·933
3	1·226	8	1·028
4	1·078	9	1·042
5	0·927	10	1·479

The scaling of the original periodogram values by means of the spectral density estimate in the appropriate frequency interval should in the absence of any discrete spectral component in the 'hole' of the first weight function provide a periodogram equivalent to that with a uniform time density (fluctuations due to sampling fluctuations in the smoothed estimates of table 8 may be allowed for if necessary, but will be neglected to a first approximation). A test of non-uniformity of the scaled periodogram, made along the same lines as above in figs. 16 and 17, did indicate significance ($P \sim 0.025$ for a one-sided confidence band) due to the sudden increase in the cumulative total from $p = 20$ onwards. However, where the departure from uniformity is due to a discrete component, especially at the start of the confidence band, this test is likely to be less sensitive than a direct test of the maximum ordinate of the scaled periodogram. Here we have on the new scale $I_{max} = 21.97$ compared with the mean value of 340 items (including this value) of 1.10. The approximate significance level by this test is

$$P = 340 \exp(-21.97/1.1) < 10^{-6},$$

a significance extreme enough to render any closer approximation to its value unnecessary. It should also be noted that the first one or two estimates in table 8 contain slight bias due to the discrete component at zero, and it would be advisable to eliminate this (cf. the previous example) before any more precise discussion of the continuous spectrum; but the neglect of such a bias somewhat undervalues the evidence for a periodic component.

To illustrate the recovery of further information in the neighbourhood of the peak value I_{20} by the method given at the end of §9·22 the 34 individual A_p and B_p values are shown in fig. 18. With one periodogram intensity much larger than the others, the line may be adequately fitted with this observation alone. The 'rapid method' outlined then results in the estimate 1.849 (36 d.f.) given in table 8 being modified to

$$\frac{33.29 + 16.23}{18 + \frac{1}{2}(15)} = 1.942 \ (51 \, \text{d.f.}).$$

The more complete method gave

$$\hat{\epsilon} = 0.368 \ (\text{period } 1700/20.368 = 83.46 \text{ seconds})$$

and a final estimate for the spectral density of 1.830 (66 d.f.).

Adaptation to discrete time. It will have been noticed from the expressions at the beginning of this section that, although the number of events in formula (6) is finite, the theoretical formulation is in continuous time. Computational methods available for the spectral analysis of *discrete* time-series, and in particular the use of 'fast Fourier transform' methods (see Cooley and Tukey,

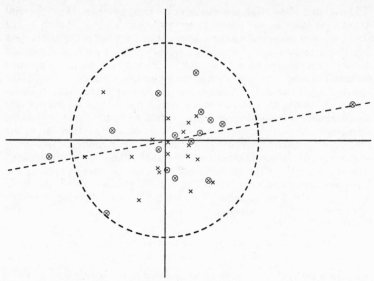

Fig. 18. Mixed spectrum. Thirty-four individual A_p and B_p plotted as abscissae and ordinates. Ringed points refer to the central 16 values, and the dotted circular boundary denotes the $P = 0·01$ significance level for the individual points. (Reprinted from *J. R. Statist. Soc.* A **130** (1967), 468.)

1965), are thus not immediately available. An ingenious modification of the above spectral analysis of point processes has, however, been introduced by French and Holden (1971). A similar modification for two-dimensional processes (see § 9.41) is also obviously possible.

In place of the point process $dN(t)$, we consider the new process

$$X(t) = \int_{-\infty}^{\infty} \frac{\sin\left[(t-\tau)\pi\right]}{(t-\tau)}\, dN(\tau),$$

which may be sampled at unit time intervals. The spectral function of the resulting process X_j, from the principles of § 6.2, is equivalent to that of the original process, with a cut-off of frequencies $|\omega| > \pi$, which in any case are not wanted when sampling at unit intervals because of frequency aliasing difficulties. The frequency cut-off point may be selected by appropriate choice of the unit of time.

The spectral analysis of line processes. Line processes in a plane and their spectra were defined in § 6.52, though it was noted at the end of that section that other approaches might be more convenient in particular contexts. If a set of lines of constant length l were to be analysed, it could be specified by the coordinates (x, y) of the mid-points, together with the orientation θ (0 to π) of the line with a fixed direction. If l were infinite, the specification can be in terms of the equation for each line

$$x \cos \theta + y \sin \theta = p,$$

where p is the perpendicular distance to the origin ($-\infty$ to ∞). Notice that in each case the problem has been transformed to the analysis of a *point* process, though one specified in a hybrid space of one or two Cartesian axes and an angle variable θ. The spectral function consists of a *series* of Fourier coefficients corresponding to θ, each coefficient being a spectral function for p or x, y (see Bartlett, 1965).

In the different type of problem mentioned at the end of § 6·52 which is an extension of the type of traffic data already discussed in § 9·23, the velocities of the vehicles passing a fixed point might have been recorded as well as the times. For such data it seems convenient to consider the point process $dN(t)$ where each point t_r has an associated quantitative variable U_r. In particular, we may consider the extended process

$$dM(t) = dN(t)\,[1 + \xi U'_r]$$

where $U'_r = U_r - E\{U_r\} \sim U_r - \overline{U}_r$, and compare the observed spectrum of $U'_r dN(t)$ and the cross-spectrum of $dN(t)$ and $U'_r dN(t)$ with those for any theoretical model, in addition to the spectrum of $dN(t)$. For a detailed analysis of data of this last kind see also Bartlett (1965), where such a spectral analysis was shown to be consistent with the clustering model used for the data of table 7, together with a correlation of velocities of vehicles in the same cluster.

Random sampling of continuous processes. A somewhat analogous process to $U'_r dN(t)$ above arises from the random sampling of a process $X(t)$ in continuous time, leading to the process $X(t)\,dN(t)$, where $dN(t)$ is a Poisson point process with uniform spectrum. In the spectral analysis of a stationary continuous series $X(t)$, this derived process has a possible advantage over a sample at equidistant points, in avoiding any 'aliasing' problem (cf. remarks after equation (13), § 6·11).

Thus if $E\{X(t)\} = 0$ and

$$J(\omega) = \sqrt{\left(\frac{2}{T}\right)} \int_0^T e^{i\omega t} X(t) \, dN(t),$$

then, for $\lambda = n/T$, and T large,

$$E\{JJ^*\} \sim 2\pi\sigma^2 f_+(\omega) \lambda^2 + 2\lambda\sigma^2$$

(see Gaster and Roberts, 1977).

9·3 Multivariate autoregressive series

Discussion so far in this chapter has been confined to the analysis of the correlation structure of single series, but the simultaneous relations among two or more series may have to be considered. For example, econometric models may consist of a number of relations with unknown coefficients which have to be estimated, and correspondingly, the cross-correlations between series will have to be investigated. The problems raised are similar in principle to those for a single series, but tend to become more complex, and are not considered here in complete detail.

As an example of the more complicated formulae, we quote the covariance formula for the sample cross-correlation $R_{12}(s)$ between $X_1(t)$ and $X_2(t+s)$ for two real normal sequences $X_1(t)$, $X_2(t)$:

$$
\begin{aligned}
\text{cov} \, & (R_{12}(s), R_{12}(s+t)) \\
& \sim \frac{1}{n-s} \sum_{-\infty}^{\infty} \{\rho_{11}(v)\rho_{22}(v+t) + \rho_{21}(v)\rho_{12}(v+t+2s) \\
& + \rho_{12}(s)\rho_{12}(s+t)\,[\rho_{12}^2(v) + \tfrac{1}{2}\rho_{11}^2(v) + \tfrac{1}{2}\rho_{22}^2(v)] \\
& - \rho_{12}(s)\,[\rho_{11}(v)\rho_{12}(v+s+t) + \rho_{21}(v)\rho_{22}(v+s+t)] \\
& - \rho_{12}(s+t)\,[\rho_{11}(v)\rho_{12}(v+s) + \rho_{21}(v)\rho_{22}(v+s)]\}.
\end{aligned}
\tag{1}
$$

If the sequences are not normal, we must also include in this formula an expression involving fourth-order cumulants; even for linear processes this extra term does not necessarily vanish.

If in (1) we put $X_1(t) \equiv X_2(t)$, it reduces to formula (8), § 9·1. If, alternatively, $X_1(t)$ and $X_2(t)$ are sequences independent of each other, we obtain

$$\left.\begin{aligned} \operatorname{var}\{R_{12}(s)\} &\sim \frac{1}{n-s} \sum_{-\infty}^{\infty} \rho_{11}(v)\,\rho_{22}(v), \\ \operatorname{cov}\{R_{12}(s), R_{12}(s+t)\} &\sim \frac{1}{n-s} \sum_{-\infty}^{\infty} \rho_{11}(v)\,\rho_{22}(v+t). \end{aligned}\right\} \tag{2}$$

Corresponding formulae hold for continuous processes, with integration replacing summation.

The estimation problem requires a careful formulation of the equations comprising any multivariate autoregressive model, if consistent estimates of uniquely defined coefficients are to be obtainable. The set of equations must in general be considered together, with a specification of the assumed relation between the residuals of the different series. For example, with two coupled series specified by the model (cf. § 5·2, equation (32))

$$\left.\begin{aligned} X_t + a_{11} X_{t-1} + a_{12} Y_{t-1} &= U_t, \\ Y_t + a_{21} X_{t-1} + a_{22} Y_{t-1} &= V_t, \end{aligned}\right\} \tag{3}$$

while the least-squares estimates of a_{11} and a_{12}, say, may obviously be obtained from the first equation alone if U_t and V_t are uncorrelated, in the correlated case the least-squares estimates are defined as the asymptotic maximum likelihood estimates on the assumption of normal processes, and involve minimizing the quadratic expression in the two sets of residuals occurring in the logarithm of the likelihood function. Fortunately, however, in the usual case when none of the coefficients is assumed zero *a priori*, it is readily shown that the least-squares estimates of all the coefficients obtained from the separate equations also satisfy these maximum-likelihood equations, and are consequently still the optimum estimates. The asymptotic standard errors of these least-squares estimates may also be derived by methods similar to those previously used.

Similar methods apply in the case of continuous processes. The estimation problem for one or more series will be found further discussed in econometric literature, at least in the case of discrete time (see, for example, the monograph *Statistical Inference in Dynamic Economic Models*, edited by Koopmans (1950), in which some attention is given to the uniqueness or 'identification' problem; cf. also Wold, 1953).

The methods for testing the goodness of fit of the correlogram may also be extended to multivariate series. Consider a set of simultaneous autoregressive sequences, written in vector and matrix notation,

$$\mathbf{H}_t \mathbf{X}_t \equiv [\mathbf{I} + \mathbf{A}_1 E_t^{-1} + \ldots + \mathbf{A}_p E_t^{-p}] \mathbf{X}_t = \mathbf{U}_t, \tag{4}$$

where $\mathbf{U}_t$ is independent of $\mathbf{U}_{t-\tau}$ $(\tau \neq 0)$. It will be convenient to write equation (4) equivalently in 'tensor' notation, with the summation convention for any repeated suffices,

$$H_{ij}(t) X_j(t) = U_i(t). \tag{5}$$

$\mathbf{H}_t \equiv H_{ij}(t)$ is also a difference operator on the time t, as indicated by (4). In the notation of (5), we write (as the processes are real, we revert to the convention for covariance matrices used in § 5·2)

$$C_{jk}(\tau) = \frac{1}{n-\tau} \sum_{u=1}^{n-\tau} X_j(u) X_k(u+\tau),$$

$$E\{X_r(u+\tau) X_j(u+t)\} = V_{rj}(t-\tau). \tag{6}$$

For comparing (4) with some wider class of model, likelihood criteria can be directly obtained on the assumption of normal residuals as in the univariate case. We shall, however, derive the extension of the test discussed in detail in § 9·11, which, though asymptotically equivalent to the likelihood criterion, depended on the basic autoregressive model only. We have

$$E\{H_{ik}(\tau) C_{jk}(\tau)\} = E\left\{\frac{1}{n-\tau} \sum_{u=1}^{n-\tau} X_j(u) U_i(u+\tau)\right\} = 0 \quad (\tau > 0), \tag{7}$$

$$H_{gr}(\tau) H_{hj}(t) V_{rj}(t-\tau) = E\{U_g(t) U_h(\tau)\}$$
$$= W_{gh} \delta(t-\tau), \tag{8}$$

say, where $\delta(0) = 1$, $\delta(t-\tau) = 0$ $(t \neq \tau)$. Further

$$E\{H_{ik}(\tau) C_{jk}(\tau) H_{pq}(t) C_{rq}(t)\}$$
$$= E\left\{\frac{1}{(n-t)(n-\tau)} \sum_{u=1}^{n-t} \sum_{v=1}^{n-\tau} X_j(v) U_i(v+\tau) X_r(u) U_p(u+t)\right\}$$

$$= \frac{1}{n-\tau} E\{U_i U_p\} E\{X_r(u) X_j(u+t-\tau)\}$$

$$= \frac{1}{n-\tau} W_{ip} V_{rj}(t-\tau) \quad (t \geqslant \tau > 0). \tag{9}$$

From (8) and (9),

$$E\{H_{gr}(-t) H_{pq}(t) C_{rq}(t) \cdot H_{hj}(-\tau) H_{ik}(\tau) C_{jk}(\tau)\}$$

$$\sim \frac{1}{n-\tau} W_{ip} W_{gh} \delta(t-\tau) \quad (t \geqslant \tau > 0). \tag{10}$$

This is the appropriate generalization of $H_{-t} H_t C_t$ in the uni-variate case, but we still have to generalize the more useful form $H_t^2 C_t$, and this is not so immediate, as in general $V_{rj}(t-\tau) \neq V_{rj}(\tau-t)$. We must look for a new operator $G_{ij}(t)$ such that

$$G_{gr}(t) G_{hj}(\tau) V_{rj}(t-\tau) = T_{gh} \delta(t-\tau), \tag{11}$$

where we expect $G_{ij}(t)$ to be associated with the series (5) reversed in time, and to be of 'length' p. We shall then have as an alternative to (10)

$$E\{G_{gr}(t) H_{pq}(t) C_{rq}(t) \cdot G_{hj}(\tau) H_{ik}(\tau) C_{jk}(\tau)\}$$

$$= \frac{1}{n-\tau} W_{ip} T_{gh} \delta(t-\tau) \quad (t \geqslant \tau > p). \tag{12}$$

We need only find $G_{ij}(t)$ for $p = 1$, for we saw in §5·2 that the more general case can always be regarded as a degenerate case of a first-order equation in more variables. In this case, reverting to matrix notation,

$$\mathbf{H}_t = \mathbf{I} + \mathbf{A} E_t^{-1},$$

$$\mathbf{H}_t \mathbf{V}'_{t-\tau} = \mathbf{H}_t E\{\mathbf{X}_t \mathbf{X}'_\tau\}$$

$$= \mathbf{V}'_{t-\tau} + \mathbf{A} \mathbf{V}'_{t-\tau-1}$$

$$= E\{\mathbf{U}_t \mathbf{X}'_\tau\} = 0 \quad (t > \tau). \tag{13}$$

From (13) we have (cf. equation (38), §5·2)

$$\mathbf{V}'_1 + \mathbf{A} \mathbf{V}_0 = 0. \tag{14}$$

Suppose we define similarly

$$\mathbf{G}_t \equiv \mathbf{I} + \mathbf{B} E_t^{-1}$$

$$= \mathbf{I} - \mathbf{V}_1 \mathbf{V}_0^{-1} E_t^{-1}. \tag{15}$$

Then we shall find that G_t has the required properties, and that

$$G_{-t}X_t = Z_t$$

has similar properties to the original equation, with time reversed. (The operator G_t is only identical with the original operator H_t if the covariance matrix $V_{t-\tau}$ is symmetric. It is interesting to note that this condition is made a basic postulate in non-equilibrium thermodynamics; see de Groot (1951).)

We have $\qquad E\{X_{t+\tau}Z'_t\} = E\{X_{t+\tau}X'_tG'_{-t}\}$

(where the operator G'_{-t} acts on X'_t)

$$= V'_\tau - V'_{\tau-1}V_0^{-1}V'_1 = 0, \qquad (16)$$

from the result $V'_\tau = (-A)^\tau V_0$, and also

$$E\{Z_\tau Z'_t\} = E\{G_{-\tau}X_\tau . X'_tG'_{-t}\}$$
$$= E\{(X_\tau - V_1V_0^{-1}X_{\tau+1})(X'_t - X'_{t+1}V_0^{-1}V'_1)\}$$
$$= V_{t-\tau} - V_1V_0^{-1}V_{t-\tau-1} - V_{t-\tau+1}V_0^{-1}V'_1 + V_1V_0^{-1}V_{t-\tau}V_0^{-1}V'_1$$
$$= 0 \quad (t \neq \tau). \qquad (17)$$

For $t = \tau$, the matrix T introduced in (11) is given by

$$G_{-\tau}V_{t-\tau}G'_{-t} \equiv E\{Z_\tau Z'_t\} = V_0 + BV'_1$$
$$= [G_{-t}V'_t]_{t=0}, \qquad (18)$$

whereas $\qquad W \equiv E\{U_tU'_t\} = V_0 + AV_1$
$$= [H_tV'_t]_{t=0}. \qquad (19)$$

It should be noted that even if W is diagonal, T is not necessarily so.

The result (12) has similar advantages over (10) as in the univariate case, when estimation of the coefficients is necessary. For if after estimation we use $h_{pq}(t)$ in place of $H_{pq}(t)$, etc.,

$$(n-t)^{\frac{1}{2}}[g_{kr}(t)h_{pq}(t) - G_{kr}(t)H_{pq}(t)]C_{rq}(t)$$
$$= (n-t)^{\frac{1}{2}}[g_{kr}(t) - G_{kr}(t)]h_{pq}(t)C_{rq}(t)$$
$$+ (n-t)^{\frac{1}{2}}G_{kr}(t)[h_{pq}(t) - H_{pq}(t)]C_{rq}(t)$$
$$\sim 0 \quad (t > p), \qquad (20)$$

because $\qquad h_{pq}(t)C_{rq}(t) \sim H_{pq}(t)V_{rq}(t) = 0 \quad (t > 0),$

and $\qquad G_{kr}(t)C_{rq}(t) \sim G_{kr}(t)V_{rq}(t) = 0,$

from (16).

As T_{ij} is not in general diagonal even if W_{ij} is, it is desirable in (12) to transform the linear forms in C_{ij} further so that they are

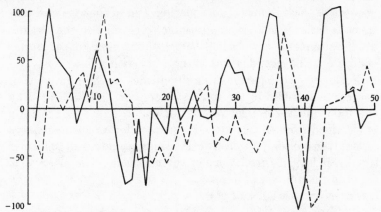

Fig. 19.   Realization of a bivariate autoregressive Markov process.
(See equations (21); X_t is the unbroken line in the figure.)

uncorrelated also for $t = \tau$. This is relatively simple, when it is
remembered that the latent roots and latent vector components
of a product $W_{ip} T_{gh}$ are given by $\lambda_r \mu_s$ and $x_{rj} y_{sk}$, where λ_r, x_{rj} are
the latent roots and vector components of W_{ip}, and μ_s, y_{sk} those of
T_{gh}. Uncorrelated forms may then be obtained by taking linear
combinations of the original forms, with the latent vector
components as coefficients. The asymptotic normality of these
forms is established by extensions of the theorems for the
univariate case.

The extension of these methods to m-dependent and continuous multivariate series has been indicated by Bartlett and
Rajalakshman (1953). Their discussion includes also a numerical
illustration of the above test on the bivariate artificial series

$$\left.\begin{aligned}
X_t - 0 \cdot 6 X_{t-1} + 0 \cdot 5 Y_{t-1} = U_t, \\
Y_t - 0 \cdot 4 X_{t-1} - 0 \cdot 5 Y_{t-1} = V_t.
\end{aligned}\right\} \tag{21}$$

The beginning of this series (with U_t and V_t uniform and independent of each other) is shown in fig. 19, and indicates how such
bivariate series, even of the first order, can exhibit oscillatory
behaviour.

In the paper just referred to, two further points were emphasized in the course of the numerical check on the artificial
series constructed from series (21). The first is that the totalling
of the individual χ^2 items (cf. §9·13) does not check the *ortho-*

gonality of these items before squaring, and in the case of simultaneous series it is at least advisable to check also the absence of simultaneous correlation between the final linear forms computed. The second is that the rapidity with which the asymptotic theory becomes available diminishes as the number k of degrees of freedom in the total χ^2 increases, as is indicated by the following theoretical results for the variance of the total χ^2 obtained from the $H(t)H(-t)$ operator for an autoregressive series given *a priori* (this test we have seen is equivalent to testing the direct lagged correlations of the residuals). The mean χ^2 is equal to the number of items k.

One series: Variance of $\chi^2 =$

$$2k + \frac{4k^2}{n}\left\{2 + \gamma_2\right\}\left\{1 + O\left(\frac{1}{k}, \frac{1}{n}\right)\right\}.$$

Two series: Variance of $\chi^2 =$

$$8k' + \frac{22k'^2}{n}\left\{2 + \gamma_2\right\}\left\{1 + O\left(\frac{1}{k}, \frac{1}{n}\right)\right\},$$

where $k = 4k'$ (k' being comparable to k for a single series, indicating the order of the maximum correlation lag considered), and γ_2 is the coefficient of kurtosis κ_4/κ_2^2 (depending on the departure from the normal or Gaussian distribution in the fourth moment) in the distribution of the residuals.

Spectral analysis. This is covered in principle by the theory of § 6·5, where examples of spectral matrices were given. Thus the remark in that section that the theory of a vector process $\mathbf{X}_t$ may be related to that of the scalar series $\boldsymbol{\lambda}'\mathbf{X}_t$ facilitates the interpretation of the spectral functions, and examination of their sampling errors.

9·4 Multidimensional series

The statistical analysis of multidimensional (and in particular two-dimensional) processes must also of course be related to the correlational and spectral theory of § 6·5, but raises rather more difficulties than multivariate series, if only because comprehensive sets of figures for spatial variation in some characteristic such as, say, soil nitrogen content, or thickness of a sheet of paper, are more laborious to compile and analyse. Moreover, as noted in § 6·51, the absence of the natural ordering properties relevant for one-dimensional series tends to make theoretical models for such multidimensional data less familiar and often more complex.

General empirical analyses of two-dimensional data, whether via correlations or spectral functions, are sufficiently straightforward extensions of one-dimensional techniques to require little further specific explanation. Any orthogonality properties used in the analyses will be direct products of the corresponding one-dimensional properties. An example of the spectral analysis of a two-dimensional point process is referred to in § 9·41. In the remainder of the present section we will therefore confine our attention to special features which may be required in theoretical models for such two-dimensional data; in particular, to isotropy. We shall, remembering that, even if observations are discrete, the model should relate to the intrinsic topographic properties of the material under study, consider only 'continuous parameter' models.

The mathematical equivalence of the classes of autocorrelation functions and characteristic functions extends to more than one dimension (see equation (1) of § 6·51), and enables us to check the admissibility of any theoretical autocorrelation function. For isotropic functions, a general formula (applicable in fact in any number of dimensions) noted by Matern (1960) for the correlation function, which must be a function $\rho(r)$ of the radial distance r, is

$$\rho(r) = \int_{-\infty}^{\infty} e^{-a^2 r^2} \, dH(a), \tag{1}$$

where $H(a)$ is a univariate distribution function. For example, if $H(a)$ corresponds to a χ^2-type distribution for a^2,

$$\rho(r) = (1 + r^2/b^2)^{-s}. \tag{2}$$

As the function in (2) is also a probability density function if multiplied by the appropriate constant, the corresponding characteristic function viz.

$$2(\tfrac{1}{2}br)^\alpha K_\alpha(br)/\Gamma(\alpha), \tag{3}$$

where K is the modified Bessel function of the second kind, is also a valid $\rho(r)$. Two particular cases are $\alpha = \frac{1}{2}$, giving the exponential function noted in § 6·51, and $\alpha = 1$, giving

$$\rho(r) = br K_1(br). \tag{4}$$

This last formula has been proposed by Whittle (1954) in the two-dimensional case, because of its relation with isotropic stochastic models analogous to the stochastic differential equation for a Markov process. Thus the formula $f(\omega) \propto h(\omega) h^*(\omega)$ of § 6·3 becomes in two dimensions, for the isotropic differential operator $\partial^2/\partial x^2 + \partial^2/\partial y^2 - \kappa^2$,

$$f(\omega_x, \omega_y) \propto (\kappa^2 + \omega_x^2 + \omega_y^2)^{-2}, \tag{5}$$

whence (4) follows. The function $\rho(r)$ drops less quickly than the exponential function, and $1 - \rho(r)$ is of order $r^2 \log r$ at the origin.

Further possible models are suggested by introducing the time into the differential operator considered. Thus consider the diffusion operator (with a damping term α)

$$\frac{\partial}{\partial t} + \alpha - \frac{1}{2} \nabla^2,$$

where in two dimensions $\nabla^2 \equiv \dfrac{\partial^2}{\partial x^2} + \dfrac{\partial^2}{\partial y^2}$ (and in three dimensions $\nabla^2 \equiv \dfrac{\partial^2}{\partial x^2} + \dfrac{\partial^2}{\partial y^2} + \dfrac{\partial^2}{\partial z^2}$). Then the simultaneous spectral density function in ω_t and $\boldsymbol{\omega} \equiv \omega_x, \omega_y$ (and ω_z) must be, for a 'noise' input with constant spectrum,

$$f(\omega_t, \boldsymbol{\omega}) \propto h(\omega_t, \boldsymbol{\omega}) h^*(\omega_t, \boldsymbol{\omega}),$$

where $\qquad h(\omega_t, \boldsymbol{\omega}) = 1/(\alpha + i\omega_t + \frac{1}{2}\omega^2),$

ω^2 being $\omega_x^2 + \omega_y^2$ ($+ \omega_z^2$). The marginal spectrum in $\boldsymbol{\omega}$ is

$$\int_{-\infty}^{\infty} f(\omega_t, \boldsymbol{\omega}) \, d\omega_t \propto 1/(\alpha + \frac{1}{2}\omega^2). \tag{6}$$

Its inverse is not, unless $\alpha > 0$, convergent in two dimensions but in three dimensions is, even for $\alpha = 0$, when it corresponds to an autocovariance function in $\mathbf{r} \equiv x, y, z$ of the form

$$w(\mathbf{r}) \propto 1/r \quad (r > 0),$$

a model suggested by Whittle (1962) to be relevant for autocovariance functions in agriculture.

Multidimensional functions also arise even with one-dimensional series in the study of higher-order moments of non-normal series (see Rosenblatt and Van Ness, 1965; cf. also § 6.4).

9·41 The analysis of multidimensional point processes.

The analysis of a point process may have to cope with the various extensions already encountered with stationary processes in quantitative variables. One of these, as noted in § 9·4, is to two or three dimensional processes; and an advantage of spectral analysis technique is its immediate availability in more than one dimension, though some care may be necessary in planning the analysis, even on an electronic computer, if excessive time spent on the analysis is to be avoided. Let us examine the two-dimensional problem, applicable, for example, to the study of spatial pattern in a plant species.

We extend formula (1) of § 9·23 to

$$J(p, q) = \sqrt{\left(\frac{2}{\mu A}\right)} \sum_{s=1}^{n} e^{i(x_s \omega_p + y_s \eta_q)}, \qquad (1)$$

where there are n points with Cartesian co-ordinates (x_s, y_s), and (ω, η) are the corresponding 'frequency' variables. The points are contained in a well-defined sample area A, which will be assumed to be rectangular with dimensions $L_x L_y = A$.

The calculations may be standardized, as recommended in the one-dimensional case, by setting $\mu A' = n$, $x'_s = n x_s / L_x$, $y'_s = n y_s / L y$, and choosing integral p, q, where $\omega_p = 2\pi p / n$, $\eta_q = 2\pi q / n$. The same factor 2 is retained in the multiplier $2/n$, so that $E\{JJ^*\} \to 2$ as $\omega \to \infty$, corresponding to JJ^* having two degrees of freedom.

Example. Examples are as yet hardly available in the literature†, but the analysis of a 'stand' of Japanese black pine saplings (from fig. 8 given by Makoto Numata, 1961) is briefly reported below. These particular observations (65 in number) taken over a square 5 × 5 m. appeared roughly random, as is indicated by the counts in individual squares of side 1 m. (table 9).

Table 9

2	1	5	3	3
3	0	2	4	6
3	1	3	3	3
4	2	1	3	3
3	1	2	3	1

† See, however, Bartlett (1964).

Table 10

$p \rightarrow$				
q	135·55	132·65	120·98	138·08
$\downarrow$	112·74	109·59	100·48	112·53
	131·67	108·59	166·04	119·50
	126·71	138·66	123·90	139·82

The spectral analysis was programmed† to provide cumulative totals of the periodogram intensity $I_{p,q}$ over blocks 8×8 in $\omega_p \times \eta_q$, with the additional recording of individual $I_{p,q}$ values exceeding 8 (as against a mean of about 2); this was to avoid the excessive tabulation of all $I_{p,q}$ values. The range of p and q was 1 to 32 (about $\frac{1}{2}n$).

Cumulative totals are shown in table 10.

These values appear fairly constant, apart perhaps from the value 166·04, which moreover contains the largest individual $I_{p,q}$ values, viz. $I_{20,20} = 18·62$, $I_{24,21} = 14·33$, $I_{18,24} = 13·63$. However, the number of $I_{p,q}$ values exceeding 8 was 19, compared with an expectation of 18·7 on the hypothesis of a uniform spectrum, and even the maximum value 18·62 has a significance level of $P = 0·09$, which is not very anomalous. A statistical test of homogeneity of the values in table 10 (which would be χ^2's with 128 degrees of freedom on the null hypothesis) also gives no significant evidence of departure from this hypothesis. It should be noticed that the complete range of p, q should also cover the axes $p = 0$ and $q = 0$ (excluding, say, $p = q = 0$), and *either* negative p *or* negative q.

9·42   The analysis of multidimensional lattice models.

The distinction between simultaneous and conditional lattice models was emphasized in § 6·54 and the properties of conditional nearest-neighbour models developed. Examples of the fitting of these conditional models in the case of the autonormal and autologistic models are referred to below.

Example 1, the autonormal model. An autocorrelation analysis of the Mercer and Hall data, consisting of 500 grain yields from plots 11 ft by 10·82 ft arranged in a 20×25 rectangle, was first given by Whittle (1954) and his 'simultaneous model' (6) of § 6·54 was among those fitted to the deviations from the mean. The example is not ideal for fitting a 'stationary' lattice model, both because of a possible trend across the area and because the data strictly arise from a variable integrated over the plot area. However, these limitations were noted by both Whittle (1954)

† I am indebted to D. Walley for the programming of the spectral analysis of point processes reported here and in § 9·23.

and Besag (974), and the latter illustrated the fitting of the autonormal model with regression relation (8), § 6·54. The values of β_1 and β_2 he obtained were 0·368 and 0·107 respectively by maximum likelihood estimation. From the exponential form (13), § 6·54, for the autonormal and autologistic models, maximum likelihood estimation for these models is equivalent to equating the observed and theoretical mean and nearest-neighbour correlations. The form of the spectral function (12), § 6·54, determines the latter as

$$r_{10} = I_{10}/I_{00}, \quad \text{or} \quad r_{01} = I_{01}/I_{00},$$

where
$$I_{jk} = \int_{-\pi}^{\pi} \int_{-\pi}^{\pi} \frac{\cos(j\omega_1)\cos(k\omega_2)\,d\omega_1\,d\omega_2}{1 - 2(\beta_1\cos\omega_1 + \beta_2\cos\omega_2)}. \tag{1}$$

Example 2, the autologistic model. Besag (1974) has also illustrated the fitting of the autologistic model to some observations of Dr J. T. Gleaves on the presence/absence of *Plantago lanceolata* in a long rectangular area of 10×940 cells, each 2 cm by 2 cm, at Treloggan, Flintshire. Here again the data have the same limitation that they refer to areas rather than lattice points. There is the further technical difficulty that we cannot fit by maximum likelihood in the absence of a general theoretical solution for the nearest-neighbour correlation (unlike the linear autonormal case, the correlations are affected by the value of the mean, which is dependent on α). Besag, however, has proposed an ingenious coding method of fitting this model (available also for the autonormal model, though less necessary in that case). If the sites are marked alternatively × and . as in the diagram below, we may consider the likelihood function for the values at the × sites, conditional on the values at the . sites (or vice versa). These values are conditionally independent, and the fitting of the model is comparatively straightforward.

```
×  .  ×  .  ×  .  ×
.  ×  .  ×  .  ×  .
×  .  ×  .  ×  .  ×
.  ×  .  ×  .  ×  .
```

The method is not fully efficient, but if the mean of the two estimates is taken, the efficiency should be adequate. Besag's estimates (for values 0, 1 in the model) are noted in the table.

	α	β
Coding pattern (1)	− 2·254	0·724
Coding pattern (2)	− 2·141	0·748
Mean	− 2·198	0·736

These estimates were for the *isotropic* model

$$p_{rs} = \frac{\exp\{(\alpha + \beta y_{rs})\,x_{rs}\}}{1 + \exp(\alpha + \beta y_{rs})}, \tag{2}$$

where $y_{rs} = x_{r-1,\,s} + x_{r+1,\,s} + x_{r,\,s-1} + x_{r,\,s+1}$. The conditional likelihood function, as noted above, is simply

$$\Pi_i \left[\frac{\exp\{(\alpha + \beta y_i)\,x_i\}}{1 + \exp(\alpha + \beta y_i)} \right] = \Pi_i p_i, \text{ say,} \tag{3}$$

for the chosen sites i, whence the estimation equations are

$$\frac{\partial L}{\partial \alpha} = \Sigma_i q_i(2x_i - 1) = 0, \quad \frac{\partial L}{\partial \beta} = \Sigma_i y_i q_i(2x_i - 1) = 0.$$

From the form (3) for the conditional likelihood function, the χ^2 technique of § 8·22 is now also available for testing the goodness of fit of model (2), though of course only for one coding scheme at a time.

BIBLIOGRAPHY

[The references for each chapter are listed separately, section by section. References to a particular author may be traced through the general Index.]

CHAPTER 1

§1·1   BARTLETT, M. S. (1947). *Stochastic Processes* (mimeographed notes of a course given at the University of North Carolina in the Fall Quarter, 1946).

1·2   CRAMÉR, H. (1937). *Random Variables and Probability Distribu-*
and   *tions.* Cambridge.

1·21   CRAMÉR, H. (1946). *Mathematical Methods of Statistics.* Princeton.

KOLMOGOROV, A. (1933). Grundbegriffe der Wahrscheinlichkeitsrechnung. *Ergebn. Math.* 2, no. 3.

1·3   BLANC-LAPIERRE, A. and FORTET, R. (1953). *Théorie des Fonctions Aléatoires.* Paris.

DOOB, J. (1937). Stochastic processes depending on a continuous parameter. *Trans. Amer. Math. Soc.* 42, 107.

DOOB, J. (1953). *Stochastic Processes.* New York.

LÉVY, P. (1948). *Processus Stochastiques et Mouvement Brownien.* Paris.

1·31   BOCHNER, S. (1947). Stochastic processes. *Ann. Math.* 48, 1014.

HOPF, E. (1952). Statistical hydromechanics and functional calculus. *J. Ration. Mech. Anal.* 1, no. 1, p. 87.

KENDALL, D. G. (1949). Stochastic processes and population growth. *J. R. Statist. Soc.* B, 11, 230.

LE CAM, L. (1947). Un instrument d'étude des fonctions aléatoires: la fonctionnelle caractéristique. *C.R. Acad. Sci., Paris,* 224, 710.

CHAPTER 2

§2·1   DOOB, J. (1953). See 1·3 above.

FELLER, W. (1950). *An Introduction to Probability Theory and its Applications,* 1. New York.

SAMUELSON, P. (1948). Exact distribution of continuous variables in sequential analysis. *Econometrica,* 16, no. 2, p. 191.

USPENSKY, J. V. (1937). *Introduction to Mathematical Probability.* New York.

WALD, A. (1947). *Sequential Analysis.* New York.

2·11   BARTLETT, M. S. (1949). Some evolutionary stochastic processes. *J. R. Statist. Soc.* B, 11, 211.

FELLER, W. (1941). On the integral equation of renewal theory. *Ann. Math. Statist.* 12, 243.

LOTKA, A. J. (1939). A contribution to the theory of self-renewing aggregates, with special reference to industrial replacement. *Ann. Math. Statist.* 10, 1.

OWEN, A. R. G. (1949). The theory of genetical recombination. I. Long-chromosome arms. *Proc. Roy. Soc.* B, 136, 67.

2·2 BARTLETT, M. S. (1937). Deviations from expected frequencies in the theory of inbreeding. *J. Genet.* **35**, 83.

BUSH, R. R. and MOSTELLER, F. (1951). A mathematical model for simple learning. *Psychol. Rev.* **58**, no. 5, p. 313.

FISHER, R. A. (1949). *The Theory of Inbreeding.* Edinburgh.

FRÉCHET, M. (1937-8). Recherches théoriques modernes sur la théorie des probabilités. *Traité du Calcul des Probabilités* (ed. Borel), **1**, no. 3. Paris.

ROMANOVSKY, V. I. (1949). *Discrete Markov Chains* (Russian). Moscow-Leningrad.

2·21 DOEBLIN, J. (1939). Sur deux problèmes de M. Kolmogoroff concernant les chaines dénombrables. *Bull. Soc. math. Fr.* **66**, 1.

DOOB, J. L. (1942). Topics in the theory of Markoff chains. *Trans. Amer. Math. Soc.* **52**, 37.

DOOB, J. L. (1945). Markoff chains—denumerable case. *Trans. Amer. Math. Soc.* **59**, 455.

FELLER, W. (1950). *An Introduction to Probability Theory and its Applications.* New York.

FRAZER, R. A., DUNCAN, W. J. and COLLAR, A. R. (1946). *Elementary Matrices.* Cambridge.

FRÉCHET, M. (1937-8). See 2·2 above.

KOLMOGOROV, A. (1936). Anfangsgründe de Theorie der Markoffschen Ketten mit unendlich vielen möglichen Zuständen. *Rec. Math., Moscou (Mat. Sbornik),* **1** (43), 607.

MILNE-THOMSON, L. M. (1933). *The Calculus of Finite Differences.* London.

2·22 FRAZER, R. A., DUNCAN, W. J. and COLLAR, A. R. (1946). See 2·21 above.

MONTROLL, E. W. (1947). On the theory of Markoff chains. *Ann. Math. Statist.* **18**, 18.

2·23 BELLMAN, R. (1957). On a generalization of the fundamental identity of Wald. *Proc. Camb. Phil. Soc.* **53**, 257.

MILLER, H. D. (1962). Absorption probabilities for sums of random variables defined on a finite Markov chain. *Proc. Camb. Phil. Soc.* **58**, 286.

PHATARFOD, R. M. (1965). Sequential analysis of dependent observations. *Biometrika,* **52**, 257.

TWEEDIE, M. G. K. (1960). Generalization of Wald's fundamental identity of sequential analysis to Markov chains. *Proc. Camb. Phil. Soc.* **56**, 205.

2·3 BIENAYMÉ, I. J. (1845). De la loi de multiplication et de la durée des familles. *Soc. Philomath. Paris Extraits* Ser. 5, 37.

BARTLETT, M. S. (1949). See 2·11 above.

BARTLETT, M. S. (1951). The dual recurrence relation for multiplicative processes. *Proc. Camb. Phil. Soc.* **47**, 821.

FISHER, R. A. (1930). *The Genetical Theory of Natural Selection.* Oxford.

GALTON, F. (1889). *Natural Inheritance.* London.

GOOD, I. J. (1949). The number of individuals in a cascade process. *Proc. Camb. Phil. Soc.* **45**, 360.

HALDANE, J. B. S. (1949). Some statistical problems arising in genetics. *J. R. Statist. Soc.* B, **11**, 1.

HARRIS, T. E. (1948). Branching processes. *Ann. Math. Statist.* **19**, 474.

KENDALL, D. G. (1949). Stochastic processes and population growth. *J. R. Statist. Soc.* B, **11**, 230.

LOTKA, A. J. (1931). The extinction of families. *J. Wash. Acad. Sci.* **21**, 377, 453.

STEFFENSEN, J. F. (1930). Om Sandsynligheden for at Afkommet uddør. *Mat. Tidsskr.* B, **19**.

WOODWARD, P. M. (1947). A statistical theory of cascade multiplication. *Proc. Camb. Phil. Soc.* **44**, 404.

YAGLOM, A. M. (1947). Certain limit theorems of the theory of branching random processes (Russian). *C.R. Acad. Sci. U.R.S.S.* (N.S.), **56**, 783.

2·31 ATHREYA, K. B. and KARLIN, S. (1971). Branching processes with random environments. *Ann. Math. Statist.* **42**, 1499 and 1843.

SMITH, W. L. and WILKINSON, W. E. (1969). On branching processes with random environments. *Ann. Math. Statist.* **40**, 814.

WILKINSON, W. E. (1969). On calculating extinction probabilities for branching processes with random environments. *J. Appl. Prob.* **6**, 478.

CHAPTER 3

§3·1 BACHELIER, L. (1900). Théorie de la spéculation. *Ann. Sci. Éc. norm. sup., Paris*, series 3, **17**, 21.

BARTLETT, M. S. (1945). The large sample theory of sequential tests. *Proc. Camb. Phil. Soc.* **42**, 239.

BARTLETT, M. S. (1953). Stochastic processes or the statistics of change. *Appl. Statist.* **2**, 44.

CHANDRASEKHAR, S. (1943). Stochastic problems in physics and astronomy. *Rev. Mod. Phys.* **15**, 1.

CRAMÉR, H. (1937). *Random Variables and Probability Distributions.* Cambridge.

CURTISS, J. (1949). *Sampling Methods Applied to Differential and Difference Equations* (mimeographed report, National Bureau of Standards U.S.A.).

HUYGHENS, C. (1654). *De ratiociniis in ludo aleae.*

KHINTCHINE, A. (1933). Asymptotische Gesetze der Wahrschienlichkeitsrechnung. *Ergebn. Math.* **2**, no. 4.

NATIONAL BUREAU OF STANDARDS U.S.A. (1951). *Monte Carlo Method.* Applied Mathematics Series, no. 12, Washington, D.C.

3·2 FELLER, W. (1950). *An Introduction to Probability Theory and its Applications,* **1**. New York.

KOLMOGOROV, A. (1931). Über die analytische Methoden in der Wahrscheinlichkeitsrechnung. *Math. Ann.* **104**, 415.

LEDERMANN, W. (1950). On the asymptotic probability distribution for certain Markoff processes. *Proc. Camb. Phil. Soc.* **46**, 581.

LUNDBERG, O. (1940). *On Random Processes and their Application to Sickness Accident Statistics.* Uppsala: Almqvist and Wiksells.

MCKENDRICK, A. G. (1914). Studies on the theory of continuous probabilities with special reference to its bearing on natural phenomena of a progressive nature. *Proc. Lond. Math. Soc.* (2), **13**, 401.

3·3 BARNETT, V. D. (1965). Wald's identity and absorption probabilities for two-dimensional random walks. *Proc. Camb. Phil. Soc.* **61**, 747.

BARTLETT, M. S. (1953). Recurrence and first passage times. *Proc. Camb. Phil. Soc.* **49**, 263.

CHUNG, K. L. (1960). *Markov Chains and Stationary Transition Probabilities.* Berlin.

DYNKIN, E. B. (1960). *Theory of Markov processes.* (Translated from the Russian by D. E. Brown.) Oxford.

FELLER, W. (1949). Fluctuation theory of recurrent events. *Trans. Amer. Math. Soc.* **67**, 98.

FELLER, W. (1950). See 3·2 above.

HARRIS, T. E. (1952). First passage and recurrence distributions. *Trans. Amer. Math. Soc.* **73**, 471.

KEILSON, J. (1963). The first passage-time density for homogeneous skip-free walks on the continuum. *Ann. Math. Statist.* **34**, 1003.

WHITTLE, P. (1964). Stochastic processes in several dimensions. *Bull. Inst. Int. Statist.* **40** (Book 2), 974.

3·31 BLACKWELL, D. (1948). A renewal theorem. *Duke Math. J.* **15**, 145.

ERDÖS, P., FELLER, W. and POLLARD, H. (1949). A theorem on power series. *Bull. Amer. Math. Soc.* **55**, 201.

FELLER, W. (1950). See 3·2 above.

KOLMOGOROV, A. (1936). Anfangsgründe der Markoffschen Ketten mit unendlich vielen möglichen Zuständen. *Rec. Math., Moscou* (*Mat. Sbornik*), N.S., **1**, 607.

SMITH, W. L. (1954). Asymptotic renewal theorems. *Proc. Roy. Soc. Edinb.* **64**, 9.

3·32 BARTLETT, M. S. (1953). See 3·3 above.

3·4 ARLEY, N. (1943). *On the theory of stochastic processes and their application to the theory of cosmic radiation.* Copenhagen.

BARTLETT, M. S. (1951). The dual recurrence relation for multiplicative processes. *Proc. Camb. Phil. Soc.* **47**, 821.

CONSAEL, R. (1950). Sur quelques points de la théorie des processus stochastiques. *Bull. Acad. Roy. Belg.* (Cl. Sci.), series 5, **36**, 870.

KENDALL, D. G. (1948). On the generalized 'birth-and-death' process. *Ann. Math. Statist.* **19**, 1.

KENDALL, D. G. (1950). An artificial realisation of a simple 'birth-and-death' process. *J. R. Statist. Soc.* B, **12**, 116.

KOLMOGOROV, A. and DMITRIEV, N. A. (1947). Branching stochastic processes. *C.R. Acad. Sci. U.R.S.S.* (N.S.), **56**, 5.

RAMAKRISHNAN, A. (1951). Some simple stochastic processes. *J. R. Statist. Soc.* B, **13**, 131.

3·41 BARTLETT, M. S. (1949). Some evolutionary stochastic processes. *J. R. Statist. Soc.* B, **11**, 211.

CHANDRASEKHAR, S. (1943). See 3·1 above.

ROTHSCHILD, LORD (1953). A new method of measuring the activity of spermatozoa. *J. Exp. Biol.* **30**, 178.

3·42 BARTLETT, M. S. and KENDALL, D. G. (1951). On the use of the characteristic functional in the analysis of some stochastic processes occurring in physics and biology. *Proc. Camb. Phil. Soc.* **47**, 65.

BHABHA, H. J. (1950). On the stochastic theory of continuous parametric systems and its application to electron cascades. *Proc. Roy. Soc.* A, **202**, 301.

BHABHA, H. J. and RAMAKRISHNAN, A. (1950). On the mean square deviation of the number of electrons and quanta in the cascade theory. *Proc. Indian Acad. Sci.* **32**, 141.

HOPF, E. (1952). Statistical hydromechanics and functional calculus. *J. Ration. Mech. Anal.* **1**, 87.

JÁNOSSY, L. (1950). Note on the fluctuation problem of cascades. *Proc. Phys. Soc. Lond.* A, **63**, 241.

KENDALL, D. G. (1949). Stochastic processes and population growth. *J. R. Statist. Soc.* B, **11**, 230.

RAMAKRISHNAN, A. (1950). Stochastic processes relating to particles distributed in a continuous infinity of states. *Proc. Camb. Phil. Soc.* **46**, 595.

RAMAKRISHNAN, A. (1952). A note on Jánossy's mathematical model of a nucleon cascade. *Proc. Camb. Phil. Soc.* **48**, 451.

RAMAKRISHNAN, A. (1953). Stochastic processes associated with random divisions of a line. *Proc. Camb. Phil. Soc.* **49**, 473.

STEVENS, W. L. (1937–8). Significance of grouping. *Ann. Eugen., Lond.*, **8**, 57.

3·5 BARTLETT, M. S. (1949). See 3·41 above.

COX, D. R. and MILLER, H. D. (1965). *The Theory of Stochastic Processes.* London.

DOBRUŠIN, R. L. (1952). On conditions of regularity of stationary Markov processes with a denumerable set of possible states (Russian). *Progr. Math. Sci., Moscow* (N.S.), **7**, no. 6 (52), p. 185.

FELLER, W. (1940). On the integro-differential equations of

purely discontinuous Markoff processes. *Trans. Amer. Math. Soc.* **48**, 488.

KARLIN, S. and McGREGOR, J. (1957). The differential equations of birth and death processes and the Stieltjes moment problem. *Trans. Amer. Math. Soc.* **85**, 489.

KEILSON, J. (1963). On transient behaviour in diffusion and birth-death processes. *Research Report No.* 380, *Sylvania Electronic Systems* (Mass.).

KENDALL, D. G. (1958). Integral representations for Markov transition probabilities. *Bull. Amer. Math. Soc.* **64**, 358.

KOLMOGOROV, A. (1931). See 3·2 above.

LEDERMANN, W. and REUTER, G. E. H. (1953). Spectral theory for the differential equations of simple birth and death processes. *Phil. Trans.* A, **246**, 321.

MOYAL, J. E. (1949). Stochastic processes and statistical physics. *J. R. Statist. Soc.* B, **11**, 150.

MOYAL, J. E. (1957). Discontinuous Markoff processes. *Acta Math.* **98**, 221.

PALM, C. (1943). Intensitätsschwankungen im Fernsprechverkehr. *Ericsson Tech.* no. 44.

REUTER, G. E. H. and LEDERMANN, W. (1953). On the differential equations for the transition probabilities of Markov processes with enumerably many states. *Proc. Camb. Phil. Soc.* **49**, 247.

3·51 BARTLETT, M. S. (1975). The paradox of probability in physics, in *Probability, Statistics and Time*. London.

CANE, V. R. (1967). Random walks and physical processes. *Bull. Int. Statist. Inst.* **42** (Book 1), 622.

KENDALL, D. G. (1974). Pole-seeking Brownian motion and bird navigation. *J. R. Statist. Soc.* B, **36**, 365.

CHAPTER 4

§4·1 BARTLETT, M. S. (1945). The large sample theory of sequential tests. *Proc. Camb. Phil. Soc.* **42**, 239.

KOLMOGOROV, A. (1933). Sulla determinazione empirica di una legge di distribuzione. *G. Ist. ital. Attuari*, **4**, 1.

SEGERDAHL, C.-o. (1939). *On Homogeneous Random Processes and Collective Risk Theory*. Uppsala.

WALD, A. (1947). *Sequential analysis.* New York.

4·2 BARTLETT, M. S. (1951). The dual recurrence relation for multiplicative processes. *Proc. Camb. Phil. Soc.* **47**, 821.

SMITH, W. L. (1954). Asymptotic renewal theorems. *Proc. Roy. Soc. Edinb.* **64**, 9.

4·21 BROCKMEYER, E., HALSTRØM, H. L. and JENSEN, A. (1948). *The Life and Works of A. K. Erlang.* Copenhagen.

COX, D. R. and SMITH, W. L. (1961). *Queues.* London.

FELLER, W. (1950). *An Introduction to Probability Theory and its Applications*, 1. New York.

KENDALL, D. G. (1951). Some problems in the theory of queues. *J. R. Statist. Soc.* B, **13**, 151.

KENDALL, D. G. (1953). Stochastic processes occurring in the theory of queues and their analysis by the method of the imbedded Markov chain. *Ann. Math. Statist.* **24**, 338. (These two papers by D. G. Kendall contain useful bibliographies on the theory of queues.)

KHINTCHINE, A. (1932). Mathematisches über die Erwartung vor ein em öffentlichen Schaffer (Russian; German summary). *Mat. Sbornik*, **39**, 73.

KINGMAN, J. F. C. (1962). The use of Spitzer's identity in the investigation of the busy period and other quantities in the queue *GI/G/*1. *J. Austral. Math. Soc.* **2**, 345.

LINDLEY, D. V. (1952). The theory of queues with a single server. *Proc. Camb. Phil. Soc.* **48**, 277.

POLLACZEK, F. (1930). Über eine Aufgabe der Wahrscheinlichkeitstheorie. *Math. Z.* **32**, 64 and ·729.

SMITH, W. L. (1953). Distribution of queuing times. *Proc. Camb. Phil. Soc.* **49**, 449.

4·22 MORAN, P. A. P. (1959). *Theory of Storage*. London.

PHATARFOD, R. M. (1963). Application of methods in sequential analysis to dam theory. *Ann. Math. Statist.* **34**, 1588.

PRABHU, N. U. (1964). Time dependent results in storage theory. *J. App. Prob.* **1**, 1.

WEESAKUL, B. (1961). The random walk between a reflecting and an absorbing barrier. *Ann. Math. Statist.* **32**, 765.

4·3 BARTLETT, M. S. (1947). *Stochastic Processes* (mimeographed notes of a course given at the University of North Carolina in the Fall Quarter, 1946).

BARTLETT, M. S. (1951). See 4·2 above.

BARTLETT, M. S. and KENDALL, D. G. (1951). On the use of the characteristic functional in the analysis of some stochastic processes occurring in physics and biology. *Proc. Camb. Phil. Soc.* **47**, 65.

FELLER, W. (1939). Die Grundlagen der Volterraschen Theorie des Kampfes ums Dasein wahrscheinlichkeitstheoretischer Behandlung. *Acta biotheor., Leiden*, **5**, 11.

KENDALL, D. G. (1949). Stochastic processes and population growth. *J. R. Statist. Soc.* B, **11**, 230.

4·31 ARMITAGE, P. (1951). The statistical theory of bacterial populations subject to mutation. *J. R. Statist. Soc.* B, **14**, 1.

BELLMAN, R. and HARRIS, T. E. (1948). On the theory of age-dependent stochastic branching processes. *Proc. Nat. Acad. Sci., Wash.*, **34**, 601.

HARRIS, T. E. (1951). Some mathematical models for branching processes. *Proc. Second Berkeley Symposium on Mathematical Statistics and Probability*. California.

KELLY, C. D. and RAHN, O. (1932). The growth rate of individual bacterial cells. *J. Bacteriol.* **23**, 147.

KENDALL, D. G. (1948). On the role of variable generation time in the development of a stochastic birth process. *Biometrika*, **35**, 316.

KENDALL, D. G. (1952). Les processus stochastiques de croissance en biologie. *Ann. Inst. Poincaré*, **13**, (fasc. 1), 43.

LEA, D. E. and COULSON, C. A. (1949). The distribution of the number of mutants in bacterial populations. *J. Genet.* **49**, 264.

LURIA, S. E. and DELBRÜCK, M. (1943). Mutations of bacteria from virus sensitivity to virus resistance. *Genetics*, **28**, 491.

4·32 FELLER, W. (1951). Diffusion processes in genetics. *Proc. Second Berkeley Symposium on Mathematical Statistics and Probability*. California.

FISHER, R. A. (1930). *The Genetical Theory of Natural Selection*. Oxford.

MALÉCOT, G. (1948). *Les Mathématiques de l'Hérédité*. Paris.

MALÉCOT, G. (1952). Les processus stochastiques et la méthode des fonctions generatrices ou caracteristiques. *Publ. Inst. Statist.* (*Univ. de Paris*), **1**, (fasc. 3).

MORAN, P. A. P. (1962). *Statistical Processes of Evolutionary Theory*. Oxford.

SEWALL WRIGHT (1951). The genetical structure of populations. *Ann. Eugen., Lond.*, **15**, 323.

4·4 BAILEY, N. T. J. (1950). A simple stochastic epidemic. *Biometrika*, **37**, 193.

BAILEY, N. T. J. (1953). The total size of a general stochastic epidemic. *Biometrika*, **40**, 177.

BARTLETT, M. S. (1947). See 4·3 above.

BARTLETT, M. S. (1949). Some evolutionary stochastic processes. *J. R. Statist. Soc. B*, **11**, 211.

BARTLETT, M. S. (1953). Stochastic processes or the statistics of change. *Appl. Statist.* **2**, 44.

BARTLETT, M. S. (1954). Processus stochastiques ponctuels. *Ann. Inst. Poincaré*, **14**, (fasc. 1), 35.

BARTLETT, M. S. (1960). *Stochastic Population Models in Ecology and Epidemiology*. London.

DANIELS, H. E. (1977). The advancing wave in a spatial birth process. *J. Appl. Prob.* **14**.

FISHER, R. A. (1937). The wave of advance of advantageous genes. *Ann. Eugen., Lond.*, **7**, 355.

GANI, J. (1965). On a partial differential equation of epidemic theory I. *Biometrica*, **52**, 617.

KERMACK, W. O. and McKENDRICK, A. G. (1927–33). Contributions to the mathematical theory of epidemics. *Proc. Roy. Soc. A*, **115**, 700; **138**, 55; **141**, 94.

McKENDRICK, A. G. (1926). Applications of mathematics to medical problems. *Proc. Edin. Math. Soc.* **44**, 98.

MOLLISON, D. (1977). Spatial contact models for ecological and epidemiological spread. *J. R. Statist. Soc. B*, **39**.

RAPOPORT, A. (1951). Nets with distance bias. *Bull. Math. Biophys.* **13**, 85.

SERFLING, R. E. (1952). Historical review of epidemic theory. *Hum. Biol.* **24**, 145.

SISKIND, V. (1965). A solution of the general stochastic epidemic. *Biometrica*, **52**, 613.

SOPER, H. E. (1929). Interpretation of periodicity in disease-prevalence. *J. R. Statist. Soc.* B, **92**, 34.

WHITTLE, P. (1953). Certain non-linear models of population and epidemic theory. *Skand. Aktuar.* **14**, 211.

WILSON, E. B. and WORCESTER, J. (1945). Damping of epidemic waves *and* The spread of an epidemic. *Proc. Nat. Acad. Sci.*, *Wash.*, **31**, 294 and 327.

CHAPTER 5

§5·1 DOOB, J. L. (1953). *Stochastic Processes.* New York.

FRÉCHET, M. (1937–8). Recherches théoriques modernes sur la théorie des probabilités. *Traité du Calcul des Probabilités* (ed. Borel), **1**, fasc. 3. Paris.

LOÈVE, M. (1945). Sur les fonctions aléatoires stationnaires de second ordre. *Rev. sci.*, *Paris*, **83**, 297.

MOYAL, J. E. (1949). Stochastic processes and statistical physics. *J. R. Statist. Soc.* B, **11**, 150.

5·11 KARHUNEN, K. (1947). Über lineare Methoden in der Wahrscheinlichkeitsrechnung. *Ann. Acad. Sci. fenn.* ser. A, **37**.

LOÈVE, M. (1945). See 5·1 above.

SLUTSKY, E. (1928). Sur les fonctions éventuelles continues, intégrables et dérivables dans le sens stochastique. *C.R. Acad. Sci.*, *Paris*, **187**, 370.

5·2 BARTLETT, M. S. (1947). See 4·3 above.

EDWARDS, D. A. and MOYAL, J. E, (1955). Stochastic differential equations. *Proc. Camb. Phil. Soc.* **51**, 663.

MOYAL, J. E. (1949). See 5·1 above.

5·21 BASS, J. (1945). Les fonctions aléatoires et leur interpretation mécanique. *Rev. sci.*, *Paris*, **83**, 3.

CHANDRASEKHAR, S. (1943). Stochastic problems in physics and astronomy. *Rev. Mod. Phys.* **15**, 1.

GREEN, H. S. (1952). *Molecular Theory of Fluids.* Amsterdam.

MOYAL, J. E. (1949). See 5.1 above.

5·22 BARTLETT, M. S. (1961). Equations for stochastic path integrals. *Proc. Camb. Phil. Soc.* **57**, 568.

COUTEUR, K. J. le (1949). Contribution (p. 277) to discussion at Symposium on Stochastic Processes. *J. Roy. Statist. Soc.* B, **11**, 150.

KENDALL, D. G. (1948). See 3·4 above.

WAUGH, W. A. O'N. (1958). Conditional Markov processes. *Biometrika*, **45**, 241.

5·3 BARBOUR, A. D. (1972). The principle of the diffusion of arbitrary constants. *J. Appl. Prob.* **9**, 519.

BARTLETT, M. S. (1960). See 4·4 above.

DANIELS, H. E. (1960). Solutions of Green's type for stochastic processes. *J. R. Statist. Soc.* B, **22**, 376.

MCNEIL, D. R. and SCHACH, S. (1973). Central limit analogues for Markov population processes. *J. R. Statist. Soc.* B, **35**, 1.

WHITTLE, P. (1957). On the use of the normal approximation in the treatment of stochastic processes. *J. R. Statist. Soc.* B, **19**, 268.

CHAPTER 6

§6·1 BOCHNER, S. (1936–7). *Lectures on Fourier Analysis.* Princeton.

6·11 DOOB, J. L. (1944). The elementary Gaussian processes. *Ann. Math. Statist.* **15**, 229.

DOOB, J. L. (1953). *Stochastic Processes.* New York.

KHINTCHINE, A. (1934). Korrelationstheorie der stationären stochastischen Prozesse. *Math. Ann.* **109**, 604.

LOÈVE, M. (1945). Sur les fonctions aléatoires stationnaires de second ordre. *Rev. sci., Paris*, **83**, 297.

WIENER, N. (1930). Generalized harmonic analysis. *Acta Math.* **55**, 117.

WOLD, H. (1938). *Analysis of Stationary Time-Series.* Uppsala.

6·13 BARTLETT, M. S. (1963). The spectral analysis of point processes. *J. Roy. Statist. Soc.* B, **25**, 264.

6·2 BLANC-LAPIERRE, A. and FORTET, R. (1946). Deux notes: Sur la décomposition spectrale des fonctions aléatoires stationnaires d'ordre deux. *C.R. Acad. Sci., Paris*, **222**, 467 and 713.

CRAMÉR, H. (1937). *Random Variables and Probability Distributions.* Cambridge.

CRAMÉR, H. (1942). On harmonic analysis in certain functional spaces. *Ark. mat. Astr. Fys.* **28**, no. 12.

KOLMOGOROV, A. (1941). Stationary sequences in Hilbert space (Russian). *Bull. math. Univ. Moscou*, **2**, no. 6.

LÉVY, P. (1948). *Processus Stochastiques et Mouvement Brownien.* Paris.

LOÈVE, M. (1945). See 6·11 above.

SLUTSKY, E. (1938). Sur les fonctions aléatoires presques periodiques et sur la décomposition des fonctions aléatoires stationnaires en composantes. *Actualités Sci. industr.* **738**, 35.

6·21 KHINTCHINE, A. (1934). See 6·11 above.

6·3 BARTLETT, M. S. (1946). On the theoretical specification and sampling properties of autocorrelated time-series. *J. R. Statist. Soc.* (Suppl.), **7**, 211.

KARHUNEN, K. (1947). Über lineare Methoden in der Wahrscheinlichkeitsrechnung. *Ann. Acad. Sci. fenn.* ser. A, **37**.

PALEY, R. E. A. C. and WIENER, N. (1934). *Fourier Transforms in the Complex Domain.* New York.

6·31 COX, D. R. (1962). *Renewal Theory.* London.

6·4 BLANC-LAPIERRE, A. and FORTET, R. (1947). Analyse harmonique des fonctions aléatoires et caractère stationnaire. *C.R. Acad. Sci., Paris,* **225**, 1119.

6·41 BARTLETT, M. S. (1953). Recurrence and first passage times. *Proc. Camb. Phil. Soc.* **49**, 263.

RICE, S. O. (1944–5). Mathematical analysis of random noise. *Bell Syst. Tech. J.* **23**, 282; **24**, 46.

SMOLUCHOWSKI, M. v. (1915). Molekulartheoretische Studien über Umkehr thermodynamisch irreversibler Vorgänge und über Wiederkehr abnormaler Zustände. *S.B. Akad. Wiss. Wien* (2a), **124**, 339.

6·5 BARTLETT, M. S. (1947). *Stochastic Processes* (mimeographed notes of a course given at the University of North Carolina in the Fall Quarter, 1946).

CRAMÉR, H. (1940). On the theory of stationary random processes. *Ann. Math.* **41**, 215.

QUENOUILLE, M. H. (1949). Problems in plane sampling. *Ann. Math. Statist.* **20**, 355.

6·51 BASS, J. and AGOSTINI, L. (1950). *Les Théories de la Turbulence.* Paris.

BATCHELOR, G. K. (1953). *The Theory of Homogeneous Turbulence.* Cambridge.

MOYAL, J. E. (1952). The spectra of turbulence in a compressible fluid; eddy turbulence and random noise. *Proc. Camb. Phil. Soc.* **48**, 329.

ROBERTSON, H. P. (1940). The invariant theory of isotropic turbulence. *Proc. Camb. Phil. Soc.* **36**, 209.

6·52 BARTLETT, M. S. (1957). Some problems associated with random velocity. *Publ. Inst. Statist. Univ. Paris,* **6**, 261.

BARTLETT, M.S. (1965) The spectral analysis of line processes. *Proc. 5th Berkeley Spmposium on Mathematical Statistics and Probability.*

LIGHTHILL, M. J. and WHITHAM, G. B. (1955). On kinematic waves II. A theory of traffic flow on long crowded roads. *Proc. Roy. Soc. Lond.* A, **229**, 317.

MILLER, A. J. (1962). Road traffic flow considered as a stochastic process. *Proc. Camb. Phil. Soc.* **58**, 312.

NEWELL, G. F. (1955). Mathematical models of freely flowing highway traffic. *Opns. Res.* **3**, 176.

PRIGOGINE, I., HERMAN, R. and ANDERSON, R. (1963). Further developments in the Boltzman-like theory of traffic flow. *Proc. 2nd. Int. Symp. on Theory of Road Traffic Flow.*

6·53 HANNAN, E. J. (1960). *Time Series Analysis.* London.

JONES, R. H. (1963a). Stochastic processes on a sphere. *Ann. Math. Statist.* **34**, 213.

JONES, R. H. (1963b). Stochastic processes on a sphere as applied to meteorological 500-millibar forecasts. *Proc. Symposium on Time Series Analysis* (ed. M. Rosenblatt). New York.

6·54 BARTLETT, M. S. (1974a). The statistical analysis of spatial pattern. *Adv. Appl. Prob.* **6**, 336.

BARTLETT, M. S. (1974b). Physical nearest-neighbour models and non-linear time-series III. *J. Appl. Prob.* **11**, 715.

BESAG, J. E. (1972). Nearest neighbour systems and the auto-logistic model for binary data. *J. R. Statist. Soc.* B, **34**, 74.

BESAG, J. E. (1974). Spatial interaction and the statistical analysis of lattice systems. *J. R. Statist. Soc.* B, **36**, 192.

NEWELL, G. F. and MONTROLL, E. W. (1953). On the theory of the Ising model of ferromagnetism. *Rev. Mod. Phys.* **25**, 353.

ONSAGER, L. (1944). Crystal statistics. I. A two-dimensional model with an order–disorder transition. *Phys. Rev.* **63**, 117.

WHITTLE, P. (1954). On stationary processes in the plane. *Biometrica*, **41**, 434.

CHAPTER 7

§7·1 WIENER, N. (1949). *The Extrapolation, Interpolation and Smoothing of Stationary Time-Series, with Engineering Applications.* New York.

7·11 KOLMOGOROV, A. (1941). Interpolation und Extrapolation von stationären zufälligen Folgen. *Bull. Acad. Sci. U.R.S.S.*, Ser. Math., **5**, 3.

PALEY, R. E. A. C. and WIENER, N. (1934). *Fourier Transforms in the Complex Domain.* New York.

WIENER, N. (1949). See 7·1 above.

WOLD, H. (1938). *Analysis of Stationary Time-Series.* Uppsala.

7·12 BOX, G. E. P. and JENKINS, G. M. (1962). Some statistical aspects of adaptive optimization and control. *J. Roy. Statist. Soc.* B, **24**, 297.

ROBINSON, E. A. (1962). *Random Wavelets and Cybernetic Systems.* London.

WHITTLE, P. (1963). *Prediction and Regulation.* London.

7·2 BARNARD, G. A. (1951). The theory of information. *J. Roy.*
and *Statist. Soc.* B, **13**, 46.

7·21 BRILLOUIN, L. (1956). *Science and Information Theory.* New York.

FEINSTEIN, A. (1958). *Foundations of Information Theory.* New York.

GOOD, I. J. (1950). *Probability and the Weighing of Evidence.* London.

LAWSON, J. L. and UHLENBECK, G. E. (1950). *Threshold Signals,* no. 24, M.I.T. Radiation Laboratory Series.

McMILLAN, B. (1953). The basic theorems of information theory. *Ann. Math. Statist.* **24**, 196.

SHANNON, C. E. (1948). A mathematical theory of communication. *Bell Syst. Tech. J.* **27**, 379 and 623.

SHANNON, C. E. (1949). Communication in the presence of noise. *Proc. Inst. Radio Engrs, N.Y.*, **37**, 10.

WHITTAKER, J. M. (1935). *Interpolatory Function Theory.* Cambridge.

BIBLIOGRAPHY   377

CHAPTER 8

§8·1 BARTLETT, M. S. (1950). The statistical approach to the analysis of time series. *Symposium on Information Theory* (mimeographed Proceedings, Ministry of Supply, London).

BARTLETT, M. S. (1953). Approximate confidence intervals. *Biometrika*, **49**, 12.

CRAMÉR, H. (1946). *Mathematical Methods of Statistics*. Princeton.

KENDALL, M. G. (1946). *The Advanced Theory of Statistics*, **2**. London.

8·11 GRENANDER, U. (1950). Stochastic processes and statistical inference. *Ark. Mat.* **1**, 195.

8·2 BARTLETT, M. S. (1951). The frequency goodness of fit test for and probability chains. *Proc. Camb. Phil. Soc.* **47**, 86.

8·21 CHANDRASEKHAR, S. (1943). Stochastic problems in physics and astronomy. *Rev. Mod. Phys.* **15**, 1.

PATANKAR, V. N. (1953). The goodness of fit of frequency distributions obtained from stochastic processes. Ph.D. Thesis, University of Manchester.

8·22 BARTLETT, M. S. (1967). Inference and stochastic processes. *J. R. Statist. Soc.* A, **130**, 457.

BARTLETT, M. S. (1968). A further note on nearest neighbour models. *J. R. Statist. Soc.* A, **131**, 579.

BESAG, J. E. (1972). On the correlation structure of some two-dimensional stationary processes. *Biometrika*, **59**, 43.

WHITTLE, P. (1954). See 6·54 above.

8·3 BAILEY, N. T. J. (1953). The total size of a general stochastic epidemic. *Biometrika*, **40**, 177.

BARTLETT, M. S. (1953). See 8·1 above.

GRENANDER, U. (1950). See 8·11 above.

KENDALL, D. G. (1952). Les processus stochastiques de croissance en biologie. *Ann. Inst. Poincaré*, **13** (fasc. 1), 43.

MORAN, P. A. P. (1951). Estimation methods for evolutive processes. *J. R. Statist. Soc.* B, **13**, 141.

PATANKAR, V. N. (1953). See 8·2, 8·21 above.

PATIL, V. T. (1957). The consistency and adequacy of the Poisson–Markoff model for density fluctuations. *Biometrika*, **44**, 43.

ROTHSCHILD, LORD (1953). A new method of measuring the activity of spermatozoa. *J. Exp. Biol.* **30**, 178.

RUBEN, H. (1962). Some aspects of the emigration-immigration process. *Ann. Math. Statist.* **33**, 119.

8·31 BARTLETT, M. S., BRENNAN, JENNIFER M. and POLLOCK, J. N. (1971). Stochastic analysis of some experiments on the mating of blowflies. *Biometrics*, **27**, 725.

COX, D. R. (1972). Regression models and life tables. *J. R. Statist. Soc.* B, **34**, 187.

McQUARRIE, D. A. (1967). Stochastic approach to chemical kinetics. *J. App. Prob.* **4**, 413.

CHAPTER 9

9·1 ANDERSON, R. L. (1954). The problem of autocorrelation in regression analysis. *J. Amer. Statist. Ass.* **49**, 113. (This survey contains a useful bibliography.)

BARTLETT, M. S. (1946). On the theoretical specification and sampling properties of autocorrelated time-series. *J. R. Statist. Soc.* (Suppl.), **7**, 211.

BARTLETT, M. S. (1947). *Stochastic Processes* (mimeographed notes of a course given at the University of North Carolina in the Fall Quarter, 1946).

KENDALL, M. G. (1944). On autoregressive time-series. *Biometrika*, **33**, 105.

MANN, H. B. and WALD, A. (1943). On the statistical treatment of linear stochastic difference equations. *Econometrica*, **11**, 173.

SLUTSKY, E. (1927). The summation of random causes as the source of cyclic processes. *Problems of Economic Conditions*, ed. by the Conjuncture Institute, Moscow, **3**, no. 1 (later reprinted in *Econometrica*, **5**, 105)

YULE, G. U. (1927). On the method of investigating periodicities in disturbed series, with special reference to Wolfer's sunspot numbers. *Phil. Trans.* A, **226**, 267.

9·11 BARTLETT, M. S. and DIANANDA, P. H. (1950). Extensions of Quenouille's test for autoregressive schemes. *J. R. Statist. Soc.* B, **12**, 108.

DIANANDA, P. H. (1953). Some probability limit theorems with statistical applications. *Proc. Camb. Phil. Soc.* **49**, 239.

LEE, G. (1951). Some aspects of autoregressive time series. M.Sc. Thesis, University of Manchester.

QUENOUILLE, M. H. (1947). A large-sample test for the goodness of fit of autoregressive schemes. *J. R. Statist. Soc.* **110**, 123.

WALKER, A. M. (1952). Some properties of the asymptotic power functions of goodness-of-fit tests for linear autoregressive schemes. *J. R. Statist. Soc.* B, **14**, 117.

WALKER, A. M. (1954). The asymptotic distribution of serial correlation coefficients for linear autoregressive processes with dependent residuals. *Proc. Camb. Phil. Soc.* **50**, 60.

WHITTLE, P. (1951). *Hypothesis Testing in Time Series Analysis.* Uppsala.

WHITTLE, P. (1952). Tests of fit in time series. *Biometrika*, **39**, 309.

WOLD, H. (1949). A large-sample test of moving averages. *J. R. Statist. Soc.* B, **11**, 297.

9·12 BARTLETT, M. S. (1946). See 9·1 above.
and BARTLETT, M. S. and DIANANDA, P. H. (1950). See 9·11 above.

9·13 FISHER, R. A. and YATES, F. (1938). *Statistical Tables for Biological, Agricultural and Medical Research.* Edinburgh.

KENDALL, M. G. (1946). *Researches in Oscillatory Time-Series.* Cambridge.

PATIL, V. T. (1957). See 8·3 above.

QUENOUILLE, M. H. (1947). See 9·11 above.

YULE, G. U. (1927). See 9·1 above.

9·2 BARTLETT, M. S. (1947). See 9·1 above.

BARTLETT, M. S. (1950). Periodogram analysis and continuous spectra. *Biometrika*, 37, 1.

GRENANDER, U. (1951). On empirical spectral analysis of stochastic processes. *Ark. Mat.* 1, 503.

KENDALL, M. G. (1946a). *The Advanced Theory of Statistics*, 2. London.

KENDALL, M. G. (1946b). See 9·12, 9·13 above.

LÉVY, P. (1948). *Processus Stochastiques et Mouvement Brownien*. Paris.

SARGAN, J. D. (1953). An approximate treatment of the properties of the correlogram and periodogram. *J. R. Statist. Soc.* B, 15, 140.

SLUTSKY, E. (1927). See 9·1 above.

YULE, G. U. (1927). See 9·1 above.

9·21 BARTLETT, M. S. (1963). Statistical estimation of density functions. *Sankhyā*, A, 25, 245.

BONEVA, LILIANA I., KENDALL, D. G. and STEFANOV, I. (1971). Spline transformations: three new diagnostic aids for the statistical data-analyst. *J. R. Statist. Soc.* B, 33, 1–70.

GRENANDER, U. and ROSENBLATT, M. (1953). Statistical spectral analysis of time series arising from stationary stochastic processes. *Ann. Math. Statist.* 24, 536.

GRENANDER, U. and ROSENBLATT, M. (1957). *Statistical Analysis of Stationary Time Series*. New York.

PRIESTLEY, M. B. (1962). Basic considerations in the estimation of spectra. *Technometrics*, 4, 551.

ROSENBLATT, M. (1956). Remarks on some nonparametric estimates of a density function. *Ann. Math. Statist.* 27, 832.

9·22 BARTLETT, M. S. (1954). Problèmes de l'analyse spectrale des séries temporelles stationnaires. *Publ. Inst. Statist. (Univ. de Paris)*, 3, fasc. 3, 119.

BARTLETT, M. S. (1967a). Some remarks on the analysis of time-series. *Biometrika*, 54, 25.

BARTLETT, M. S. (1967b). See 8·22 above.

GRENANDER, U. and ROSENBLATT, M. (1952). On spectral analysis of stationary time-series. *Proc. Nat. Acad. Sci., Wash.*, 38, 519.

PRIESTLEY, M. B. (1962). The analysis of stationary processes with mixed spectra. *J. Roy. Statist. Soc.* B, 24, 215 and 511.

PRIESTLEY, M. B. (1964). Estimation of the spectral density function in the presence of harmonic components. *J. Roy. Statist. Soc.* B, 26, 123.

WHITTLE, P. (1952). The simultaneous estimation of a time series harmonic components and covariance structure. *Trab. Estadistica*, 3, 43.

WHITTLE, P. (1954). Some recent contributions to the theory of stationary processes. Appendix 2 to the second edition of H. WOLD, *A Study in the Analysis of Stationary Time-Series.* Uppsala.

9·23 BARTLETT, M. S. (1963). See 6·13 above.

BARTLETT, M. S. (1965). See 6·52 above.

BARTLETT, M. S. (1967a). See 9·22 above.

BARTLETT, M. S. (1967b). See 8·22 above.

COOLEY, J. W. and TUKEY, J. W. (1965). An algorithm for the machine calculation of complex Fourier series. *Math. of comput.* 19, 297.

COX, D. R. (1955). Some statistical methods connected with series of events. *J. R. Statist. Soc.* B, 17, 129.

FRENCH, A. S. and HOLDEN, A. V. (1971). Alias-free sampling of neuronal spike trains. *Kybernetik*, 8, 165.

GASTER, M. and ROBERTS, J. B. (1977). On the spectral analysis of randomly sampled records by a direct transform. *Proc. Roy. Soc.* A 354, 27.

GREEN, H. S. (1952). See 5·21 above.

9·3 BARTLETT, M. S. and RAJALAKSHMAN, D. V. (1953). Goodness of fit tests for simultaneous autoregressive series. *J. R. Statist. Soc.* B, 15, 107.

GROOT, S. R. DE (1951). *Thermodynamics of Irreversible Processes.* Amsterdam.

KOOPMANS, T. C. (editor) (1950). *Statistical Inference in Dynamic Economic Models.* New York.

WHITTLE, P. (1953). The analysis of multiple stationary time series. *J. R. Statist. Soc.* B, 15, 125.

WOLD, H. (1953). *Demand Analysis.* New York.

9·4 MATERN, B. (1960). *Spatial Variation.* Stockholm.

ROSENBLATT, M. and VAN NESS, J. S. (1965). Estimation of the bispectrum. *Ann. Math Statist.* 36, 1120.

WHITTLE, P. (1954). See 6·54 above.

WHITTLE, P. (1962). Topographic correlation, power-law covariance functions and diffusion. *Biometrika*, 49, 305.

9·41 BARTLETT, M. S. (1964). The spectral analysis of two dimensional point processes. *Biometrika* 51, 299.

MAKOTO NUMATA (1961). Forest vegetation in the vicinity of Choshi—coastal flora and vegetation at Choshi, Chiba Prefecture IV. *Bull. Choshi Marine Laboratory*, Chiba University, 3, 28.

9·42 BARTLETT, M. S. (1976). *The statistical analysis of spatial pattern.* London.

BESAG, J. E. (1974). See 6·54 above.

WHITTLE, P. (1954). See 6.54 above.

GLOSSARY OF STOCHASTIC PROCESSES

[Definitions are to be found in the pages listed; for complete references,
see Index; for a general definition of stochastic processes, see p. 1]

INDEX